# 2nd International Symposium on Laser-Ultrasonics – Science, Technology and Applications 2010



Journal of Physics: Conference Series Volume 278

## Talence, France
## 5-8 July 2010

**Editors:**

**Thomas Dehoux** **Yannick Guillet**

ISBN: 978-1-61782-662-7
ISSN: 1742-6588

**Printed from e-media with permission by:**

Curran Associates, Inc.
57 Morehouse Lane
Red Hook, NY 12571

**Some format issues inherent in the e-media version may also appear in this print version.**



Printed by Curran Associates, Inc. (2011)

For permission requests, please contact the Institute of Physics
at the address below.

Institute of Physics
Dirac House, Temple Back
Bristol BS1 6BE UK

Phone: 44 1 17 929 7481
Fax: 44 1 17 920 0979

techtracking@iop.org

**Additional copies of this publication are available from:**

Curran Associates, Inc.
57 Morehouse Lane
Red Hook, NY 12571 USA
Phone: 845-758-0400
Fax: 845-758-2634
Email: curran@proceedings.com
Web: www.proceedings.com

# TABLE OF CONTENTS

## MATERIAL EVALUATION AND CHARACTERIZATION



## FLAW AND DEFECT



## PICOSECOND LASER ULTRASONICS

## THERMOMECHANICS



## OPTICAL DETECTION AND GENERATION OF ULTRASOUND



## SIGNAL PROCESSING, MODELING AND IMAGING

## INDUSTRIAL APPLICATIONS, TECHNIQUES AND INSTRUMENTATION



## BIOMEDICAL APPLICATIONS



## SHOCK WAVES

Journal of Physics: Conference Series **278** (2011) 011001 doi:10.1088/1742-6596/278/1/011001

LU2010

**Foreword**

In the year 2010 we celebrated the 50th birthday of the laser, and laser-ultrasonics is only slightly younger. For the last few decades, laser-ultrasonics has attracted the attention of an increasing number of academics and engineers. It has motivated research in many directions, often in pluridisciplinary fields; and in parallel, having gained maturity, it has attracted more and more interest from industry.

Over the years laser-ultrasonics has been the topic of several sessions in different conferences dedicated to ultrasonics, optics, photo-acoustics, phonon scattering, material characterization, and non-destructive evaluation and control. With the aim or creating a fruitful forum for the exchange of ideas, national periodic meetings dedicated to laser-ultrasonics have been organized in Asia, North America and in Europe. For example, from 1999, we began holding convivial meetings in France, in which contributions by PhD students are strongly encouraged. Discussions at an international level were initiated a few years ago, examining the opportunity for creating a dedicated international conference assembling researchers and end-users with an interest in the field. Under Jean-Pierre Monchalin's instigation, the first International Symposium on Laser-Ultrasonics took place in Montreal in July 2008, and was a great success. We were very honoured to be in charge of the organization of the second meeting of the conference in Bordeaux in July 2010.

This second symposium assembled about 150 attendees from 30 countries. The program included 90 oral contributions, with 17 invited and plenary talks, and 40 contributions with posters. Despite extending the conference from three to four days, the organization of two parallel sessions was necessary. Thanks to the contributions of the participants, the scientific and organizing committees could construct an intensive and attractive scientific program.

The conference tour was a sunny visit to Saint-Emilion vineyard and the gala dinner took place in a wine chateau in the medieval village. We hope this experience will remain in the attendees' memories as a pleasant and convivial time.

We would like to express our thanks to the members of the organizing committee, the scientific committee, and all our generous sponsors, either institutions or companies, for their help in making this event possible. We would also like to thank the scientists involved for their confidence in our organization, and for their contributions.

As unanimously decided by the attendees, the next symposium will be held in Japan, most likely in 2012.

The papers published in this volume of the *Journal of Physics: Conference Series* provide a collection of state-of-the-art and recent advances in research and applications of laser ultrasonics as presented at this second Symposium.

Bertrand Audoin
Conference Chair

Thomas Dehoux
Conference proceedings co-editor

Yannick Guillet
Conference proceedings co-editor

Journal of Physics: Conference Series **278** (2011) 011001 doi:10.1088/1742-6596/278/1/011001

## Scientific committee

| | |
|---|---|
| *Bertrand Audoin* | Université Bordeaux 1, France |
| *Alain Blouin* | NRC, Canada |
| *Christ Glorieux* | Katholieke Universiteit Leuven, Belgium |
| *Vitalyi Gusev* | Université du Maine, France |
| *Peter Hess* | University of Heidelberg, Germany |
| *Alexander A Karabutov* | Moscow State University, Russia |
| *Sridhar Krishnaswamy* | Northwestern University, USA |
| *Humphrey Maris* | Brown University, USA |
| *Jean-Pierre Monchalin* | NRC, Canada |
| *Todd Murray* | Boston University, USA |
| *Makoto Ochiai* | Toshiba Corp, Japan |
| *Menglu Qian* | Tongji University, China |
| *Daniel Royer* | Ecole Supérieure de Physique et de Chimie, France |
| *Hervé Trétout* | Dassault Aviation, France |
| *Hubert Voillaume* | EADS, France |
| *Oliver B. Wright* | Hokkaido University, Japan |
| *Shu-yi Zhang* | Nanjing University, China |

## Organizing committee

| | |
|---|---|
| *Bertrand Audoin* | Chair |
| *Sandrine Guit* | Local administrator |
| *Christophe Aristégui* | |
| *Thomas Dehoux* | |
| *Evelyne Dewayse* | |
| *Yannick Guillet* | |
| *Armelle Guilloux* | |

All papers published in this volume of *Journal of Physics: Conference Series* have been peer reviewed through processes administered by the proceedings Editors. Reviews were conducted by expert referees to the professional and scientific standards expected of a proceedings journal published by IOP Publishing.

# Surface Brillouin scattering measurement of the elastic constants of single crystal $InAs_{0.91}Sb_{0.09}$

**L M Kotane[1], J D Comins[1], A G Every[1] and J R Botha[2]**

[1]Materials Physics Research Institute, School of Physics, University of the Witwatersrand, Johannesburg, Wits 2050, South Africa
[2]Department of Physics, Nelson Mandela Metropolitan University, Port Elizabeth, South Africa

E-mail: Lesias.Kotane@wits.ac.za

**Abstract.** Surface Brillouin scattering of light has been used to measure the angular dependence of the Rayleigh surface acoustic wave (SAW), pseudo surface acoustic wave (PSAW) and longitudinal lateral wave (LLW) speeds in a (100)-oriented single crystal of the ternary semiconductor alloy $InAs_{0.91}Sb_{0.09}$. The wave speed measurements have been used to determine the room temperature values of the elastic constants $C_{11}$, $C_{12}$ and $C_{44}$ of the alloy. A simple and robust fitting procedure has been implemented for recovering the elastic constants, in which the merit function is constructed from explicit secular functions that determine the surface and lateral wave speeds in the [001] and [011] crystallographic directions. In the fitting, relatively larger weighting factors have been assigned to the SAW and PSAW data because of the greater precision with which the surface modes can be measured as compared with the lateral wave.

## 1. Introduction

The III-V ternary semiconductor alloy, $InAs_{1-x}Sb_x$, is important in the development of long-wavelength optoelectronic devices due to its room temperature band gap ranging from about 0.17 to 0.35 eV [1]. The alloy with a composition x = 0.09 is of importance as it has a room temperature energy gap corresponding to a wavelength of about 4.2 $\mu$m which is within the infrared wavelength range 3-5 $\mu$m, the transmission window in the atmosphere, where long-range detection is of interest [2, 3].

In this paper we present the results of a study which makes use of the surface Brillouin scattering (SBS) technique to measure the room temperature elastic constants of the $InAs_{0.91}Sb_{0.09}$ ternary semiconductor alloy. The elastic constants are essential in furthering the understanding of the mechanical properties of devices that are fabricated from such alloys. The utilised SBS technique is a nondestructive measurement that exploits light scattering in the analysis of the characteristics of surface acoustic waves (SAWs).

## 2. Experimental method

For opaque materials such as $InAs_{0.91}Sb_{0.09}$, where only very limited light penetration occurs, laser light incident on the surface is inelastically scattered by surface acoustic waves via the so-called surface ripple mechanism [4, 5]. The backscattering geometry used in our experiments leads via wavevector conservation to the relation $k_{//} = 2k_i \sin\theta_i$, where $k_{//}$ is the component of the wavevector parallel to the surface, $\theta_i$ is the incident angle and $k_i$ is the wavevector of the incident light. The phase velocity $V$ of the SAW is related to the spectral frequency shift$(\omega_s - \omega_i)$ by $V = (\omega_s - \omega_i)/2k_i \sin\theta_i$.

Published under licence by IOP Publishing Ltd

The rotation of the sample about the azimuthal angle $\varphi$ selects different SAW propagation directions. The phase velocities of the (SAWs) are then used to determine the elastic constants of the material.

Undoped samples with a mirror-like finish were grown by atmospheric pressure metalorganic vapor phase epitaxy (MOVPE-grown) on an InAs substrate. The samples were measured to be about 2.7 $\mu$m thick, thick enough to be regarded as bulk material in SBS, by the Nomarski differential interference method [6] and were lattice matched to the (100) orientation of the InAs substrate.

X-ray diffraction measurements were carried out on the specimen to determine its lattice constant and hence its density. The measurement was done at room temperature using a Philips PW1050 diffractometer (Cu K$\alpha$ radiation at 40kV and 20 mA). From the X-ray diffraction measurements, the lattice constant was determined to be 6.0811 Å. The density of this cubic zinc-blende structure was then calculated from the lattice constant and the total atomic weight in a unit cell and was found to be 5.729 g/cm$^3$.

The experimental arrangement used for SBS measurements is described in Comins [5]. In brief, measurements were done at room temperature using an argon-ion laser of wavelength 514.5 nm operated in a single axial mode for illumination. The laser light was focused onto the sample by a 120 mm focal length lens and the scattered light was collected with the same lens in the backscattering geometry. The scattered light was analysed by a Sandercock-type (3+3)-pass tandem Fabry-Pérot interferometer fitted with a JRS Scientific Instruments vibration isolation system. The light was detected by an EG&G Optoelectronics Canada SPCM-200-PQ detector, which is a silicon avalanche photodiode device. The scattering angle $\theta_s = \theta_i =$ 71° used in the measurements is close to a maximum for the p-p scattering (polarisation in the sagittal plane) cross section.

## 3. Results and discussion

The values of the elastic constants $C_{11}$, $C_{12}$ and $C_{44}$ were recovered from the SBS measurements taken in the [001] and the [011] directions of the (100) surface of the $InAs_{0.91}Sb_{0.09}$ specimen. Our measurements resulted in 4 velocities and with 3 parameters (i.e. $C_{11}$, $C_{12}$ and $C_{44}$) to be determined – this is an over determined problem that is solved by a fitting procedure. One approach would be to do a least squares fit to the 4 velocities. We have, however, adopted a much simpler approach to the fitting, which avoids the necessity of solving the cubic equations for the SAW and PSAW velocities. We do this by taking the merit function to be $\chi^2 = (h_1^2 + h_2^2) + w\left(h_3^2 + h_4^2\right)$, where $w$ is the relative weighting factor for the longitudinal velocities as compared with the Rayleigh SAW measurements. $h_1$ and $h_3$ are secular functions that determine the surface and longitudinal lateral wave speeds in the [001] crystallographic direction and are defined as follows

$$h_1 = \rho V_{RSAW}^2 - \left( \frac{C_{11}}{C_{44}} \times \frac{C_{44} - \rho V_{RSAW}^2}{C_{11} - \rho V_{RSAW}^2} \right)^{\frac{1}{2}} \times \left( C_{11} - \frac{C_{12}^2}{C_{11}} - \rho V_{RSAW}^2 \right)$$

$$h_3 = \rho V_L^2 - C_{11}\,.$$

Similarly, expressions for secular functions $h_2$ and $h_4$ that determine the surface and longitudinal lateral wave speeds in the [011] crystallographic direction are defined as follows

$$h_2 = \rho V_{PSAW}^2 - \left\{ \frac{C_{11}}{C_{44}} \times \frac{C_{44} - \rho V_{PSAW}^2}{\left[\frac{1}{2}(C_{11} + C_{12} + 2C_{44})\right] - \rho V_{PSAW}^2} \right\}^{\frac{1}{2}} \times \left( \frac{1}{2}(C_{11} + C_{12} + 2C_{44}) - \frac{C_{12}^2}{C_{11}} - \rho V_{PSAW}^2 \right)$$

$$h_4 = \rho V_L^2 - \frac{1}{2}(C_{11} + C_{12} + 2C_{44}).$$

A perfect fit is obtained when all the secular functions are equal to zero. The quantities $h_3$ and $h_4$ correspond to the deviation of the squared longitudinal velocities from their measured values and $h_1$ and $h_2$ are not very different from being the deviation of calculations from squared measured SAW velocities [7].

In the calculation of the elastic constants, the SAW velocities were given a bigger weighting than the longitudinal velocities. This is largely due to the possibility of locating the centre of the SAW peaks by fitting Lorentzian functions whilst on the other hand the longitudinal thresholds presented themselves more as somewhat rounded valleys rather than sharp dips. This can be clearly seen in figure 1. Given this factor of a greater uncertainty in measuring the longitudinal velocities, a weighting factor of $w = 0.25$ was assigned to the longitudinal velocities in the calculation of the merit function $\chi^2$. A detailed computational approach is found in Every et al. [7], also providing information on elasto-dynamic Green's function method in the calculation of the surface dynamics.

**Table 1.** Measured and calculated velocities (with a $V_L$ weighting of 0.25) for the (100) surface of the ternary semiconductor alloy $InAs_{0.91}Sb_{0.09}$.

| | **[001]** $V_{SAW}$ **(m/s)** | **[011]** $V_{PSAW}$ **(m/s)** | **[001]** $V_L$ **(m/s)** | **[011]** $V_L$ **(m/s)** |
|---|---|---|---|---|
| **Measured** | 2011 | 2139 | 3483 | 4179 |
| **Calculated** | 1984 | 2156 | 3587 | 4087 |
| **% deviation** | -1.34 | 0.79 | 2.99 | -2.20 |

On minimisation of $\chi^2$ the fitted values of the elastic constants were found to be:
$C_{11}$ = 73.7 GPa, $C_{12}$ = 40.4 GPa, $C_{44}$ = 38.7 GPa.

The SBS spectra were calculated from the determined values of the elastic constants $C_{11}$, $C_{12}$ and $C_{44}$ and compared with spectra measured in the [001] and the [011] directions based on the assumption that the scattering is entirely by the ripple mechanism. The comparison, illustrated in figures 1 and 2, shows a good agreement between measured and calculated spectra particularly for the Rayleigh SAW modes. The uncertainty in locating exact positions of the longitudinal minima is evident when comparing the "valleys" of the measured spectra to the calculated "dips".

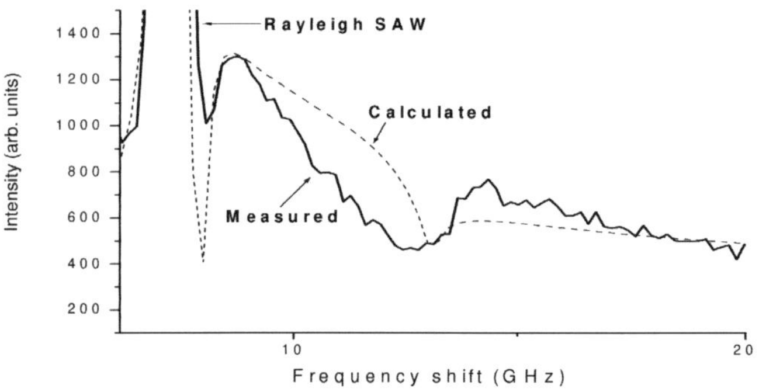


**Figure 1.** The measured and calculated SBS spectra for the [001] direction on the (100) surface of $InAs_{0.91}Sb_{0.09}$.

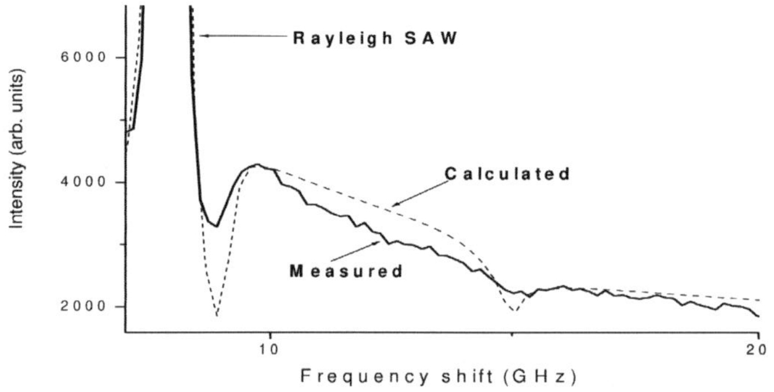


**Figure 2.** The measured and calculated SBS spectra for the [011] direction on the (100) surface of $InAs_{0.91}Sb_{0.09}$.

## 4. Conclusion

Using a simple and robust fitting procedure employing a merit function, the elastic constants of single crystal $InAs_{0.91}Sb_{0.09}$ have been determined from velocity measurements of the Rayleigh SAW, pseudo-SAW and the longitudinal threshold of the Lamb continuum observed in the SBS spectra corresponding to [001] and the [011] directions on the (100) surface.

## References

[1] Sze S M 1981 *Physics of Semiconductor Devices* (New York; John Wiley)
[2] Biefeld R M 1986 *J. Cryst. Growth* **75** 255
[3] Baisitse T R, Forbes A, Katumba G, Botha J R and Engelbrecht J A A 2008 *Phys. Status Solidi* (c) **5** 573
[4] Beghi M, Every A G and Zinin P V 2004 Chapter 10 *Ultrasonic Nondestructive Evaluation: Engineering and Biological Material Characterization*, ed Kundu T, (Boca Raton; CRC Press)
[5] Comins J D 2001 Chapter 15 *Handbook of Elastic Properties of Solids, Liquids and Gases. Volume I: Dynamic Methods for Measuring the Elastic Properties of Solids*, edited by Levy et al (San Diego; Academic Press)
[6] Krug T, Botha L, Shamba P, Baisitse T R, Venter A, Engelbrecht J A A and Botha J R 2007 *J. Cryst. Growth* **298** 163
[7] Every A G, Kotane L M and Comins J D 2010 *Phys. Rev. B* **81** 224303

# Thickness gauging of thin layers by laser ultrasonics and neural network

**F Lefèvre, F Jenot, M Ouaftouh, M Duquennoy, M Ourak**

Institut d'Electronique de Microélectronique et de Nanotechnologie Département Opto-Acousto-Electronique (UMR CNRS 8520), Université de Valenciennes, Le Mont Houy, 59313 Valenciennes Cedex 09, France.

Fabien.lefevre@univ-valenciennes.fr

**Abstract**. Non destructive testing has been performed on a thin indium layer deposited on a two inches silicon wafer. Guided waves were generated and studied using a laser ultrasonic setup, and a two-dimensional Fourier transform technique was employed to obtain the dispersion curves. The inverse problem, in other words the determination of the layer thickness and the elastic constants of the substrate, has been solved by means of a feedforward neural network. These parameters were then evaluated simultaneously, the dispersion curves being entirely fitted. The experimental results show a good agreement with the theoretical model. This inversion method was found to be prompt and easy to automate.

## 1. Introduction

Silicon wafers are widely used in microelectronics as substrates to receive microelectronics devices. In recent years, growing interest and concern for thin layer deposition has been observed. Numerous methods have been developed to characterize such films using, for example, nanoindentation [1], x-ray reflectivity [2] or laser ultrasonics [3,4].

In this paper, guided waves are probed to study an indium layer deposited on a silicon substrate. Experimental dispersion curves are obtained in a non contact and non destructive way, and the inverse problem is solved by a feed-forward neural network [5]. The experimental dispersion curves are fitted on several modes which allows us to find out simultaneously the layer thickness and the elastic characteristics of the silicon substrate.

## 2. Lamb wave propagation in anisotropic bilayered structures

The transfer matrix method [6] has been used to describe the propagation of guided waves in a bilayered platelike structure. Such a structure is depicted in figure 1, where the exponents (1) and (2) refer to the different layers, whereas the indices 1, 2, and 3 refer to the different directions.

In this method, for each layer, the displacements and stresses at the same surface are expressed as a vector P. The relationship between the $P_+$ vector at the top layer surface and the $P_-$ vector at the bottom surface of the same layer is given by:

$$P_+^{(k)} = A^{(k)} \cdot P_-^{(k)} \quad (1)$$

where $A^{(k)}$ is the transfer matrix of the anisotropic layer k. The use of continuity conditions for the displacements and the stresses at the interface between the two layers leads to the expression of the global transfer matrix $A$. This matrix links the surface vector $P_{-}^{(1)}$ of the first layer to the surface vector $P_{+}^{(2)}$ of the second layer:

$$A = A^{(1)} \cdot A^{(2)} \tag{2}$$

Using free boundary conditions to $P_{-}^{(1)}$ and $P_{+}^{(2)}$ allows us to obtain an equation which solutions can exist if and only if:

$$\begin{vmatrix} A_{41} & A_{42} & A_{43} \\ A_{51} & A_{52} & A_{53} \\ A_{61} & A_{62} & A_{63} \end{vmatrix} = 0 \tag{3}$$

This problem has been solved in the case of a 4 μm thick indium layer deposited on a 2 inches silicon wafer. The formulation employed to express the components of the different transfer matrices was the one of Nayfeh [7]. The silicon having a diamond cubic structure, which implies an anisotropic behaviour, the resolution has been made in the (100) direction. The results are presented into dispersion curves in figure 2.

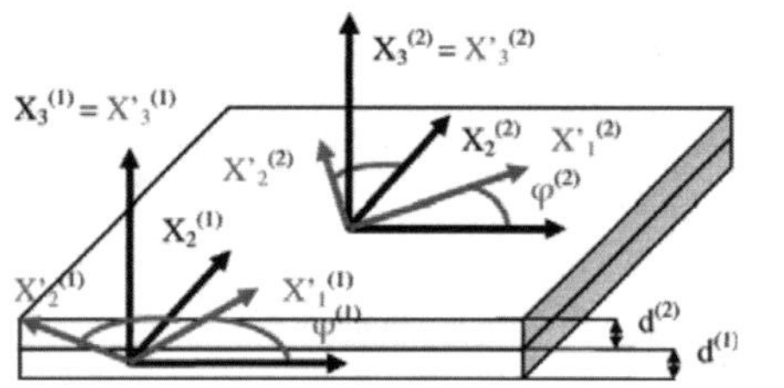


**Figure 1.** The coordinate systems of a plate composed of two anisotropic layers of thickness $d^{(1)}$ and $d^{(2)}$.

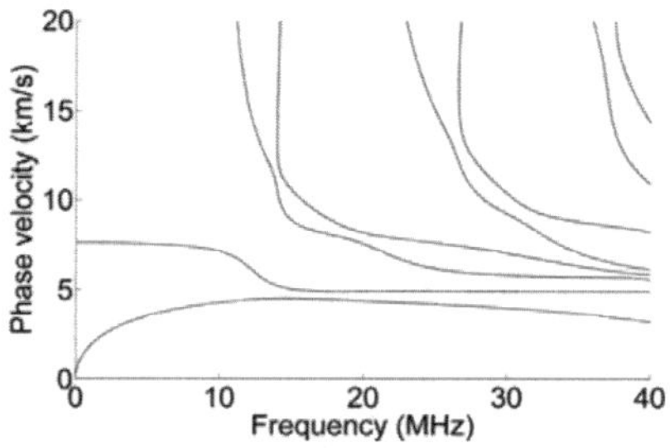


**Figure 2.** Theoretical dispersion curves for a 275 μm thick silicon wafer coated with a 4 μm thick indium layer.

## 3. Experimental setup

The system used to generate the acoustic waves is schematically shown in Figure 3. A 10 ns duration Q-switched Nd-YAG laser pulse of 532 nm wavelength was focused at the sample surface as a line source of about 5 mm length and 0.5 mm width [8]. The energy per pulse was around 6 mJ, which allowed us to work in the thermoelastic mode.

The normal displacement of each Lamb mode was detected by a Mach-Zehnder type interferometer with a power of 100 mW and a large bandwidth (200 kHz to 45 MHz). The received signals were sampled and averaged by a digital oscilloscope before acquisition. Each recorded signal corresponded to an average of sixteen laser shots in order to improve the signal-to-noise ratio. Motorized motion tables allowed us to move the laser line source to perform measurements at different distances between source and detector.

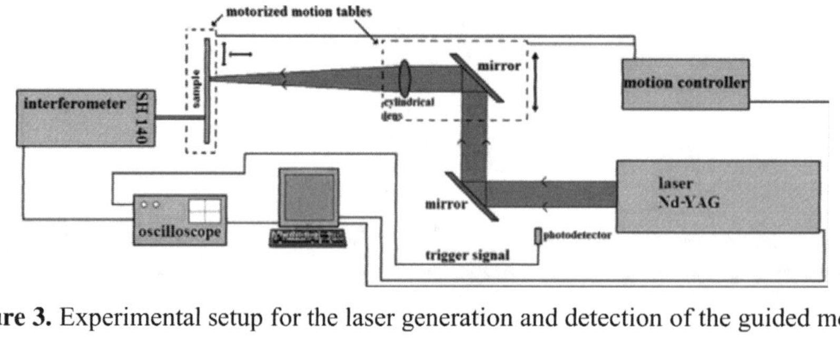


**Figure 3.** Experimental setup for the laser generation and detection of the guided modes.

## 4. Neural networks

The sample's dispersion curves are characterized by several modes. Classical fitting methods are consequently difficult to implement and time consuming in this case. A more global method, using feed-forward neural networks, has been investigated.

Feed-forward neural networks (FF networks) are widely used to approximate functions. They are known by many different names, such as "multilayer perceptrons" [9]. Figure 4 illustrates a one-hidden-layer FF network with inputs $X_1, \ldots, X_n$ and output $Y$. Each arrow in the figure symbolizes a parameter in the network which is divided into layers. The input layer consists of just the inputs to the network. Then follows a hidden layer, which consists of any number of neurons or hidden units placed in parallel. Each neuron performs a weighted summation of the inputs, which then passes a nonlinear activation function $\sigma$, also called the neuron function.

Mathematically the functionality of a hidden neuron can be described by $F$, such as:

$$F = \sigma\left(\sum_{j=1}^{n} X_j \cdot w_j + b_j\right) \tag{4}$$

where the weights $w_j$ ,$b_j$ are symbolized with the arrows feeding into the neuron.

The network output is formed by another weighted summation of the outputs of the neurons in the hidden layer. This summation on the output is called the output layer. The neurons in the hidden layer of the network in figure 4 are similar in structure to those of the perceptron with the exception that their activation functions can be any differential function. The output of this network is given by

$$Y = \sum_{i=1}^{nh} w_i^2 \cdot \sigma\left(\sum_{j=1}^{n} X_j \cdot w_{i,j}^1 + b_{j,i}^1\right) + b^2 \tag{5}$$

where $n$ is the number of inputs and $nh$ is the number of neurons in the hidden layer. The variables $\{w^1_{i,j}, b^1_{j,i}, w^2_i, b^2\}$ are the parameters of the network model. The size of the input and output layers are defined by the number of inputs and outputs of the network respectively and, therefore, only the number of hidden neurons has to be specified when the network is defined.

Supervised learning can then be used to modify and to fix the parameters of the defined model. The network is fed with a training data, which consists of pairs of input values $x$, and desired outputs $y$. The weights are adjusted incrementally until the data satisfy the desired mapping as well as possible; that is, until $Y$, the network output, matches the desired output $y$ as closely as possible up to a maximum number of iterations.

The FF network in figure 4 is just one of the possible architectures. The architecture can be modified in various ways by changing the options.

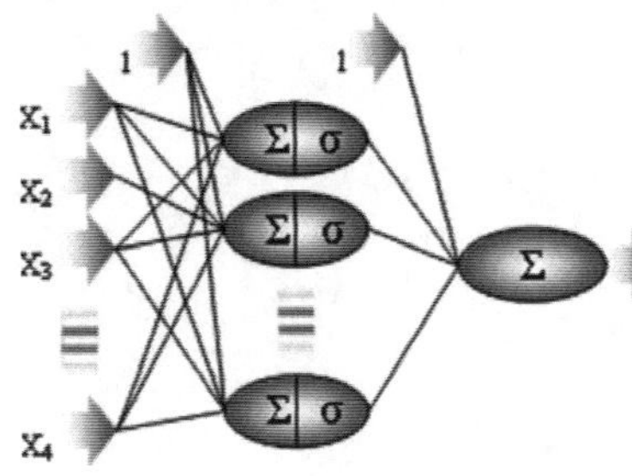


**Figure 4.** A feed-forward network with one hidden layer and one output.

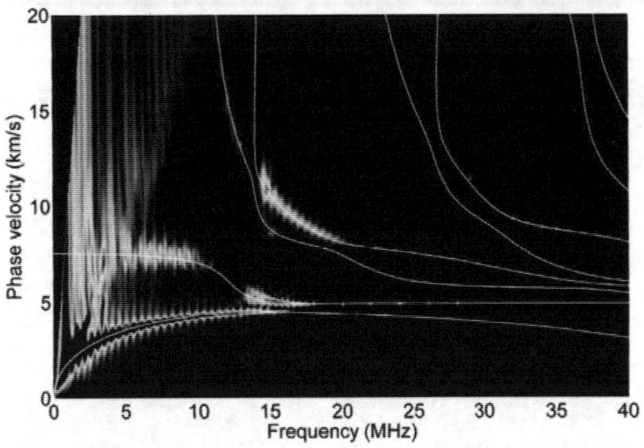


**Figure 5.** Neural network fit of the 2DFFT results for an indium layer deposited on a 275 μm thick silicon wafer.

## 5. Determination of the indium layer thickness and of the substrate's elastic properties

Measurements were achieved with the experimental setup described earlier on a two inches silicon wafer on which 4 μm of indium has been deposited. 500 signals were recorded, the distance between the emitter and the receiver being increased of 20 μm before each new recording over a total distance of 10 mm. The sampling frequency was set to 500 MHz.

A two-dimensional spectral analysis was applied to the time signals and the results are presented as contour plots of amplitude versus phase velocity and frequency in figure 5. Four two-layer feed-forward neural networks were used to estimate the structure characteristics. They were trained using the Levenberg–Marquardt algorithm [10] and with a database of 2365 theoretical dispersion curves. The input vector of the different networks is composed of 29 velocity values, the hidden layer is made of 250 neurons, and the output layer gives the indium layer thickness and the elastic characteristics of the anisotropic substrate. The thickness and density of the substrate were assumed to be respectively 275 μm and 2330 $kg/m^3$. Concerning the layer, the Young modulus was set to 11 GPa, the Poisson ratio to 0.45 and the density to 7310 $kg/m^3$.

The inversion results are given in table 1 for the four networks. The layer thickness was moreover measured by means of a high contact profilometer. Concerning the elastic properties of the substrate, the inversion results were compared to the elastic characteristics currently used in the literature.

A good agreement between the inversion solutions and the structure characteristics is observed. Moreover, the fit of the 2DFFT results in figure 5 is found to be satisfying for the different propagating modes.

**Table 1.** Results of the network inversion for the indium layer deposited on a 2 in. silicon wafer.

| | Layer thickness (μm) | $C_{11}$ (GPa) | $C_{12}$ (GPa) | $C_{44}$ (GPa) |
|---|---|---|---|---|
| Structure characteristics | 4.11 | 165.6 | 63.9 | 79.5 |
| Inversion solution n°1 | 4.35 | 165.0 | 67.7 | 83.5 |
| Inversion solution n°2 | 4.03 | 164.7 | 65.3 | 81.3 |
| Inversion solution n°3 | 4.14 | 166.5 | 64.5 | 81.3 |
| Inversion solution n°4 | 4.12 | 164.6 | 65.0 | 80.9 |
| **Mean of the inversion solutions** | **4.16** | **165.2** | **65.6** | **81.7** |
| **Inversion error in %** | **1.2** | **0.25** | **2.7** | **2.8** |

## Conclusion

A laser ultrasonic method for the analysis of guided waves in an anisotropic bilayered structure has been presented. Dispersion curves, with a clear identification of multimode guided waves, were obtained with a 2DFFT. Determination of the layer thickness and of the substrate elastic parameters has been made in an original way using feed-forward neural networks.

## References

[1] Chudoba T, Schwarzer N, Linss V, and Richter F 2004 *Thin Solid Films* **239** 469–470.
[2] Santucci S, la Cecilia A V, Digiacomo A, Phani R A and Lozzi L 2001 *J. Non- Cryst. Solids* **280** 228.
[3] Bonello B, Perrin B, Romatet E and Jeannet J C 1997 *Ultrasonics* **35** 223.
[4] Reverdy F and Audoin B 2001 *J. Appl. Phys.* **90** 4829.
[5] Lefèvre F, Jenot F, Ouaftouh M, Duquennoy M and Ourak 2009 *Rev. Sci. Inst.* **80** 014901.
[6] Haskell N A 1953 *Bull. Seismol. Soc. Am.* **43** 17.
[7] Nayfeh A H 1991 *J. Acoust. Soc. Am.* **89** 1521.
[8] Jenot F, Ouaftouh M, Duquennoy M and Ourak M 2005 *J. Appl. Phys.* **97** 094905.
[9] Rosenblatt F 1961 *Principles of Neurodynamics*, Spartan Washington D.C.
[10] Marquardt D 1963 SIAM *J. Appl. Math.* **11** 431.

Journal of Physics: Conference Series **278** (2011) 012003 doi:10.1088/1742-6596/278/1/012003

# Photoacoustic thermal conductivity determination of layered structures PS-Si: piezoelectric detection

**S Alekseev, D Andrusenko, R Burbelo, M Isaiev, A Kuzmich**

Taras Shevchenko Kyiv National University, Faculty of Physics, Volodimirska Str., 64, 01601 Kyiv, Ukraine

E-mail: and_@univ.kiev.ua

**Abstracts.** Based on bending vibrations model method and also construction of multilayer transducer, which allow the measurements of thermal conductivity for porous layer on Si wafer is proposed. Time dependence of photoacoustic signal under rectangular modulation of exciting light was numerical modeling. The experimental tests were done on the samples of porous silicon on Si wafers as well as for the porous silicon free standing layers, obtained under the same anodization regime. The value of thermal diffusivity of the porous silicon layer correlates well with the value of thermal diffusivity of the porous silicon free layer, determined by TDC method.

## 1. Introduction

Photoacoustic (PA) methods are successfully applied to determine the values of thermophysical parameters of porous silicon (PS) [1,2]. An important advantage of these techniques is the simplicity of its experimental realization. In particular, the problem of insuring the reliable thermal contact to the porous layer characteristic for classical methods of thermal conductivity measurements eliminates easily in PA methods.

The gas-microphone PA methods are the most frequently used to measure the thermal conductivity $k$ (thermal diffusivity $D = k/c\rho$, where c is thermal capacity, $\rho$ - density) of the PS. These methods have the solid theoretical background; they are highly sensitive to variable component of the surface temperature. We'll distinguish here the method of TDC (transmission detection configuration [1] method) and the PA spectroscopy (PAS) method [2]. The TDC method is based on the transmission of thermal waves through the sample. In the case of supported layers the TDC method allows to determinate effective parameters of entire structure and the determination of the parameters of porous layer itself comprise a problem. The PAS method is rather complicated, it is not very precise, and applies only to a narrow range of materials.

The piezoelectric method of PA signal formation has the same sensitivity the gas-microphone PA method. However, this method has no wide practical application for thermophysical measurements, due to absence of unified theory of signal generation for this method. In some particular cases at definite geometries of experimental system and different types of registered acoustic vibrations, number of rather complicated theoretical models was proposed [3, 4]. The model of excitation of low-frequency quasistatic thermoelastic bending vibrations of the system “sample-transducer” [5] should be mentioned as one of the most simple among them. For this case, the applicability of multilayer piezotransducer for measurements of thermal conductivity of highly light-absorbing samples is shown in the work [6].

In the present work we propose relatively simple and precise method based on bending vibrations model and the setup of multilayer transducer, which allow the measurements of thermal conductivity for porous layer on monolithic support. Difference of elastic properties of porous and non-porous layers, as in the case of supported PS, is crucial in this case.

## 2. Numerical model

From the qualitative considerations, based on the quasistatic deformations model at rectangular modulation of light, which excite the thermal wave in steady-state conditions, for the dependence of the electric potential $U(t)$, taken from the electrodes of the transducer when the region of localization of thermal excitation in depth $l_t$ is much smaller than thickness of the system sample-piezoelectric transducer $h$:
$l_t << h$, at approach of isotropic layers we can write an expression:

$$U(t)=a\cdot\sum_{n=1}^{k}\frac{Q(t)}{c_n\rho_n}\cdot\alpha_{Tn}\hat{E}_n\,, \tag{1}$$

where $\hat{E}_n=E_n/(1-\sigma_n)$, $E_n$ - modulus of elasticity, $\sigma_n$ - Poisson coefficient, $\alpha_{Tn}, c_n, \rho_n$ correspondingly thermal expansion coefficient, thermal capacity and density of material with number $n$, $a$ – constant value and $Q_n(t)$ - difference of heat quantity in the layer n from its stationary value in $t$ time. Thus, the potential induced on the electrodes of a piezoelectric transducer (PA signal) is proportional to the amount of thermoelastic stresses in the layers of the system laying within the field of thermal disturbances.

The ration $\left(\alpha_{T1}\cdot\hat{E}_1\cdot c_2\cdot\rho_2\right)/\left(\alpha_{T2}\cdot\hat{E}_2\cdot c_1\cdot\rho_1\right)$ << 1 for the sample where the 1st layer is porous with the porosity > 50%, and the 2nd is non-porous. That is why due to this expression in the moment when the edge of thermal agitation reaches the interface between the porous and non-porous layers, due to the redistribution of heat between the layers, significant change of rate of growth of $U(t)$ should be achieved.

In the more general case of dependence of the material parameters on the occurrence depth of the layer z under the surface of the sample, taking to the consideration that for the ac component of the temperature $\theta(z,t)$ we can write down: $\theta(z,t)\cdot \mathrm{d}z=\mathrm{d}Q(z,t)/c(z,t)\cdot\rho(z,t)$ then the expression (1) takes the form:

$$\widetilde{U}\sim\int_0^h\theta(z,t)\cdot\alpha_T(z,t)\cdot\widehat{E}(z,t)\,\mathrm{d}z\,. \tag{2}$$

It makes possible also to model the development of PA signal through the time, by calculating the temperature distribution over the thickness of the sample.

Numerical calculations of the spatial distribution of temperature in different moments of time where realized due to the finite-element method, and the dependence $U(t)$ through the numerical integration (2).

## 3. Experiment

The experiment was conducted with samples of mesoporous silicon in the form of a layer on the monocrystal substrate. Geometry of the experiment is shown in figure 1. The glass-ceramic layer 3 was added in the piezo-transformer to locate a thermal disturbance within the sample and to fulfill the conditions $l_T \ll h$.

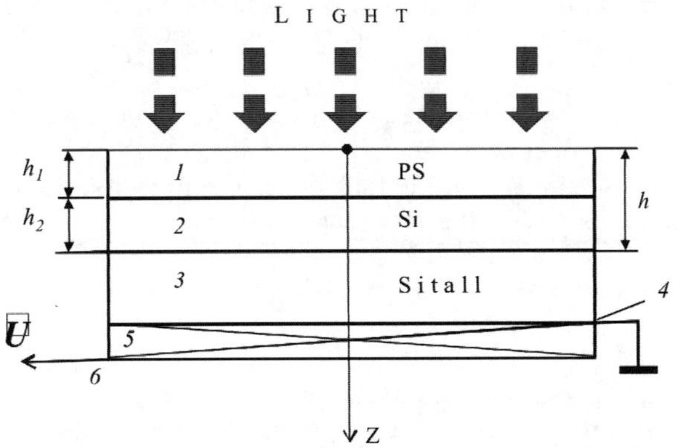


**Figure 1.** Sandwich structure comprising (1,2) composite sample where 1 - PS, 2 - Si and (3-6) piezoelectric transducer where which 3 - buffer layer, 4 and 6 - electrodes, 5 - piezoelectric ceramics.

Glass-ceramic has the low thermal conductivity, from whence thermal perturbation is almost completely localized in the sample due to the rapid decay in the glass ceramics. In addition, glass ceramics has a small coefficient of thermal expansion. As a result, the contribution of thermoelastic stresses in glass-ceramic layer (and the corresponding component in the expression 2) and the pyroelectric effect in the transducer to the PA signal can be neglected.

Thermal perturbations in the sample were excited in the samples by radiation of a blue LED operating at an electric power of 3 W , which was modulated by rectangular pulses (at an off/duty ratio of two). The intensity of light collected and uniformly distributed over the surface of the optical system was about 1mW/mm². The light flux was modulated by interrupting the pump current to the LED. The dependence $U(t)$ was reflected by the oscilloscope.

Figure 2 compares the experimental dependency, obtained by processing the oscillograms (curve 1), and the dependency obtained by numerical modeling (curve 2) for PA signal. By selecting the parameters of the material in the numerical model, we were pursuing the coincidence of the character of the time changes in calculated and experimental dependencies.

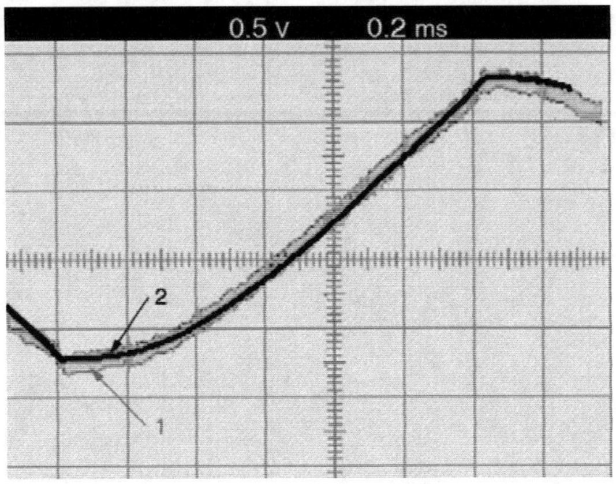


**Figure 2.** The oscillogram of a half-period of heating (shade line – 1) and the dependence resulting from the numerical simulation (solid line – 2) dependence of potential on time. Graduating mark on the time scale - 2 ms / grad.

The increase in the rate of growth of potential $U(t)$ (nearly 5 times) was experimentally observed in a certain time interval (0.4 ms for the porous layer with thickness of 50 microns with a porosity of 60%) after the start of heating.

The value of thermal diffusivity of mesoporous layer of double-layer sample obtained, was compared with the experimental value of thermal diffusivity of the porous layer, grown under similar regimes of anodization, but separated from the substrate. The porosity of this sample is lower - 55%. The measure of thermal diffusivity of the layer separated from the substrate was carried out by the TDC. The results obtained by both methods are close. The results of the first case of the TDC – $0.014 \pm 0.001$ $cm^2/s$ and of the second – $0.012 \pm 0.003$ $cm^2/s$.

## 4. Conclusion

Hence the method proposed in our work allows to precise the determination of thermal conductivity and thermal diffusivity of porous layer on the monolithic support. Although this method provides a lower accuracy than the method of TDC, it should be mentioned that it can be applied to the samples (thin layers on a substrate), where the use of other methods, including method of TDC, is problematic. It should be mentioned, that the piezoelectric registration looks most promising among the PA methods if the temperature dependencies of parameters of a material on temperature are desirable.

[1] Shen Q and Toyoda T 2003 *Review of Scientific Instruments* **74** 601
[2] Obraztsov A, Karavanskii V, Okushiy H and Watanabe H 1997 *Semiconductors* **31** 629
[3] Gusev V and Karabutov A 1991 *Laser Optoacoustics* (AIP Press) p 336
[4] Jachson W and Amer M 1980 *Appl. Phys.* **51** 3343.
[5] Blonskij J, Tkhoryk V and Shendeleva M 1996 J. *Appl. Phys.* **79** 3512
[6] Andrusenko D and Kucherov I 1999 *Tech. Phys.* **44** 1397

# Determination of elastic properties of surface layers and coatings by resonant ultrasound spectroscopy

**M Růžek[1,2], P Sedlák[1], H Seiner[1], M Landa[1]**

[1]Institute of Thermomechanics, Academy of Sciences of the Czech Republic, Dolejškova 5, Prague 8, CZ 182 00, Czech Republic

[2]Faculty of Nuclear Sciences and Physical Engineering, Czech Technical University in Prague, Trojanova 13, Prague 2, CZ 120 00, Czech Republic

E-mail: ml@it.cas.cz

**Abstract.** This paper deals with determination of in-plane elastic constants of thin layers deposited on substrates. Modified resonant ultrasound spectroscopy is used to measure resonant spectra before and after layer deposition . These two spectra are compared and changes in the position of the resonant peaks are associated with layer properties. It is shown that for thin layers either the elastic moduli or the surface mass density can be determined, providing the complementary information (the surface mass density for the determination of the moduli, the elastic moduli for the determination of the surface mass density) is known. As an experimental demonstration of this approach, the elastic moduli of diamond-like-carbon film deposited on a silicon substrate and the surface mass density of a thin spray paint on a silicon substrate are determined.

## 1. Introduction

Resonant ultrasound spectroscopy (RUS) is one of the most precise methods for experimental determination of the elastic properties of solids [1]. It has been shown recently that this method can also be used for determination of the elastic constants of thin layers [2] and [3]. In this article we present a novel approach to this technique with simplified problem modeling, and verify this approach by application to two distinctly different systems layer-substrate.

## 2. Method description

The RUS method for thin layers uses comparative measurements of the free vibration spectra of the same substrate with and without the layer, as shown schematically in figure 1(a). It is assumed that the measured frequency shifts of the individual resonant peaks are given by the properties of the layer which is added (or removed).

Generally, the frequency shifts depend on many factors but most important are the stiffness of the layer, its density and thickness. We can write the forward problem mathematically as

$$\mathbf{F}(\mathbf{Q}, \rho, h, geometry) = \Delta\omega \tag{1}$$

where $\mathbf{Q}$ is the vector of in-plane layer elastic constants (plane stress is assumed), $\rho$ and $h$ are the layer density and thickness and $\Delta\omega$ is the vector of angular frequency shifts. Function $\mathbf{F}$ cannot be constructed analytically (with exception of some simple 1D geometries) but it can

Published under licence by IOP Publishing Ltd

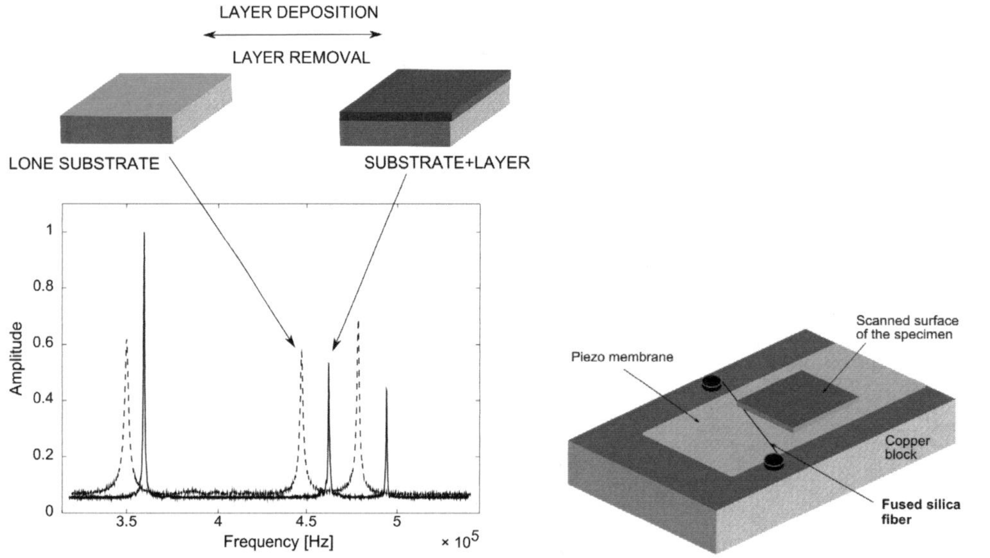


(a) Principle of RUS method on thin layers. (b) Excitation of the sample.

**Figure 1.** Principle of the method and experimental layout of the excitation.

be evaluated numerically by different methods. We use the Ritz method described in [4]. For thin layers, the problem given by Eq.1 can be approximated by the linear equation

$$\mathbf{AQ} - \mathbf{M}\rho h = \Delta\omega \tag{2}$$

where the matrix **A** describes the effect of the layer stiffness and the matrix **M** effects of the layer mass on frequency shifts $\Delta\omega$. Both effects are multiplied by stiffness coefficients **Q** and the surface density $\rho h$ of the layer. Construction of all the matrices is specified in [4]. From equation 2 we clearly see that there are two competing factors that change the resonances when the layer is added. Stiffness represented by the first term in equation 2 always tends to increase the resonant frequencies (sample gets more stiffer) and mass represented by the second term in equation 2 always shifts the resonant frequencies down. This is not surprising, though, but we must always keep in mind that in order to solve equation 2 for **Q** the mass density must be known.

Based on solution of the inverse problem defined by equation 2, we can distinguish two application approaches :

- **Stiffness determination:** The mass of the layer is known or it is negligible. In most cases, we cannot neglect the mass, especially if we are working with high frequencies. However, the mass density of the layer is not easy to measure when we are dealing with submicron thin films. Fortunately, the mass density of the layer can be usually estimated from bulk density measurement.
- **Mass determination:** The stiffness of the layer is known or negligible. Stiffness can be neglected, only if the material of the substrate is incomparably stiffer than the material of the layer. If deal with dust or paint particles and polymer layers we can assume $\mathbf{Q}\to\mathbf{0}$ and calculate surface mass density from equation 2.

Both approaches are verified experimentally later in this paper.

## 3. Experimental setup

As it was described earlier, the key part of the RUS method is the measurement of the the spectra of the free-vibrating specimen. In order to fulfill the free boundary condition, the specimen was supported on one side by a thin optical fiber and the opposite corner was laid on a piezoelectric film (PVDF) as shown in figure 1(b). The mechanical excitation of the sample was made by the piezoelectric membrane though this corner. For point detection of the dynamical response the Microsystem Analyser Polytec MSA-500 with the laser interferometer embedded into the optical microscope was used. This Doppler vibrometer was equipped by a piezo scanner, able to measure vibration at different points of the smooth surface of the sample. By repeating excitation and vibration measurement we can get resonant spectra at each scanned point of a predetermined mesh on the specimen surface and compose the shape of vibrational eigenmodes.

Samples were plate-like parallelepipeds with the measured layer on one of their faces. Typically, their dimensions were 4.5x3.5x0.3 $mm^3$. Substrates discussed in this paper were made of silicon single crystals that exhibit very good resonant properties. From experimental point of view there were several limitations which must be considered in order to have good measurements:

- **Damping:** Both the substrate and the layer should have as low damping as possible. High damping attenuates the free vibration and makes the spectral peaks less visible.
- **Thickness:** Dimensions and geometry of both the substrate and the layer should be known, most important is the reliable determination of the layer thickness.
- **Reflectivity:** The sample must have (before and after deposition) at least one reflective surface in order to use the reflected beam for laser interferometric vibration measurement.
- **Damage:** Any damage that the sample undergoes during manipulation changes the resonant spectra and hampers the measurement. Some materials get change properties during layer deposition as well (for example if it is done at high temperature or aggressive ambiance).

## 4. Measurements

### *4.1. Stiffness measurements*

were performed on diamond-like-carbon (DLC) layer on Si substrate (DLC prepared by physical vapor deposition). Silicon substrate was measured before layer deposition and then remeasured. Thickness of the layer was 375 nm (determined by AFM step) and the mass density was taken 3.5 $g/cm^3$ from bulk measurements. Layer was considered isotropic, the resulting elastic constants were: Young's modulus $E = 581 \pm 95$ GPa and bulk modulus $K = 418 \pm 547$ GPa (the insensitivity of the resonant spectra to this coefficient is a common feature of the RUS measurement).

### *4.2. Mass measurements*

were tested on different kinds of commercial undercoat paint sprays, were deposited upon the silicon substrates. Simultaneously, the mass of the paint was measured on static microbalance (accuracy 1 mg) using 64 $cm^2$ piece of paper with and without spray to determine change of the mass due to paint. In this way, mass measured by static and dynamic means could be independently compared and dynamic experiments verified. Static surface density was determined as $\rho h$=16.3 $\pm$ 0.33 $\mu g/mm^2$ while the dynamic method gave $\rho h$=16.2 $\pm$ 0.39 $\mu g/mm^2$. The paint thickness was unknown, from density considerations it was estimated around 5 microns.

In figure 2, the good agreement of the linear model given by equation 2 with the measured data can be seen.

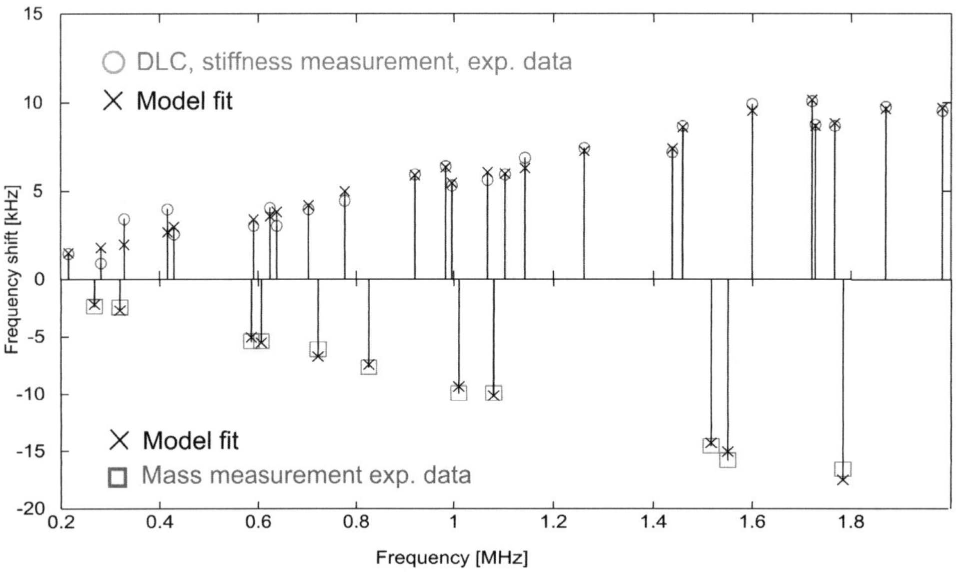


**Figure 2.** Comparison of the frequency shifts measured and fitted by linear prediction model for one DLC thin layer (green) and one spray mass measurement (red).

## 5. Conclusions

Both the stiffness and the mass measurements show good accordance with the proposed linear inverse model. Moreover, the mass measurement was validated independently by static measurement using the microbalance. The main advantage of the proposed inverse modeling is that the relations between the properties of the layer and the frequency shifts are represented by linear equations, which can be easily solved and enable us to estimate the accuracy of the measurement. There are obviously some experimental constrains. Especially, the layer thickness must be known for the stiffness measurement (for the mass measurement it is not necessary).

## Acknowledgements

The authors would like to express their thanks to dr.Tomáš Kocourek from the Institute of Physics of the Academy of Sciences of the Czech Republic for the preparation of DLC layers. This work was financially supported by the Czech Science Foundation (project No.101/09/0702) and by the research program of IT ASCR (CEZ:AV0Z20760514).

## References

[1] Migliori A and Sarrao J L 1997 *Resonant Ultrasound Spectroscopy* (New York: J.Wiley and Sons)

[2] Nakamura N, Ogi H, Nitta H, Tanei H, Fujii M, Yasui T and Hirao M 2006 Study of Elastic Anisotropy of Cu Thin Films by Resonant-Ultrasound Spectroscopy Coupled with Laser-Doppler Interferometry and Pump-Probe Photoacoustics *Japan J Appl Phys* **45** 4580-4584

[3] Nakamura N, Ogi H, Hirao M and Ono T 2005 Determination of anisotropic elastic constants of superlattice thin films by resonant-ultrasound spectroscopy *J Appl Phys* 2005 **97** 013532

[4] Růžek M, Seiner H, Sedlák P, Kruisová A and Landa M, Linearized Forward and Inverse Problem of the Resonant ULtrasound Spectroscopy for the Evaluation of Thin Surface Layers, accepted in *J Acoust Soc Am*

# Elastic characterization of Au thin films utilizing laser induced acoustic Rayleigh waves

**A Haim[1, 2], S Bar-Ad[2] and A Azoulay[1]**

1. Ultrasonic Section, NDT Department, Soreq - Nuclear Research Center, Yavne 81800, Israel.
2. School of Physics and Astronomy, Faculty of Exact Sciences, Tel-Aviv University, Tel-Aviv 69978, Israel

E-mail: arbelhai@gmail.com

**Abstract**. Wide frequency-band Rayleigh waves (~100 MHz) were utilized to characterize the elastic constants of thin Au/Cr films deposited on glass substrates. The Rayleigh waves were excited utilizing laser induced thermoelastic mechanism and detected using a knife-edge technique apparatus. The dispersion of the signals in glass substrates coated with Au/Cr was measured and fitted to theory using a non-linear regression algorithm. From the fitting, the Au films Young modulus and the film thickness were extracted. The results were analyzed with regards to AFM scans performed on the samples and independent thickness measurement done by a dektak$^3$ profiler. Results show a good agreement between the two measurements.

## 1. Introduction

Thin Au films are widely used throughout the industry in areas such as microelectronics and nanotechnology. Characterizing the elastic and mechanical properties of such films is necessary in order to optimize their fabrication processes as well as for quality-assurance. These properties have been shown to differ significantly from those of Au bulk samples [1-4]. Measurements of wide-band laser-generated ultrasonic Rayleigh waves [5] traveling along the surface are well suited for the extraction of elastic and mechanical parameters of thin coatings [6-8]. The inherent advantages of this technique over alternatives methods such as nano-indentation [1, 9] and tensile-testing [3, 10] include its being non-destructive, non-contact and require no film preparation.

The samples studied were polished glass substrates predeposited with Cr to assure good adhesion for the deposition of the Au films. Deposition was accomplished utilizing a sputtering system. Two groups consisting of five samples each were prepared, differing in either their Cr or Au thickness. The first group of samples consisted of a Cr layer with varying thicknesses ranging between 8 nm and 100 nm, and an Au layer of constant thickness of about 1.3 $\mu$m. The second group of samples consisted of a Cr layer of constant thickness of about 90 nm, and a varying Au film thickness ranging from 140 nm to 1010 nm. The films thicknesses were controlled by the sputtering time and measured using a Dektak$^3$ profiler.

The dispersion curves of surface waves traveling along the samples surface were measured for both groups of samples utilizing a laser-ultrasonic system. The elastic constants of the Au deposited layers were extracted from these dispersion curves. Additional AFM (Atomic Force Microscope) measurements were taken for all samples in order to analyze surface structure and grain size.

Published under licence by IOP Publishing Ltd

## 2. The experimental setup

Figure 1 depicts the laser-ultrasonic system used to generate and detect Rayleigh waves. The generation laser was an Nd:YAG Q-switch laser with a pulse width of about 7 ns operated in the first harmonic (1064 nm). The laser beam was focused using a cylindrical lens and a spherical doublet lens to form a narrow line-shaped pattern with dimensions of 7 mm × 20 $\mu$m. The total laser energy deposited on the sample was about 1mJ per pulse to assure thermoelastic generation [11]. Mirror M3 along with the spherical and doublet lenses were mounted on a translation stage, thereby providing control of the distance between the generation and detection of the Rayleigh waves.

The detection of the acoustic waves was done by a knife-edge technique [12, 13]. The detection laser was an Nd:YAG CW laser operated in its second harmonic (532 nm) with an output power of 350 mW. The detection beam was focused onto the sample using a spherical doublet lens having a focal length of 70 mm. The same lens was used to collect the reflected beam from the sample. The reflected beam was then diverted by the polarizing-beam-splitter where it was focused onto a fast PIN photodiode detector, while half of the beam is blocked by the knife-edge. When the acoustic wave reaches the detection spot, the reflected detection beam angle deviates to a certain extent which makes the beam power reaching the photodiode deviate as well, thereby measuring the acoustic signal. The output signals were transferred to a 250 kHz high-pass filter which eliminates the DC component, rectified using a wide-band preamplifier, and then captured using a digital oscilloscope and recorded to a PC.

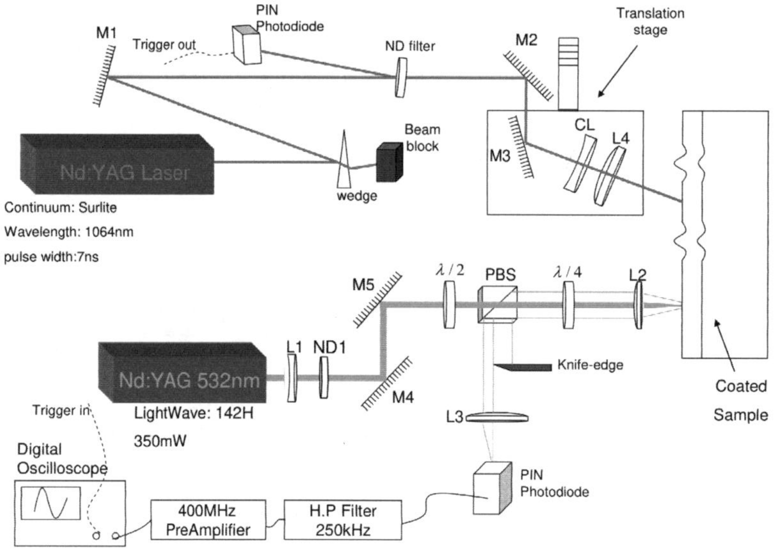


**Figure 1**. The experimental setup.

## 3. Results & discussion

Figure 2 shows an example of the measured signals and the corresponding dispersion curve from a 1.3 $\mu$m Au layer on a 8 nm Cr layer sample. The two signals in figure 2.a. were measured at different propagation distances and the dispersive nature of the signals is apparent. Figure 2.b shows the resulting dispersion curve calculated from cross-correlation [14] of the two signals, and the theoretical fitted dispersion curve obtained by a nonlinear least-square fitting algorithm. From the fitting, the Au film thickness and elastic modulus as well as the substrate elastic modulus could be obtained. Other parameters such as the film density and Poisson ratio were assumed known and taken to be those of bulk values. The Young modulus of the glass substrate was further measured by conventional ultrasonic pulse-echo technique and was found to be in excellent agreement to the one fitted.

Table 1 contains the values of the fitted parameters of the first group of samples (i.e. with different Cr layer thickness) and the Au film thickness measured using the profiler. The agreement between the directly measured and fitted Au thicknesses seems to be quite good. The results also show that the Au Young modulus was consistent between 101 GPa and 108 GPa with no apparent trend. These values are somewhat higher than the Au bulk Young modulus which is 79 GPa, however those values are supported by other researches in the literature [1], and they might be explained by a <111> texture, common in Au thin films [1, 3, 15, 16].

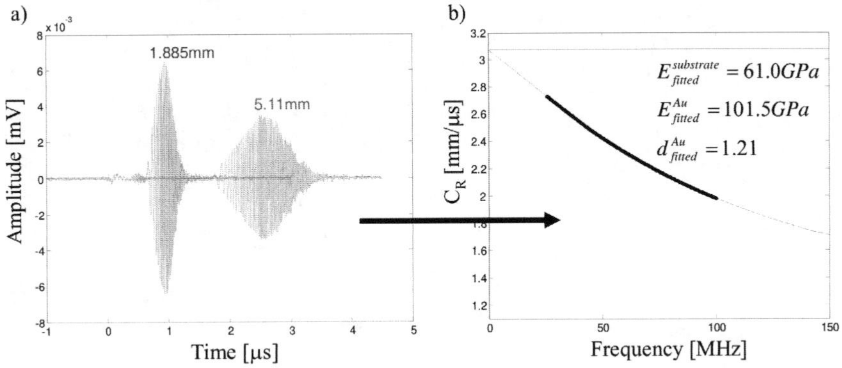


**Figure 2**. (a). Two dispersive Rayleigh wave signals measured after different propagation distances. (b). Resulting measured dispersion curve, and its fitted curve from the theoretical model.

**Table 1.** Fitted parameters (Au film thickness, Young modulus, and substrate Young modulus) and measured Au thickness vs. thickness of Cr adhesion layer

| $d^{Cr}\left[\text{nm}\right]^{a}$ | $d^{Au}_{Measured}\left[\mu\text{m}\right]^{b}$ | $d^{Au}_{Fitted}\left[\mu\text{m}\right]$ | $E^{Au}_{Fitted}\left[\text{GPa}\right]$ | $E^{Substrate}_{Fitted}\left[\text{GPa}\right]$ |
|---|---|---|---|---|
| 8 | 1.30 | 1.21 | **101.5** | 61.0 |
| 17 | 1.23 | 1.18 | **107.6** | 61.5 |
| 33 | 1.21 | 1.26 | **105.4** | 61.0 |
| 67 | 1.34 | 1.29 | **102.7** | 61.4 |
| 100 | 1.28 | 1.21 | **104.3** | 61.3 |

[a] Cr thickness was estimated from sputtering time.
[b] Measured using a Dektak$^3$ profiler

AFM measurements performed on the Au film surface revealed that the samples having different Cr thickness also had different Au film grain size without an apparent trend. This observation along with the stated results of Table 1 may indicate that the Au film Young modulus does not depend on its characteristic grain size.

Figure 3 shows the measured dispersion curves of the second group of samples, having different Au thickness and a constant Cr thickness of about 90 nm. As theory predicts, all the curves approach the substrate Rayleigh wave velocity at low frequency – $c_R^{glass} = 3.15$ mm/$\mu$s. Since the examined Au coatings were relatively thin, only one parameter could be extracted from the fitting algorithm. Therefore the film thickness was measured directly using the profiler and inserted into the fitting algorithm from which the Young modulus could be obtained. The relatively large variance in the thickness measurements lead to a large relative error in the films Young modulus, and therefore only the thicker films ($d \geq 590$ nm) Young modulus could be obtained with a reasonable accuracy (see figure 4). AFM measurements were performed for these samples as well and the results for the Au grain size are presented in figure 5. It can be seen that the characteristic grain size increases with increasing Au film thickness.

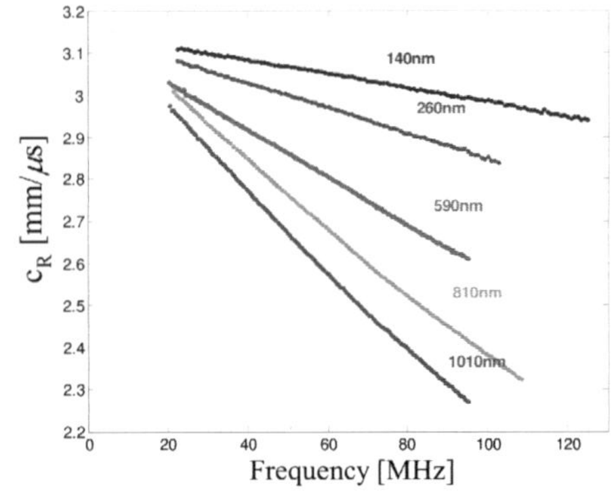


**Figure 3.** Measured dispersion curves for samples having different Au film thickness. All the samples had a constant Cr thickness of 90 nm

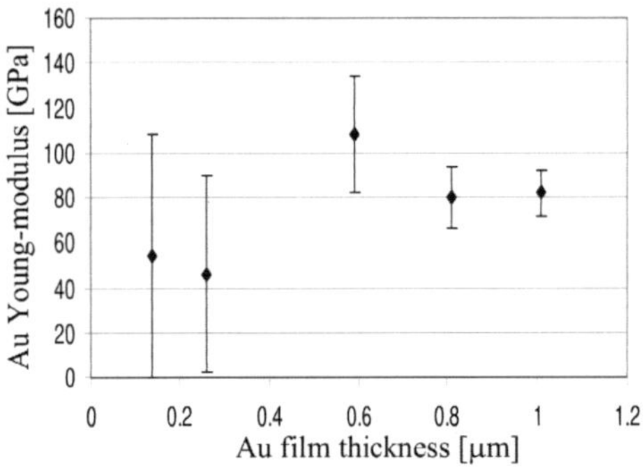


**Figure 4**. Fitted Young modulus vs. Au film thickness

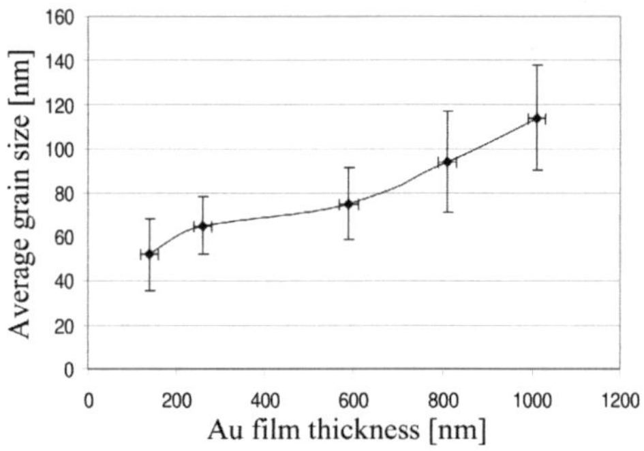


**Figure 5**. Grain size vs. Au film thickness

## References

[1] Cao Y, Allameh S, Nankivil D, Sethiaraj S, Otiti T, and Soboyejo W, 2006 *Mater. Sci. Engng. A* **427** 232-40.
[2] Baek C W, Kim Y K, Ahn Y, and Kim Y H, 2005 *Sensors and Actuators A* **117** 17-27.
[3] Varguez P, Aviles F, and Oliva A I, 2008 *Surf. Coat. Technol.* **202** 1556-63.
[4] Espinosa H D, and Prorok B C, 2003 *J. Mater. Sci.* **38** 4125-28.
[5] Victorov I A 1967 *Rayleigh and Lamb waves* (New-York: Plenum Press).
[6] Schneider D, and Schwarz T, 1997 *Surf. Coat. Technol.* **91** 136-46.
[7] Neubrand A, and Hess P, 1992 *J. Appl. Phys.* **71** 227.
[8] Lehmann G, Hess P, Weissmantel S, Reisse G, Scheible P, and Lunk A, 2002 *Appl. Phys. A* **74** 41.
[9] Xu Z H, and Rowcliffe D, 2002 *Surf. Coat. Technol.* **157** 231-37.
[10] Neugebauer C A, 1960 *J. Appl. Phys.* **31** 1096.
[11] Scruby C B and Drain L E 1990 *Laser ultrasonics* (Bristol: Adam Hilger)
[12] Noui L, and Dewhurst R J, 1993 *Ultrasonics* **31** 425-32.
[13] Monchalin J P, 1986 *IEEE Trans. Ultr. Ferroelectrics, Freq. Control UFFC-33* **5** 485-99.
[14] Aussel J D, and Monchalin J P, 1989 *Ultrasonics* **27** 165-77.
[15] Kalkman A J, Verbruggen A H, and Janssen G, 2001 *Appl. Phys. Lett.* **78** 2673.
[16] Faurie D, Castelnau O, Renault P O, Patriarche G, Brenner R, Le Bourhis E, and Goudeau P, 2006 *Surf. Coat. Technol.* **201** 4300-4

# Photoacoustic measurements on thermal properties of hydrogenated amorphous carbon films: the effect of hydrogen dilution

**Y Y Kim[1], H A Alwi[2], R Awang[2] and S Krishnaswamy[1]**

[1]Department of Mechanical Engineering, Northwestern University, Evanston IL 60208
[2]School of Applied Physics, Universiti Kebangsaan Malaysia, 43600 Bangi, Selangor, Malaysia

E-mail: s-krishnaswamy@northwestern.edu

**Abstract**. Thermal properties of hydrogenated amorphous carbon (*a*-C:H) thin-films are measured using an ultrafast optical pump-probe technique. The *a*-C:H samples were grown in a home-built direct-current (DC) plasma enhanced chemical vapor deposition (PECVD) system with varying hydrogen ($H_2$) diluents to methane ($CH_4$) flow-rate ratios. Thermal diffusivities of samples are extracted by comparing thermoreflectance measurements with numerical calculations so that thermal conductivities ($\kappa$) can be determined. Although the dependence of thermal property on $H_2$ dilution was not significant, our films show lower $\kappa$ (0.10-0.15 W/mK ± 20%) compared to the results of previous studies.

## 1. Introduction

Hydrogenated amorphous carbon (*a*-C:H) is a member of diamond-like carbon (DLC) materials that have mechanical and thermal properties suitable for applications in microelectromechanical systems (MEMS) and nanoscale devices such as thermal actuators, micro-coolers, piezoresistive sensors, and so on.[1] To design such structures it is crucial to understand their thermal characteristics especially when they are subject to large heat loads or frequent temperature changes. While previous studies [2-4] show that thermal conductivities of certain *a*-C:H thin-films are in the range of 0.20-1.4 W/mK, these properties are known to be dependent on the fabrication process that results in different configuration of atomic structures such as crystallinity, hydrogen content in the film and $sp^2$-to-$sp^3$ bonding ratios. For this reason investigations on the thermophysics of the carbon material have been vigorously performed for various regimes of *a*-C:H.

In the present study the pump-probe technique was employed to determine thermal properties of *a*-C:H thin-films grown by plasma enhanced chemical vapor deposition (PECVD) with varying hydrogen ($H_2$) diluents to methane ($CH_4$) flow-rate ratios. The ultrafast optical pump-probe technique provides a contact-free and nondestructive way to evaluate mechanical and thermal characteristics of multilayers in nanometer-scale.[5,6] The measured thermoreflectance signals were analyzed using a computational model so that thermal diffusivities of films were yielded to see if the different diluents ratio affects the material properties.

## 2. Experimental

The *a*-C:H samples were deposited on single crystal silicon (Si) wafer substrates in a home-built

**Table 1.** Film characteristics.

| Sample # | Diluents-to-methane ratio | Film Thickness (nm) | Hydrogen content (%) | $sp^2$ carbon content (%) |
|---|---|---|---|---|
| **1** | 1:1 | 386 | 8.0 | 23.8 |
| **2** | 1:2 | 470 | 4.0 | 23.8 |
| **3** | 1:3 | 490 | 3.0 | 17.7 |
| **4** | 1:4 | 560 | 5.3 | 26.1 |

direct-current (DC) PECVD system. The deposition time was fixed at 4 hours with a constant DC power of 6.7 W. The set consists of films deposited from the discharge of pure $CH_4$ diluted with $H_2$ such that the $H_2$-to-$CH_4$ flow-rate ratios varied from 1:1 to 1:4. The deposition temperature, pressure and $CH_4$ flow-rate were fixed at 100 °C, 0.2 mbar and 20 sccm, respectively.

The film thickness and Tauc gap, $E_g$ were measured from the optical transmission spectra of the films obtained using a Jasco UV-Vis-NIR 3102-PC double beam spectrophotometer scanned within a scanning range of 190 to 2500 nm. The spectra were fitted using MATLAB 7.0 developed by Suganthi *et al.*[7] Fourier Transform Infrared (FTIR) transmission spectra were done with Perkin Elmer 2000 FTIR spectrometer used in transmission mode in the range 400 to 4000 $cm^{-1}$ to determine the bonded hydrogen content (H%) of the films and to identify the $sp^2$ hybridization of C and CH bonds as well as other possible bonds that could appear due to the presence of hydrogen. The film characteristics are summarized in table 1.

After the sample fabrication, surfaces were coated with a thin layer (~30 nm) of aluminum in an e-beam evaporator (Edwards Auto 500) for the picosecond ultrasonics measurements. The DC voltage applied to a tungsten filament is 4.7 KV, and the current level was set to 25 mA to obtain a deposition rate of 0.15 nm/s. During the process the chamber pressure was maintained at $3\times10^{-6}$ Torr.

The time-resolved thermoreflectance signals were measured using the ultrafast optical pump-probe technique.[6] A femtosecond laser system (Millenia Pro and Tsunami, Spectra-physics) lasing at the beam wavelength of 780 nm and the full width at half maximum (FWHM) pulse duration of 120 fs was employed. The pulse repetition rate was 80 MHz. It was then split into the pump-beam, which is modulated at 100 kHz by an acousto-optic modulator (AOM, IntraAction Corp. AOM-405) to heat up the sample, and the probe-beam to monitor the reflectance change induced by the temperature rise. The intensity ratio of the two beams divided by a polarized-beamsplitter and a half-wave plate was 10 to 1. The transient response of thermoreflectance at a specific time-delay with respect to the application of pump-pulse was detected using a computer-controlled delay-line on which a retroreflector was installed. The maximum time-delay of the motorized stage was 1 ns, and the temporal resolution was set to 4 ps/pts in the present study. The probe-pulse was reflected by the Al-coated *a*-C:H sample mounted on an *xyz*-stage, and then collected by a photodetector (Thorlabs PDA36A) connected to a lock-in amplifier (Stanford Research SR803).

## 3. Data analysis

Curve-fittings have been performed with the measurement data to derive the thermal diffusivity ($\alpha$) following the model developed by Paddock and Eesley.[8] The one-dimensional heat conduction equation in a single-layer sample is:

$$\frac{\partial T(x,t)}{\partial t} = \alpha \frac{\partial^2 T(x,t)}{\partial x^2} + \frac{I(1-R)\beta e^{-\beta x} e^{-(t/\tau)^2}}{C} \tag{1}$$

where $T$ is the sample temperature, $x$ is the distance normal to the sample surface, $t$ is the time-delay with respect to the pump-pulse application, $C$ is the heat capacity per unit volume, $R$ is the reflectivity,

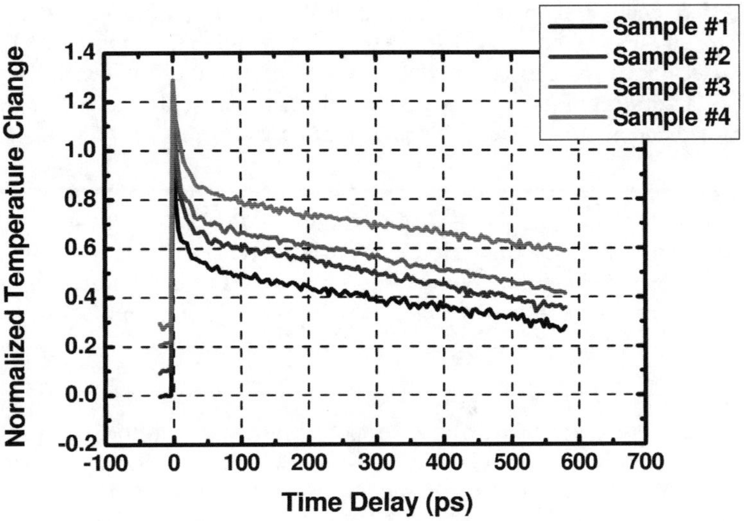


**Figure 1.** (Color online) The thermoreflectance measurement results on *a*-C:H samples.

$I$ is the pump-intensity, $\beta$ is the absorptivity per unit length, and $\tau$ is the pump-pulse width of a Gaussian shape. Therefore, the shape of normalized solution depends on $\alpha$ of the *a*-C:H film and the absorptivity of Al-coat ($\beta$). In this study the best fitting is obtained with $\beta$=105 μm$^{-1}$, which is close to the theoretical value for Al.

## 4. Results and discussion

Figure 1 shows the optical pump-probe reflectivity data from the four *a*-C:H samples. It is observed that the heat from the pump-pulse takes 9 ps approximately ( $\lambda = \sqrt{Dt}$ , where $\lambda$ is the thermal diffusion length (30 nm) and $D$ is the thermal diffusivity of Al ($9.75\times10^{-5}$ m$^2$/s)) to diffuse into the Al layer, and it propagates in the carbon material afterwards. Based upon the aforementioned methods, thermal diffusivities are extracted and thermal conductivities ($\kappa$) are also calculated using the relationship $\kappa = \alpha C$. The density of film and specific heat capacity were taken from representative values of 2.4 g/cm$^3$[3] and 1.04 J/gK,[4] respectively. Results are summarized in table 2.

While the results indicate overall thermal conductivities of 0.10-0.15 W/mK within 20% uncertainty, no significant dependence on the dilution ratio are observed. It can be explained by the general agreement that the H content has minor influence on the mechanical and thermal properties of the film,[3,4] but it is also because of the small variation of hydrogen content (3.0~8.0 %) and *sp*$^2$ carbon content (17.7~26.1 %), which do not change linearly with the flow-rate ratio. It is interesting,

**Table 2.** Thermophysical quantities of the samples studied.

| Sample # | $\alpha$ (mm$^2$/s) | $\kappa$ (W/mK) |
|---|---|---|
| **1** | 0.04 | 0.10 |
| **2** | 0.06 | 0.15 |
| **3** | 0.06 | 0.15 |
| **4** | 0.05 | 0.12 |

however, that our films show relatively lower $\kappa$ compared to those of the previous investigations. For example, the lowest conductivities in Bullen *et al.*[2] and Shamsa *et al.*[3] are 0.20 W/mK for their remote-plasma chemical vapor deposition (RPCVD)-grown sample and polymer-like *a*-C:H (PLCH), respectively. Considering the increasing trend of $\kappa$ with respect to the film density[2-4], it is inferred that the current fabrication technique produces softer films, which will be clarified by photoacoustic characterization (broadband guided-wave technique[5]) in near future.

## 5. Conclusion

Thermal properties ($\alpha$ and $\kappa$) of *a*-C:H samples prepared by DC PECVD with varying $H_2$-to-$CH_4$ flow-rate ratios are determined from the thermoreflectance measurements. Although no obvious dependence of thermophysical characteristics on the flow-rate ratio is observed, our film-growth method results in relatively lower thermal conductivities (0.10-0.15 W/mK ± 20%) compared to those of the existing studies.

**Acknowledgments**
HAA thanks MOSTI, Malaysia for the financial support during the visit to Northwestern University, and MOHE, Malaysia for research grant UKM-ST-07-FRGS0020-2010. RA thanks UKM for research grant UKM-GGPM-NBT-092-2010. The samples were fabricated during RA's Ph.D. study with Prof. S. A. Rahman at University of Malaya. SK thanks NSF for support through grant OISE-0730259.

## References

[1] Luo J K, Fu Y Q, Le H R, Williams J A, Spearing S M and Milne W I 2007 *J. Micromech. Microeng.* **17** S147-S163
[2] Bullen A J, O'Hara K E, Cahill D G, Monteiro O and Keudell A V 2000 *J. Appl. Phys.* **88** 6317-6320
[3] Shamsa M, Liu W L, Balandin A A, Casiraghi C, Milne W I and Ferrari A C 2006 *Appl. Phys. Lett.* **89** 161921
[4] Arlein J L, Palaich S E M, Daly B C, Subramonium P and Antonelli G A 2008 *J. Appl. Phys.* **104** 033508
[5] Zhang F F, Krishnaswamy S and Lilley C M 2006 *Review of Progress in Quantitative Nondestructive Evaluation* vols 25A and 25B, ed D O Thompson and D E Chimenti (New York: American Institute of Physics) pp 1098-1104
[6] Kim Y Y, Alwi H A, Huang Q, Abd-Shukor R, Tsai C F, Wang H, Kim K W, Naugle D G and Krishnaswamy S 2010 *Physica C* **470** 365-368
[7] Suganthi D, Ruzela T, and Laila H M I 2004 *Determination of optical energy gap for semiconducting film* UiTM Research Report
[8] Paddock C A and Eesley G L 1986 *J. Appl. Phys.* **60** 285-290

# Measurement of texture in steel by laser-ultrasonic surface waves

**D Lévesque[1], C S Lim[2], C Padioleau[1] and A Blouin[1]**

[1]Industrial Materials Institute, National Research Council Canada, Boucherville, Quebec, Canada
[2]Instrumentation & Control Research Group, POSCO Technical Research Laboratories, Pohang City, Gyeongbuk, Korea

Email: Daniel.Levesque@cnrc-nrc.gc.ca

**Abstract.** A laser-ultrasonic approach based on the propagation of surface skimming longitudinal wave (P-wave) and Rayleigh wave (SAW) is investigated for the measurement of texture in moderately thick steel plates. The angular dependence of the P-wave and SAW velocities presents different patterns, both of amplitude of a few % attributed to texture. Moreover, the difference of the P-wave and SAW profiles is shown to be very robust against small path length changes or other detrimental effects. For measurements on large plates, a solution based on the rotation of a generation line using an axicon lens is tested. Texture variations across the width of steel plates and the influence of residual stress are discussed.

## 1. Introduction

In the continuous production line for hot-rolled steel sheet, thick input material is transformed to much thinner steel sheet in the form of coil by several processes. The stresses generated during these processes are usually transformed into energy for shaping the steel strip. But a small amount of stress usually remains in the steel sheet after the whole processing. Also, crystallographic texture change is expected to occur for similar reasons. Therefore, the control of texture and reduction of residual stress are important and necessary for high quality steel product.

Laser ultrasonics can be a good candidate for such measurements. Successful results were obtained in recent years for residual stress in aluminum using an approach based on the laser generated surface skimming longitudinal wave (LSSLW or P-wave) [1-3]. For steel, the effect of stress on the ultrasonic velocity is small, with an acoustoelastic coefficient of about -0.1 % / 100 MPa for the P-wave, i.e. four times less than in aluminum. Also, this coefficient is only about -0.01 % / 100 MPa for the Rayleigh wave (SAW).

In this paper, a laser-ultrasonic approach based on the propagation of P-wave and SAW is investigated for the measurement of texture in steel plates. The angular dependence of the velocity of these modes presents different patterns of amplitude of a few % attributed to texture. Also, the difference of the P-wave and SAW profiles is found to be very robust against small path length changes. For measurements on large plates, a solution based on the rotation of a generation line is tested. Texture variations across the width of steel plates and the influence of residual stress are discussed.

## 2. Laser ultrasonic P-wave approach

A laser-ultrasonic setup was assembled for precise velocity measurements of P-wave and SAW. A pulsed Nd:YAG laser is employed to simultaneously generate the ultrasonic modes. For the ultrasonic detection, a long pulse Nd:YAG laser is coupled to a photorefractive interferometer by optical fibers. A line-source and line-detection configuration is used to minimize wave spreading and microstructure noise. Also, a scanning system is mounted to allow measurements for different separation distances between the generation and detection locations. The setup further includes a rotation stage for the analysis of the angular dependence of velocity variations due to both texture and stresses.

For signal processing, a cross-correlation method is used between the measured signal and a reference signal for each separation distance between the generation and detection. One has the simple relation between velocity and time delay variations:

$$\frac{\Delta V}{V^0} = \frac{\Delta d}{d^0} - \frac{\Delta t}{t^0} \tag{1}$$

where $V$ is the velocity, $t$ is the time delay and $d$ is the path length. Since variations are small, the index 0 (for zero stress) is unnecessary and if the change in path length (e.g. separation distance between generation and detection line) can be neglected, one has the simple relation:

$$\frac{\Delta V}{V} \approx -\frac{\Delta t}{t} \tag{2}$$

One improvement is to use the P-wave and SAW in combination to correct for small changes in the path length with respect to equation 1 and possibly other effects. Taking the difference between velocity variations of the two modes results in the relation:

$$\frac{\Delta V_P}{V_P} - \frac{\Delta V_{SAW}}{V_{SAW}} = -\left( \frac{\Delta t_P}{t_P} - \frac{\Delta t_{SAW}}{t_{SAW}} \right) \tag{3}$$

where the subscript denotes the wave mode. In this manner, the path length change is numerically cancelled out.

Using the rotation stage, the angular dependence of the velocity or time delay in a sample is a combined effect of texture and stress [2]. The equations for P-wave and SAW with properties for steel are given by:

$$\begin{aligned} (\Delta V/V)_P(\theta) &= -10.2 10^{-6}\,\sigma_m - 1.12\,W_{400} - 12.8 10^{-6}\,\sigma_d \cos(2\theta) \\ &\quad + 2.35\,W_{420}\cos(2\theta) + 3.11 W_{440}\cos(4\theta) \\ (\Delta V/V)_{SAW}(\theta) &= -1.2 10^{-6}\,\sigma_m + 2.2\,W_{400} - 3.2 10^{-6}\,\sigma_d \cos(2\theta) \\ &\quad - 4.17 W_{420}\cos(2\theta) + 0.74\,W_{440}\cos(4\theta) \end{aligned} \tag{4}$$

where $\theta$ is the angle with respect to rolling direction, $W_{400}$, $W_{420}$ and $W_{440}$ are the texture coefficients in the Roe's notation, $\sigma_m = (\sigma_0+\sigma_{90})/2$ is the mean stress and $\sigma_d = (\sigma_0-\sigma_{90})/2$ is the difference (in MPa) between the rolling (0 deg) and transverse (90 deg) directions. With a reference signal at 0 deg in the cross-correlation, one should subtract terms corresponding to 0 deg in equation 4. When using a reference with rotation at a different location, the variation in each physical quantity with respect to the reference is obtained. Also for the P-wave / SAW combination, the expression for the difference is easily derived.

## 3. Measurement results

A series of measurements was performed using the rotation stage on samples cut across the width of a large steel plate. It is expected that the stress should be actually zero due to the small sample dimensions. Also, measurements are made after slight surface polishing. First results showed velocity variations clearly oscillating in a different manner for P-wave and SAW, with an amplitude of more than 1 % attributed to texture, much larger than the 0.1 % variation expected for a 100 MPa stress. The different profile shapes for P-wave and SAW are due to the opposite sign for the contribution of $W_{420}$ in equation 4.

Figure 1 shows the results (time delay) from simultaneously fitting the profiles of P-wave and SAW using equation 4 with parameters $W_{420}$, $W_{440}$ and a fixed separation distance of 14 mm. Notice that the reference is the signal in the rolling direction at 0 deg. The fit is quite good except for some distortion, with texture coefficients $W_{420}$ = -3.56 x $10^{-3}$ and $W_{440}$ = 3.37 x $10^{-3}$ in the valid range, with stress values forced to zero. Figure 2 shows the results of fitting the difference profile of the P-wave / SAW combination for samples at the edge and at the center. For the edge, a very good fit is obtained with texture coefficients $W_{420}$ = -3.65 x $10^{-3}$ and $W_{440}$ = 2.91 x $10^{-3}$ close to the previous values obtained. For the center, a good fit is also obtained with smaller texture coefficients $W_{420}$ = -1.46 x $10^{-3}$ and $W_{440}$ = 1.80 x $10^{-3}$. Such difference profiles clearly show less scattering than separated P-wave and SAW profiles. This is especially the case when making measurements with a rough, as received, surface finish.

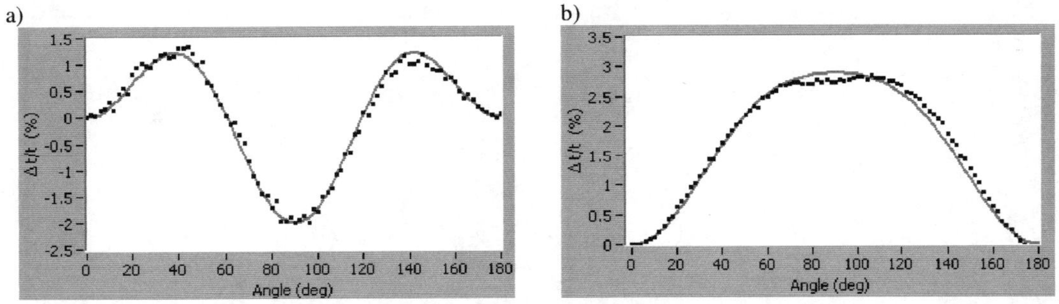


**Figure 1.** Simultaneous fit of the profiles for a) P-wave and b) SAW on a sample near the edge.

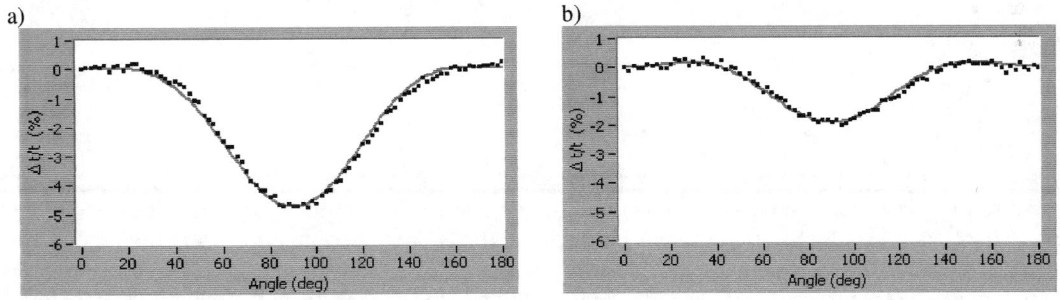


**Figure 2.** Fit of the profile of the P-wave / SAW combination for a sample a) at the edge and b) at the center.

Another analysis is to use the center as reference location and apply cross-correlation for the P-wave and SAW with corresponding signals at each angle. With this reference choice, the contribution of $W_{400}$ can be obtained as well from fit. The difference profiles of the P-wave / SAW combination for samples at both edges were obtained. For the same edge as before, a very good fit is obtained with texture coefficient variations $\Delta W_{400}$ = -0.36 x $10^{-3}$, $\Delta W_{420}$ = -2.26 x $10^{-3}$ and $\Delta W_{440}$ = 0.87 x $10^{-3}$. The changes in texture coefficients $\Delta W_{420}$, $\Delta W_{440}$ are quite consistent with the distinct values obtained from the two samples in figure 2. For the other edge, the good fit is obtained with almost the same texture coefficient $\Delta W_{400}$ = -0.35 x $10^{-3}$, but smaller values of the others, $\Delta W_{420}$ = -1.34 x $10^{-3}$ and $\Delta W_{440}$ = 0.79 x $10^{-3}$. These results indicate an asymmetry across the width of the strip.

## 4. Measurements on large samples

Using the P-wave approach on large hot-rolled samples, an optical setup was developed to perform rotation of the generation line, with a point detection at the center of the circle formed. One solution is to use an axicon lens to map the laser spot onto a circle and a mask to only have an arc rotating on the sample surface. Two large hot-rolled samples of dimensions sufficient to sustain residual stresses across the width were prepared by POSCO. From strain gauge measurements, it was found that one sample had residual stresses reaching -80 MPa (in compression) near the edges, compared to -20 MPa for the other sample.

Figure 3a shows a profile and fit of the P-wave / SAW combination near the edge on the sample with largest residual stresses, using an axicon lens of 16 mm radius and the reference signal at 0 deg. The profile shape is nearly symmetric with time delay variations of about 6 %. Such large variations should be attributed to texture. To reduce the effect of texture and possibly make more emphasis on stress, a strategy discussed before is to use a reference location far enough from the edge with almost no stress. Figure 3b shows the profile of the P-wave / SAW combination using a reference location at 175 mm from the edge. The profile is nearly symmetric, but large variations are still observed. Therefore, taking a reference location far enough from the edge does not cancel out the texture effect. According to the fit, variations of the texture coefficients between the two locations are $\Delta W_{400} = 4.00 \times 10^{-3}$, $\Delta W_{420} = -3.06 \times 10^{-3}$ and $\Delta W_{440} = 0.56 \times 10^{-3}$.

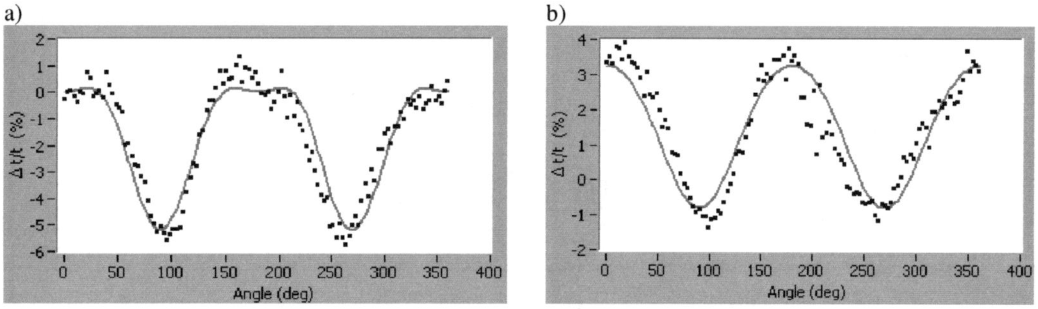


**Figure 3.** Profile of the P-wave / SAW combination near the edge on a large sample a) with the reference at 0 deg and b) using a reference location at 175 mm from the edge.

## 5. Conclusion

A laser-ultrasonic approach based on the propagation surface waves was studied for the measurement of texture in steel plates. The difference of the P-wave and SAW profiles is shown to be very robust against small path length changes or other detrimental effects. For actual measurement on large plates, a solution based on the rotation of a generation line using an axicon lens was tested. Texture variations across the width of a plate are obtained. Taking a reference far enough from the edge does not cancel out the texture effect to have variations mostly due to stress. This texture change could prevent measuring residual stress in hot-rolled steel strip with ultrasound. However, the potential of measuring the texture on-line by laser ultrasonics is demonstrated.

## References

[1] Bescond C, Monchalin J P, Lévesque D, Gilbert A, Talbot R and Ochiai M 2005 *Proc. of the SPIE - Symp. on NDE for Health Monitoring and Diagnostics* **5767** 175

[2] Bescond C, Lévesque D, Lord M, Monchalin J P, Forgues S 2006 *Review of Progress in QNDE* vol 25, ed D O Thompson and D E Chimenti (New York, AIP Conf. Proc.) p 1426

[3] Moreau A, Man C S 2006 *Review of Progress in QNDE*, vol 25, ed D O Thompson and D E Chimenti, (New York, AIP Conf. Proc.) p 1434

Journal of Physics: Conference Series **278** (2011) 012008 doi:10.1088/1742-6596/278/1/012008

# Finite element simulation of the frictional heating of cracks under excitation of intensive ultrasonic waves

**Z J Chen**[1,2,a)], **S Y Zhang**[2,b)], **J Zheng**[2], **X B Mi**[2] **and K Zheng**[2]

[1] Department of Physics, Zhejiang Normal University, Jinhua 321004, China
[2] Institute of Acoustics, Nanjing University, Nanjing 210093, China

[a)] E-mail: zjchen@zjnu.edu.cn
[b)] E-mail: zhangsy@nju.edu.cn

**Abstract.** A thermo-mechanical coupling finite element (FEM) model is built to simulate frictional heating phenomenon of a crack when the plate containing the crack is excited by intensive ultrasonic waves. In the explicit FEM model, a high-power ultrasonic transducer is included for simulating the phenomena of acoustic chaos, and piezoelectric excitation of the transducer is realized by using a piezoelectric thermal-analogy method. Moreover, the dynamical interaction between both crack faces is modelled by using contact-impact algorithm. In the simulation, Dynamical contact force and relative motion of the crack faces, temperature-rise distribution around the crack, and temperature-rise versus time curves of the crack faces, are quantitatively calculated and analyzed when the plate undergoing chaotic excitation.

## 1. Introduction

As a novel hybrid ultrasonic/infrared nondestructive testing technique, ultrasonic infrared imaging (UIR) has obtained much attention in recent years [1,2]. For explaining the mechanism of the localized heating of cracks in UIR, several studies were presented in recent years, such as, Mian et al. [3] reported a finite element (FEM) computation for the response of a fatigue damage under sonic load; Mi et al. [4] presented a theory and FEM codes to calculate the heating around a crack. However, during the above researches, the ultrasonic excitation was modeled simply by a time-dependent sinusoidal displacement load at the surfaces of the sample. In other words, the high-power ultrasonic transducer system was not included in the FEM model, which leads to the phenomena of acoustic chaos [2,5] can not been modeled reasonably. As a result, the simulation of crack heating may not be very accurate.

In this paper, a three-dimensional thermo-mechanical coupling FEM model is built up to simulate the frictional heating phenomena of a crack when the plate containing the crack is excited by intensive ultrasonic waves. In the explicit FEM model, a high-power ultrasonic transducer system is included for simulating acoustic chaos reasonably, and piezoelectric excitation of the transducer is realized by using a piezoelectric thermal-analogy method [6]. On the other hand, the contact interaction between the both crack faces is modeled by using a contact-impact algorithm based on penalty method [7]. Considering the plate undergoing sound chaotic excitation, the vibration waveform and frequency spectrum of the plate, the dynamical contact force and relative motion of the crack faces, the temperature rise distribution around the crack zone, as well as the temperature-rise versus time curve of the crack faces, are quantitatively calculated.

## 2. FEM modelling

The FEM model consists of an aluminum alloy plate and an ultrasonic transducer. The plate is of a size of 120×60×2.5 $mm^3$ and has a notch at the top side. A through-thickness crack with two crack faces is located just below the notch, where the length of the crack is about 3 mm. The boundary conditions of the plate are that all degrees of freedom of the displacements are constrained at the nodes of the left and right sides of the plate. The simulated transducer is composed of six PZT-8 piezoceramic discs, a cylinder-shaped back mass and a front mass, and a composite ultrasonic horn. For describing accurately the contact interaction between both crack faces, refined meshing is adopted in the vicinity of the crack. In the model, a static coupling force is applied to the transducer to push it against the plate surface. The corresponding FEM meshing model is shown in Figure 1.

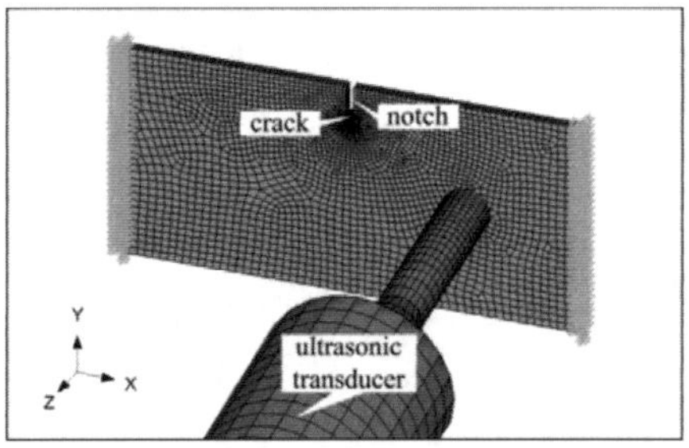


**Figure 1.** FEM modeling of plate containing a crack excited by high-power ultrasonic transducer

Owing to explicit integration algorithm has good stability and fast calculation capability, an explicit FEM code LS-DYNA is used here. During the simulation, the interface between the horn tip and the plate is assumed to be unbonded, where the surface of the horn tip is designated as the master side and the plate surface around the contact region is designated as slave side. For contact modeling of the crack faces, one face is declared as a master surface and the other one is a slave surface. In current model, penalty method is used in formulating the contact mechanisms and "surface-to-surface" contact algorithm is chosen [7]. Moreover, the Coulomb model is employed to simulate the friction process, where the friction coefficient between the crack faces is assumed to be 0.4. During the simulation, the initial temperature of the plate is assumed to be 27.0 °C.

Until now most explicit FEM programs like LS-DYNA are not possible to simulate the piezoelectric field, so the vibration excitation of piezoelectric transducer should be realized by other methods. In the work, piezoelectric thermal analogy method as described in Ref. [6] is used. According to the method, the strains due to an applied voltage field can be modeled by thermally induced strains. For the extension actuation mechanism of piezoelectric materials (the polarization axis of the piezoceramic is aligned with the z-axis), the correspondence between piezoelectric strains and thermal strains is obtained as follows:

$$\alpha_{11} = \frac{d_{31}}{t}, \alpha_{22} = \frac{d_{32}}{t}, \alpha_{33} = \frac{d_{33}}{t} \qquad (1)$$
$$\alpha_{23} = \alpha_{31} = \varepsilon_{12} = 0, \quad \Delta T = \psi_z$$

where $d_{mn}$ is the piezoelectric strain coefficient component, $\alpha_{mn}$ is thermal expansion coefficient component and $\Delta T$ is a temperature applied relative to a reference temperature, $\Psi_z$ represents the voltage difference between electrodes of each piezoelectric disc and $t$ is the thickness of each disc.

Equation 1 indicates that voltage actuation can be exactly simulated using conventional three-dimensional elastic elements with the thermal actuation rather than using piezoelectric elements. By comparing the result of harmonic analysis using thermal excitation with that using voltage excitation, it can be found that the vibration displacement of the horn tip under thermal excitation is nearly equal to that under voltage excitation, where the relative error between them is less than 3%. Therefore, in explicit transient analyses, piezoelectric transducer can be modeled by piezoelectric thermal analogy method. In this work, the resonance frequency of the longitudinal vibration mode of the transducer is 20 kHz and the vibration amplitude of the horn tip under unloading condition is 80 μm.

## 3. FEM simulation results

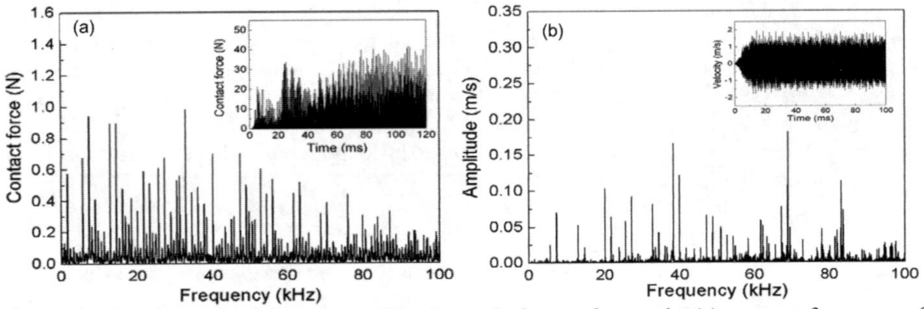


**Figure 2.** Simulation results of motions of both crack faces of a crack: (a) contact force waveform (inset) and spectrum; (b) relative velocity waveform (inset) and spectrum of two nodes.

It has been observed in the experiments [5] that a suitable static preload force may cause strong sound-chaotic vibration of the plate samples. The chaotic vibration of the plates will result in complex interaction of crack faces. Figure 2 shows the simulation results of motions of both crack faces when the plate is undergoing sound chaotic excitation (where the static coupling force is 150 N). It can be found in the inset of figure 2(a) that complex contact interaction occur in the crack faces in the process of sound chaotic excitation. The contact force waveform consists of a series of force pulses with different magnitudes, and the time intervals of the adjacent force pulses are different. Moreover, the contact force spectrum under chaotic excitation contains much rich frequency components. To produce frictional heating, in addition to have a suitable contact pressure at crack faces, it is important to find out relative motions of both crack faces. Figure 2(b) show the relative velocity of two nodes separately located at the right and left faces of the crack and 0.4 mm from crack tip. It can be found in the inset that there exists relative motion between the two crack faces. Moreover, different nodes on the same crack face have different velocity amplitudes. It also can be seen that the velocity spectrum includes many higher frequency components which helps to get higher relative velocity.

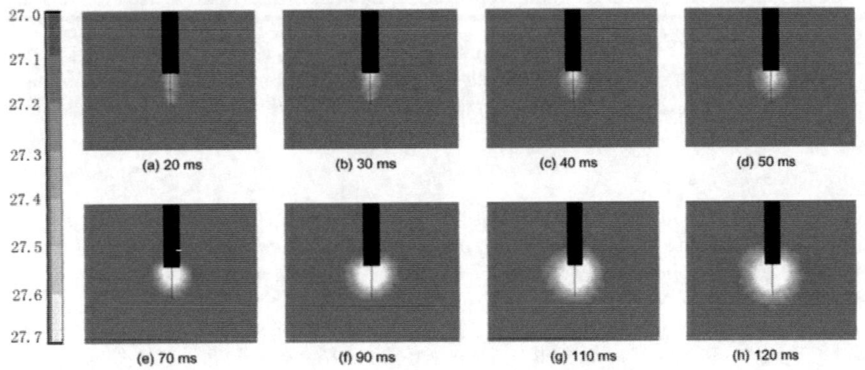


**Figure 3.** Simulation results of temperature distributions around crack under ultrasonic excitation.

The temperature distribution around the crack area under ultrasonic excitation is shown in figure 3. It can be found that frictional heating phenomena occur on the crack faces. As the time progresses, the heat energy is dissipated at crack faces and diffuses away from the crack. As a result, the area of the bright spot becomes bigger and bigger during the excitation process. When the ultrasonic pulse is turned off at time 120 ms, an obvious temperature rise in the range of 0.7 - 0.8 °C is observed in the crack region, which makes the crack detection by UIR become easier.

The temperature versus time curves for two nodes located on the crack faces are shown in figure 4, where one node is near the crack tip and the other one is at the middle region of the crack. It can be observed that the temperature rises quickly when the ultrasonic excitation is turned on. It is also clearly seen that the temperature increment is higher in the crack middle region than at the crack tip. The reason for that is the crack middle portion has larger relative velocity between both the crack faces then more thermal energies are generated. In addition, the curves in figure 4 show that the temperatures increase un-monotonously and have some fluctuation under chaotic excitation, which is induced by the intermittent contact of the crack faces.

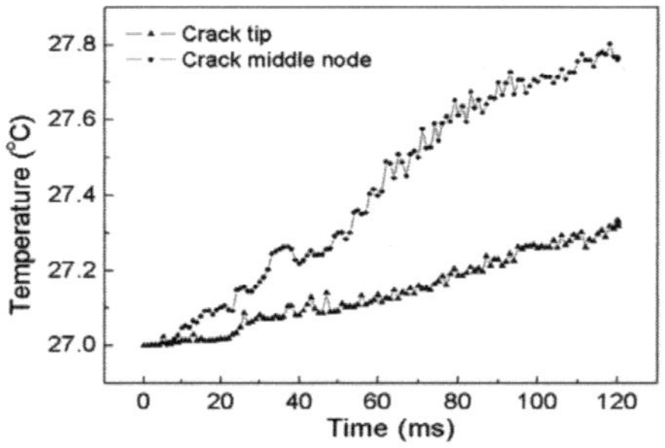


**Figure 4.** Simulation results of temperature-time curves for two nodes located at crack face.

## 4. Conclusion

The mechanism of crack detection by ultrasonic infrared imaging (UIR) has been further studied by using transient FEM simulation. Several typical conclusions are obtained as follows: (1) In the dynamic interaction process of crack under ultrasonic excitation, crack faces will produce relative motion, then contacting, slipping and separating will occur on the crack faces. As a result, the frictional heating is produced at the crack area; (2) Time dependence of contact force pattern across crack faces is complicated spatially and temporally. The waveform of the contact force consists of a series of delta pulses, and the magnitudes and time intervals of the force pulses are different; (3) Owing to the nonlinear interaction between the transducer horn and the sample, abundant frequency components appears in the frequency spectrums of contact force or relative velocity waveform between the crack faces; (4) Temperature-rise curves under sound chaotic excitation have some fluctuation, which just indicates that intermittent contact occurs at the crack faces. The computational results can explain reasonably the phenomena observed in UIR.

## References

[1] Favro L V, Han X, Ouyang Z, Sun G, Sui H and Thomas R L 2000 *Rev. Sci. Instrum.* **71** 2418
[2] Han X Y, Li W, Zeng Z, Favro L D and Thomas R L 2002 *Appl. Phys. Lett.* **81** 3188
[3] Mian A, Newaz G, Han X Y, Mahmood T and Saha C, 2004 *Compos. Sci. and Techn.* **64** 1115
[4] Mi X B and Zhang S Y 2005 *J. Phys. IV France* **125** 539
[5] Chen Z J, Zhang S Y, Zheng K and Kuo P K 2009 *J. Appl. Phys.* **106** 023507
[6] Cote F, Masson P, Mrad N and Cotoni V 2004 *Compos. Struct.* **65** 471
[7] Hallquist J O 2006 *LS-DYNA theory manual* (Livermore: LSTC)

# Laser-induced surface acoustic wave technique for precise depth measurement of stress corrosion cracking

**M Ochiai**
Power and Industrial Systems R&D Center, Toshiba Corporation
8, Shinsugita-cho, Isogo-ku, Yokohama 235-8523 Japan

E-mail: mak.ochiai@toshiba.co.jp

**Abstract.** A new method to evaluate a crack depth is introduced. Surface acoustic wave generated by Q-switched Nd:YAG laser (wavelength: 532 nm) and detected by frequency-stabilized long pulse laser (wavelength: 1,064 nm) coupled with confocal Fabry-Perot interferometer is used to evaluate a depth of surface-breaking tight crack. When the generated surface acoustic wave propagates through a crack before it is detected, only the lower frequency component is observed at the detection point due to interaction between the broadband surface acoustic wave and shallow crack. Energy of surface acoustic wave penetrates about its one wavelength into the propagation medium; it means that surface acoustic wave with higher frequency component localizes only thin layer from the surface and one with lower frequency component easily travel through cracks if it is shallow. A frequency response analysis technique, as well as an amplitude response analysis, of the surface acoustic wave is developed to quantitatively evaluate the depth of cracks. Several stress corrosion cracks introduced on type 304 stainless steel plates by immersing corrosive solution with tensile stress are prepared to verify the performance of this method. The results demonstrate the error of this depth measurement method is estimated at less than 0.3 mm.

## 1. Introduction

Initiation and growth monitoring of surface crack, such as stress corrosion cracking (SCC) and fatigue cracking is important issue for many industrial non-destructive testing (NDT) applications. In particular, due to the recent remarkable innovation on maintenance technologies [1], detection and accurate depth measurement of shallower, typically sub-mm to a few mm deep, surface cracks is important to determine efficient maintenance strategy.

Our final goal is to establish a method for sizing surface crack having a depth of about 0-3 mm. Laser-ultrasonics can offer higher spatial and temporal resolution than conventional contact type transducer since it features a very small generation and detection spot size, say less than 1mm, and very broad generation and detection response. Using the laser-generated and -detected surface acoustic wave (SAW), some pioneer studies to detect and size surface cracks in metal have been performed since 1980's [2]. In particular, feasible concepts based on the use of frequency response of transmitted SAW have been already introduced [3][4] and they demonstrated good results through some experiments with artificial slots. In this paper, we firstly carefully observe behaviours of laser-induced SAW transmitted through a surface crack, not only artificial slots but also SCCs, to obtain deeper understandings of the phenomena. Then we discuss a method to obtain accurate crack depth based on the understandings.

## 2. Experimental Setup

An experimental setup for laser-ultrasonic testing is schematically shown in figure 1. A Q-switched Nd:YAG laser having a wavelength of 532 nm, a pulse duration of 5-10 ns and a pulse energy of typically 30 mJ is used as the generation source of ultrasound. When a laser pulse from the generation laser is launched onto the surface of the test piece, longitudinal wave, shear wave and SAW are excited at the same moment through the ablative interaction process between the laser pulse and the materials.

Another Nd:YAG laser having a wavelength of 1,064nm, a pulse duration of 90 μs and a pulse energy of 2.5 mJ is used for the detection of ultrasound. The detection laser is irradiated at the point 15 mm away from the generation laser spot.

A slot or a crack formed on the test piece (details are described later) is arranged between the generation and detection spots. The generated SAW transmits through the crack and is detected as micro surface displacements by the detection laser coupled with a confocal Fabry-Perot interferometer (CFPI). The CFPI has a cavity length of 0.35 m and finess of about 22.

The ultrasonic signal from the photodetector in the interferometer is converted, processed and stored in the signal processor.

Regarding test pieces (TPs), 12 slots, each of which has a dimension of 30 mm long, 0.2 mm wide and from 0.2 to 2.4 mm deep, are machined on type 304 stainless steel plates. Also, 2 long SCCs (TP#SCC03 and TP#SCC06) are introduced on the other type 304 stainless steel plates. The cross section is observed as figure 2 after a series of laser-ultrasonic experiments and the actual average depth of SCCs are measured as 0.5 mm and 1.8 mm. For each SCC, 10 measurement points are defined with a interval of 5 mm along the SCC.

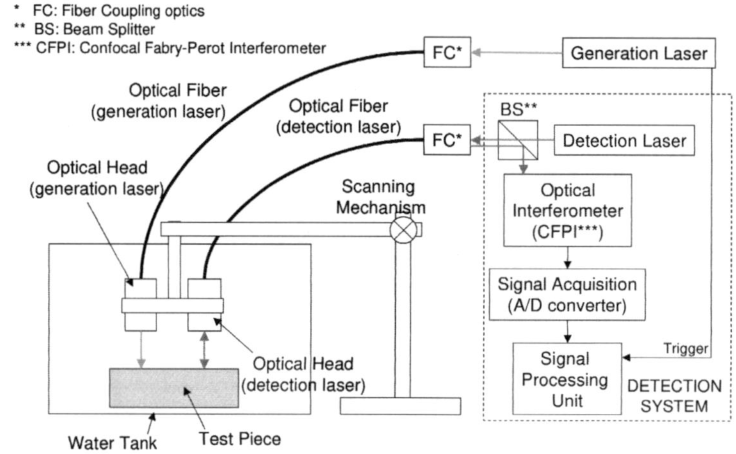


**Figure 1.** Experimental setup for laser-ultrasonic testing

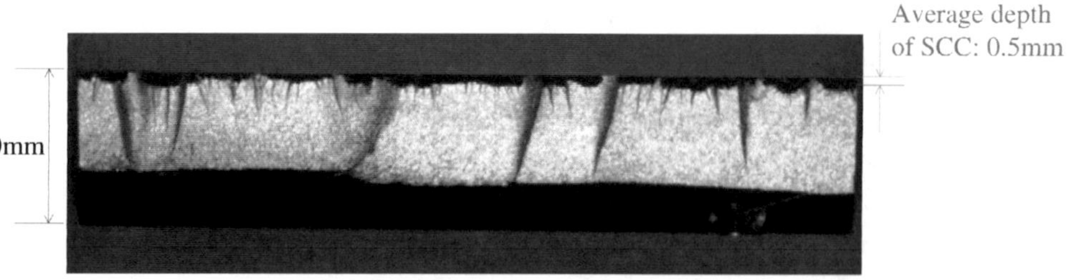


**Figure 2.** Typical cross-section observation result (TP#SCC03)

## 3. Results and Discussions

### *3.1 Amplitude change of transmitted SAW*

Acquired raw waveforms with the slots are shown in figure 3. They obviously indicate that the amplitude of transmitted SAW drastically decreases with the increase of slot depth. Signal amplitudes against the slot depth are plotted in figure 4(a). Based on the plots in figure 4(a), we obtain a fitting equation:

$$y = \exp(-1.6x) \qquad (1).$$

By using equation (1), we calculate slot depths from the observed signal amplitudes. Also, 20 data from the SCC TPs (10 data from 2 SCC TP) are calculated by the same manner. The result is shown in figure 4(b). Important note that the behaviours observed on the slot TPs and SCC TPs are mostly the same. One more important point is obvious trend that measured depths increase with the increase of the actual depths is observed but lack of linearity is also observed. Underestimation in the deeper crack depth region can be explained as "saturation", which means that the energy of SAW cannot reach to the region any more. Instead, data show rather a logarithmic behaviour drawn by the dotted line in figure 4(b). Therefore the difference should be systematic and we should establish a model to explain this phenomenon.

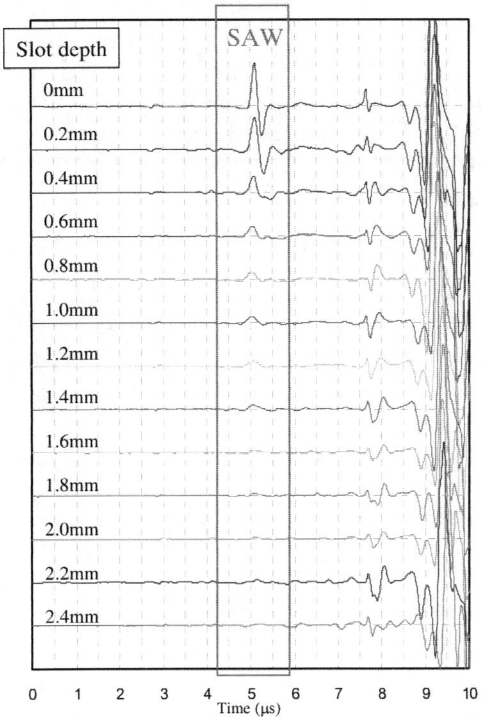


**Figure 3.** Acquired raw waveforms with the slots

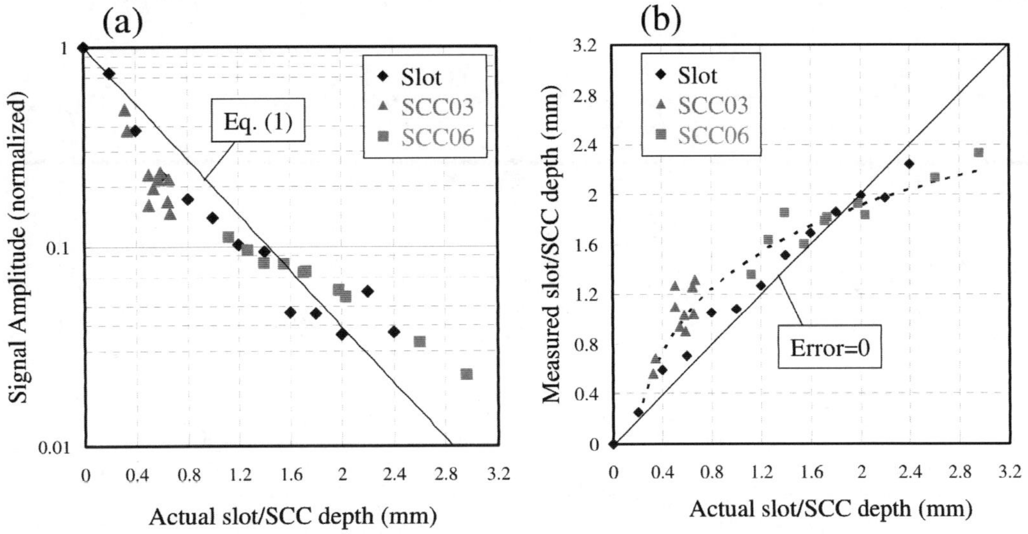


**Figure 4.** (a) Amplitude change of SAW against crack depth and (b) Relation between actual and measured crack depth

### *3.2 Frequency response of transmitted SAW*

Frequency responses of the transmitted SAW through the slots are shown in figure 5. As shown in the figure 5, the peak frequency shifts toward the lower frequency region and the peak amplitude decreases with the increase of slot depth. The cut-off (first node) frequencies against the slot depth as well as the SCC depth are plotted in figure 6. Important note that the behaviours observed in the frequency domain on the slot TPs and SCC TPs are mostly the same again. One more note that the plot with slot is divide into 2 branches denoted with the white and black diamonds in the figure 6. This caused by the "dent" around the peak frequency given by 0.4-1.2 mm depth slots in the frequency response curve, figure 5.

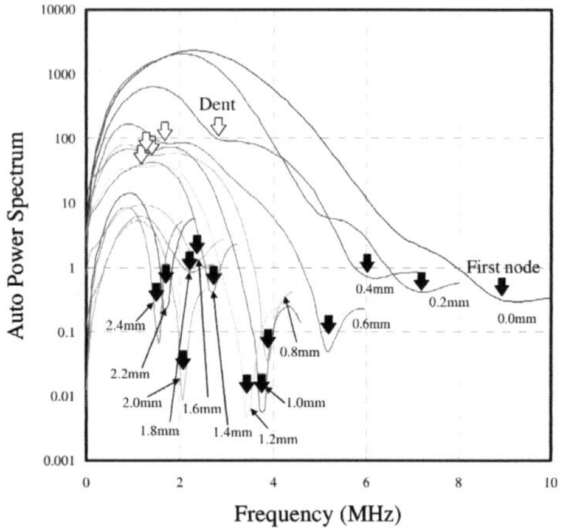


**Figure 5.** Frequency response of transmitted SAW with slots

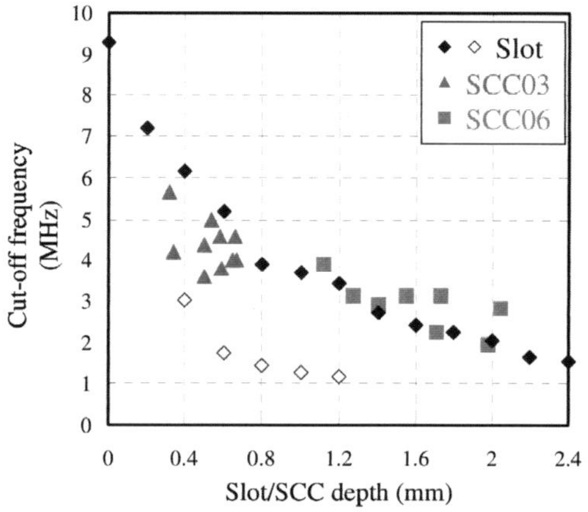


**Figure 6.** Cut-off frequency of transmitted SAW against slot/SCC depth

### *3.3 Combination use of amplitude change and frequency response*

Through the discussion in the section 3.1 and 3.2, we found that both amplitude and frequency response of the transmitted SAW gives value information of crack depth. The amplitude of SAW is

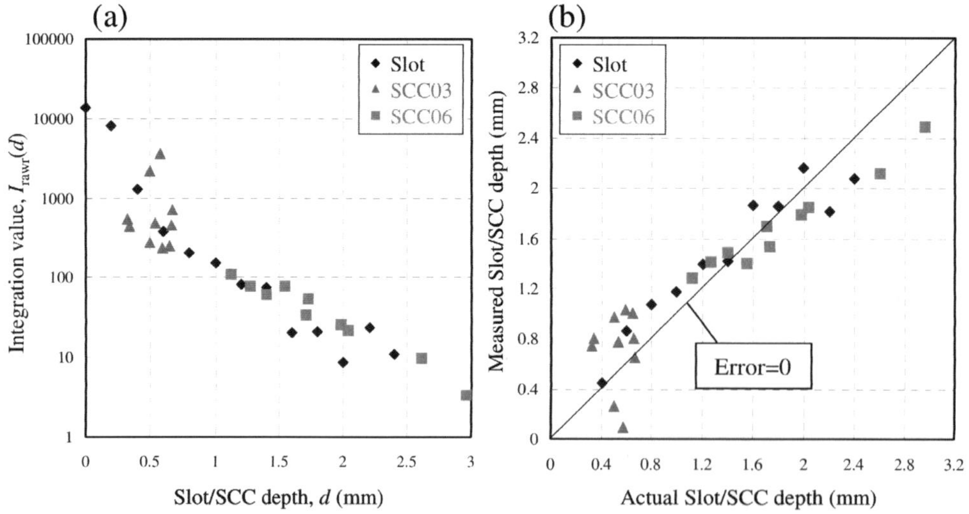


**Figure 7.** (a) Integration value of SAW against slot/crack depth and (b) Relation between actual and measured crack depth evaluated by $I(d)$

exponentially decreasing with the increase of slot/crack depth; however the linearity is slightly lost in particular in SCC data. Also, the cut-off frequency decreases with the increase of slot/crack depth but some discrepancy observed around the crack depth of 0.4-1.2 mm maybe due to the small dent around the peak in the frequency profile. Therefore, we propose that the combination use of amplitude and frequency information. This idea is realized by introducing an integration value, $I(d)$ ;

$$I(d) = \int_{f_m}^{f_M} f \cdot P_{SAW}^{d}(f)df \qquad (2).$$

Here, $f$ is frequency, $P^{d}_{SAW}(f)$ is a frequency profile of transmitted SAW at the slot/crack depth of $d$, $f_m$ and $f_M$ are minimum and maximum frequency of region of interest, respectively. In this study, $f_m$ and $f_M$ are chosen as 1 MHz and 8 MHz.

Figure 7(a) shows the calculation results of $I(d)$ for all slot/crack data against actual depth. The fitting equation obtained from figure 7(a) gives the final relation between the measured and actual depth of slot/crack as shown in figure 7(b). The statistical analysis on the error between the measured depth (plots in figure 7(b)) and actual depth (plain line in figure 7(b)) is also shown in table 1.

**Table 1** Statistical analysis result of slot/crack depth measurement

| | Slot (mm) | SCC (mm) |
|---|---|---|
| Error average | -0.00 | -0.02 |
| Maximum error | 0.39 | 0.49 |
| RMS error | 0.27 | 0.32 |

## 4. Conclusion

Based on the experimental data including SCC and their analysis results, we confirm that the amplitude of SAW is exponentially decreasing with the increase of slot/crack depth; however the linearity is slightly lost in particular in SCC data. Also, the cut-off frequency decreases with the increase of slot/crack depth but some discrepancy observed around the crack depth of 0.4-1.2 mm maybe due to the small dent around the peak in the frequency profile. Based on these careful observation, we suggested a new method for evaluating a depth of surface crack using both frequency response and amplitude change. The results obtained in slot samples and SCC samples show that the error average of this depth measurement method is 0.0 mm and its RMS error is 0.3 mm in the range between 0.0-3.0 mm.

## References

[1] Sano Y, Kimura M, Sato K and Obata M 2000 *Proceedings of 8th International Conference on Nuclear Engineering* **ICONE-8441** 191

[2] Cooper J A, Crosbie R A, Dewhurst R J, McKie A D W and Palmer S B 1986 *IEEE Transactions on Ultrasonics, Ferroelectrics, and Frequency Control* **UFFC-33** 462

[3] Scala C M and Bowles S J 2000 *Review of Progress in Quantitative Nondestructive Evaluation* **19** 327

[4] Ochiai M, Butsuen T, Miura T, Kuroda H, Soramoto S and Kanemoto S 2001 *J. At. Energy Soc. Japan* **43** 275 in Japanese

Journal of Physics: Conference Series **278** (2011) 012010 doi:10.1088/1742-6596/278/1/012010

# Laser ultrasonic investigation of laminate disbonding

**Alexander Karabutov**[1], **Oleg Sapozhnikov**[2], **Alexandr Karabutov (Jr.)**[2]

[1]International Laser Center, Moscow State University, Leninskie Gory, 119991 Moscow, Russian Federation
[2]M.V. Lomonosov Moscow State University, Physics Department, Leninkie Gory, 119991 Moscow, Russian Federation

E-mail: akarabutov@gmail.com

**Abstract.** Contact laser-ultrasonic evaluation (CLUE) provides ultrasonic testing with sharp probe ultrasonic pulse. It makes it possible to enhance spatial resolution at relatively lower frequency band than piezoelectric ultrasonic testing. The advantage of CLUE is the possibility to distinguish the acoustic wave reflection at the soft or rigid boundary. This feature can be effectively used for the investigation of disbonding in laminate structures. The sample consisting of three aluminum sheets glued with polymer layers is tested with CLUE. The results of back-reflection coefficient measurement are compared with the numerical simulation. The signatures from regular area and area with disbonded layer are compared. The sensitivity and resolution of testing is discussed.

## 1. Introduction

Composite materials are very important for industry (for making panels and other heavy loaded sets for airplanes). Production of such materials is a complicated process, which needs permanent control. Composite parts have to be controlled during exploitation. Ultrasonic investigation could give a lot of information about mechanical properties and texture of such material.

Generally, composite materials are layered structures that consist of different materials. Transmission and reflection of ultrasonic waves in laminated structures is a complex process. Theoretical foundations of propagates of ultrasonic wave in layered structures can be found in [1]. The size of components that form a composite structure varies over a wide range. Due to this fact probe ultrasonic pulse should have wide bandwidth. Practically, one needs frequency band of 10 octaves to investigate the texture of a composite. Traditional ultrasonic transducers could not emit signal with such a wide band. Fortunately, laser ultrasonics could provides signals with sufficient bandwidth [2].

Experimental and numerical investigation of transmitted acoustic fields with narrow band was made in the paper [3]. Similar work for broadband signals was performed in [4].

In this work layered structures are analyzed by the laser-ultrasonic method in pulse-echo mode.

## 2. Theoretical model

Laser ultrasonic method uses short monopolar acoustic videopulses generated during absorption of laser pulses. Temporal waveform of theses laser-generated pulses is well investigated in

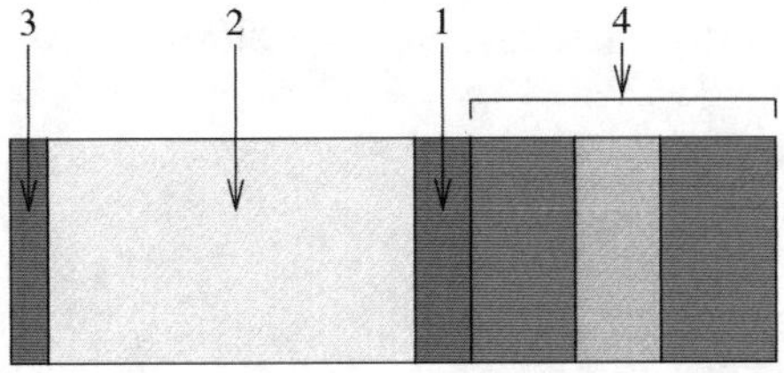


**Figure 1.** Scheme of ultrasonic control of composite materials. 1 - generator, 2 - acoustic conductor, 3 - receiver, 4 - object of control

optoacoustic [2]. Peak pressure and duration of the acoustic pulse depends on duration of the laser pulse and light absorption factor of absorbing medium ("generator"). Thanks to possibility of using short laser pulses, the acoustic pulse duration can be less than a nanosecond.

Scheme of the optoacoustic transducer shown in figure 1. A laser pulse falls on the the left side of a light absorbing layer (generator, 1) through an optically transparent acoustic conductor block (2). The opposite side of the generator is contacting an object of control (4). The generator and conductor are acoustically matched with each other. Acoustic pulse is emitted as a result of the light pulse absorption in the generator. At the other side of the acoustic conductor, a piezoelectric receiver is placed (3). It is made from a thin piezoelectric PVDF film and due to that has a wide frequency bandwidth. When light pulse is absorbed in the generator, two acoustic waves are excited that propagate in opposite directions. First of them prorogates to the left-hand side (see figure 1) through the conductor (2) to the receiver (3). The other wave propagates initially to the right, then reflects from all layers of the object and prorogates to the receiver, like the first wave.

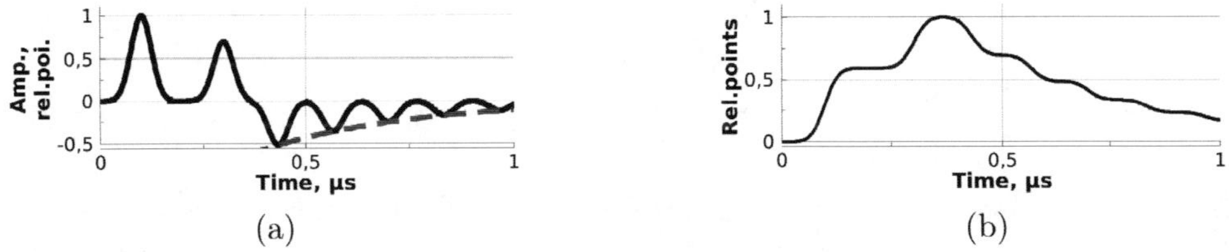


**Figure 2.** Modeling results for the signal (a) reflected from a layer contacting air (solid line). Its integral are shown on (b).

Let us consider a situation when an acoustic videopulse is reflected from one layer with its back side contacting air (this happens when delamination occurs in the structure). Modeling of this situation was made with the help of transfer matrix method [3, 1]. Results are shown in figure 2(a) by solid line. The probe acoustic pulse is purely positive, with Gaussian waveform. Its duration is 70 ns, it is shifted by 0.1 $\mu$s for the presentation convenience. Reflected signal from the boundary between the generator and the object comes to the receiver after this first pulse. It also has positive polarity and its delay reflects prorogation time inside the generator. Other pulses reflect from the back soft boundary of the layer, which explains their negative polarity.

In real materials essential role is played by the attenuation in object and acoustic conductor. This limits the possibility of evaluating acoustic impedance ratio from the levels of reflected signals, especially at high frequencies that more readily attenuate. It is therefore difficult to find coefficient of attenuation and acoustic impedance of layers by analysing reflected signals. On the other hand, low frequency signals are less attenuative. In order to use them, we decided to analyze time-integral of the received signals, which enhances low-frequency components.

Time-intergal $I(t)$ of the received signal shown in figure 2(a) is plotted in figure 2(b). Its value increases until arrival of the first signal. After integrating this first signal, the value of $I$

reaches some constant level $A_0$. Then reflection from the object arrives, and $I$ continues to rise until reaching level $A_1 = (1+R_{10})A_0$. Here $R_{10}$ is reflection coefficient related to the impedance of the investigated material:

$$R_{10} = \frac{A_1 - A_0}{A_0}, \qquad Z_1 = Z_0\frac{1+R_{10}}{1-R_{10}}. \tag{1}$$

Let us consider a structure that has two layers. We will suppose that the proximal layer properties ($Z_1$) are known (e.g. obtained from the measurements described above). We will now calculate impedance of the distal layer ($Z_2$). Its impedance could be calculated from the levels of flat portions of the integral signal $I(t)$ and acoustic impedance of the second (distal) layer:

$$M = \frac{A_2 - A_1}{A_1 - A_0} = \frac{T_{01}R_{21}T_{10}}{R_{10}}, \qquad Z_2 = Z_1\frac{4Z_1Z_0 + M(Z_1^2 - Z_0^2)}{4Z_1Z_0 - M(Z_1^2 - Z_0^2)}, \tag{2}$$

where $A_2$ and $A_1$ are values of $I(t)$ that correspond to reflections from the proximal and distal sides of the first (known) layer.

Similarly, structures with larger number of layers can be analyzed. However, this approach in practice is limited by the diffraction effect, which results in significant decrease of low-frequency components. Due to that the described approach could be applied only to a few layers.

## 3. Experimental setup

Scheme of laser-ultrasonic transducer is shown in figure 1 and it was described in the theoretical part.

A short light pulse of 10 ns duration is emitted by a Q-switched laser. The pulse energy 60 $\mu$J. The pulse is delivered to the generator through an optical fiber. Electrical signal of the acoustic receiver is digitized by an analog-to-digital converter (frequency of sampling is 100 MHz, 12 bits, analog bandwidth is 70 MHz). Filtering and deconvolution of the signals were made on a computer.

Probe ultrasonic signal has smooth waveform. Its half-amplitude duration is approximately 50 ns. Lateral distribution of acoustic beam is the same as for the optical one, and has approximately Gaussian shape. Because the absorbing layer is flat, the acoustic wavefront is plane. Diameter of the beam is approximately 3 mm.

## 4. Objects

Composite material, which was used as an object, consists of three aluminum sheets glued by a polymer. Thickness of the aluminum sheets is approximately 0.42 mm, and the glue layers have thickness of 0.08 mm. This sample has a delamination in one of the glue layers. The sample was investigated in three sites. In the first site "A" there was a delamination between the first and the second aluminum sheets, for the second site "B" it was between the second and the third sheets. In the third site "C" there was no delamination.

## 5. Experimental results

The CLUE signals in the three sites of the composite A-C are presented in figure 3(a): A - delamination between the first and the second aluminum sheets (solid line), B - delamination between the second and the third aluminum sheets (dashed line), and C - without delamination (dotted line).

The probe pulse and signal reflected from the front surface of the first layer have the same waveform for all three sites. Pulse that follows is the reflection from the rear surface of the first layer. Pulse from rear surface of the first layer has negative polarity, that corresponds to soft acoustic boundary. Signals from sites B and C have smaller amplitude of the reflection from

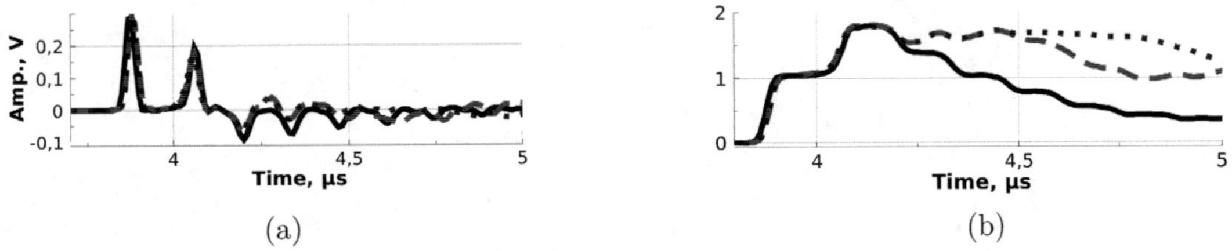


**Figure 3.** Signals (a) and there integrals (b) in site "A" (solid line), "B" (dashed line) and "C" (dotted line).

this boundary, as compared to that for the site A. It means that impedances of the following layer for the sites B and C are greater than for the site A. Ultrasonic paths through the first and the second aluminum layer for the sites B and C are the same. The difference in these signals is manifested after time of arrival of the pulse reflected from the front surface of the second aluminum layer ($t > 4.45\ \mu s$).

Integrals of the signals show a difference between reflected signals in sites A-C (figure 3(b)) very clearly. First rising slope corresponds to the partial displacement in the probe signal. The following rising slope corresponds to the partial displacement in the signal reflected from front surface of the first aluminum layer. Using measured time-of-flight in each layer and coefficient of reflection at each boundary one can calculate the ultrasonic wave velocity and acoustic impedance of layers. The result of such calculation for the first layer gives $(1.8\pm0.1)$ g/($\mu$s mm$^2$). Sound velocity can be calculated by using thickness of the layer and time of reflection signals from the boundaries. Thickness of the first layer is $(373 \pm 10)$ $\mu$m and sound velocity is $(6.5 \pm 0.6)$ mm/$\mu$s. Density is calculated from impedance and sound velocity and its value for the first layer is $(2.8 \pm 0.4)$ g/cm$^3$. As one can see, aluminum has a close value of parameters. Impedance of next layer is calculated by using formula (2) and it is equal to $(3.4 \pm 0.2)$ g/($\mu$s mm$^2$) (thickness of this layer is $(94 \pm 10)$ $\mu$m, velocity - $(2.9 \pm 0.6)$ mm/$\mu$s, density - $(1.2 \pm 0.3)$ g/cm$^3$). Low transmition coefficient of interface aluminum-glue is a one of a few reasons of errors in determining layer parameters.

## 6. Conclusions

Contact laser-ultrasonic testing is applied to the investigation of layered structures. The temporal trace of reflected ultrasonic pulse provides the information on the time-of-flight and coefficient of reflection for each peculiar layer. The results of theoretical treatment, numerical simulation and experimental investigation are in good agreement for each particular layer. Contact laser-ultrasonic testing makes it possible to measure the ultrasonic wave velocity and density of several upper layers. Contact laser-ultrasonic testing can be effectively explored for the detection of delamination in a layered structure.

## Acknowledgments

We want to thank Vitaliy Soustin, Igor Kudinov, Vladimir Solomatin, and Sergey Gorbunov. Work was partly supported by RFBR 08-02-00368, RFBR 10-02-01468 and ISTC 3691.

## References

[1] Brekhovskikh L and Godin O 1992 and 1999 *Acoustics of layered media II: point sources and bounded beams* (Berlin: Springer) 2nd ed
[2] Karabutov A and Gusev V 1993 *Laser Optoacoustics* (AIP Press)
[3] Panomarev A, Bulatitskii S and Sapozhnikov O 2007 *Acoust. Phys.* **53** 127–135
[4] Karabutov A, Kozushko V, Pelivanov I and Podymova N 2000 *Acoust. Phys.* **46** 510–515

Journal of Physics: Conference Series **278** (2011) 012011 doi:10.1088/1742-6596/278/1/012011

# Exploring surface wave interaction with angled defects in the near and far field

**B Dutton, A R Clough, M H Rosli and R S Edwards**

Department of Physics, University of Warwick, Coventry, CV4 7AL, UK

Email: b.dutton@warwick.ac.uk

**Abstract**. This paper explores some effects that occur when using laser ultrasound to scan defective samples. Surface defects can often propagate at an angle to the surface; however, for calibration, slots machined normal to the surface of the sample are typically used. Several interesting angle-dependent effects are observed when Rayleigh waves interact with angled surface defects, and are explored here using measurements and models for a scanning laser detector (SLD) or scanning laser line source (SLLS) scanned across the defect. Reflection and transmission coefficients are calculated for different crack angles and lengths. Additionally, interesting angle-dependent effects are observed in the Rayleigh wave amplitude and frequency enhancements in the near field when using SLD or SLLS.

## 1. Introduction

Ultrasound has proven to be a suitable technique for characterising surface cracks, particularly by analysing the reflection or transmission of Rayleigh waves, which have both in-plane (IP) and out-of-plane (OP) components [1-3]. Previous work on slot defects machined normal to the surface has shown that the transmission of broadband Rayleigh waves can be analysed in both the time and frequency regimes and used as a measure of crack length [4,5]. Real cracks, however, can propagate at an angle to the surface normal; for example rolling contact fatigue (RCF) in rails which propagates at around 25° to the surface. For this reason, our work considers a range of crack angles and lengths and the effect they have on surface waves, using both experimental and modelling techniques. Far field effects, such as the reflection and transmission coefficients of a Rayleigh wave following interaction with a crack, are explored for a wide range of angles and lengths. Enhancements in the near-field have been previously observed with a scanning laser detection (SLD) or scanning laser line source (SLLS) for limited number of angles [2,3,6], and this work considers these enhancements for a wider range of crack angles.

## 2. Model and experimental details

A 3D finite element method model using PZFlex was used to explore the interaction of ultrasound with surface cracks. The model used a loading force derived from a laser pulse duration of 10 ns, applied onto the sample in the form of dipoles [7,8] for SLD calculations. For the SLLS technique, to avoid complication of the boundary conditions as the laser generation passes over the crack and the crack walls are free to expand, our approach used the heat-mechanical bridge mode in PZFlex, with the temporal and spatial temperature profile of the laser line source considered [7,9]. This creates a material expansion which in turn generates ultrasound waves on the material. For SLD and SLLS model samples, all surfaces apart from the top and the crack faces were assigned to be 'Absorbing', simulating a larger sample, with symmetry applied to improve model efficiency. Figure 1 shows the location of generation and detection points for the SLD case with the detection point scanned across the crack, while for SLLS scans the generation and detection positions are switched and the generation line source scanned.

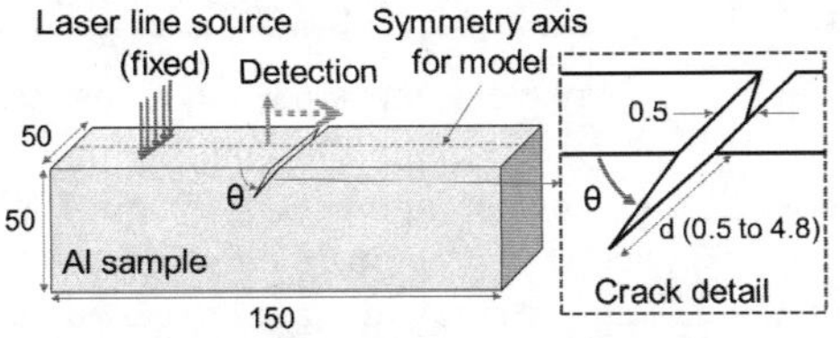


**Figure 1.** Model and experimental sample details. The SLD technique is shown and for SLLS the source and detection were reversed. Dimensions in mm.

Experimental scans were performed on aluminium samples with dimensions 150x150x50mm, using a non-contact laser-ultrasound system. The dimensions were chosen such that bulk acoustic waves (BAW) would not interfere with the Rayleigh wave. A Nd:YAG laser with 10ns rise time was used to generate surface waves in the thermoelastic regime. The laser beam was focused into a line of approximate dimensions 6 mm by 300 μm, to enhance generated ultrasound in a direction which is perpendicular to the line, in addition to increasing the frequency content of the generated surface waves [10,11]. A two-wave mixer interferometer from IOS was used to detect OP displacements on unpolished aluminium surfaces. Fabricated aluminium samples with crack angles from 30° to 150° relative to the surface, and a fixed length d of 2 mm, were used. Generated models had defect angles ranging from 10° to 170° and lengths from 1 to 4 mm.

## 3. Far field: reflection and transmission coefficients

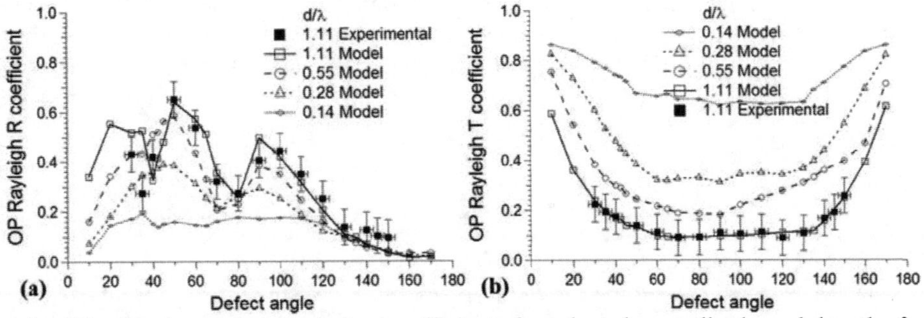


**Figure 2.** OP reflection and transmission coefficients for selected normalised crack lengths from 0.14 to 1.11, where (a) shows a general increase of reflection coefficient as d/λ increases, while (b) shows a decrease of transmission coefficient as d/λ increases.

We first explore the effect of crack angle and length on Rayleigh reflection and transmission coefficients, using both model and experimental data. OP reflection coefficients were calculated from A-Scan data as the amplitude ratio of reflected Rayleigh over incident Rayleigh wave. Figure 2a shows the model (line and open squares) and experimental (filled squares) reflection coefficients for a crack length to central wavelength ratio of $d/\lambda = 1.11$, where λ is the central wavelength of the broadband incident pulse, showing very good agreement. The same figure also shows reflection coefficients for smaller d/λ of 0.55, 0.28 and 0.14, showing a progressive decrease of amplitude, as expected when considering the broadband nature of the pulse and the fact that the majority of the energy in the Rayleigh wave is contained within one wavelength of the sample surface. Corresponding calculations were performed for the transmission coefficients using model and experimental data, with the coefficient also measured more than three wavelengths past the crack opening. Figure 2b shows the model and experimental transmission coefficients for $d/\lambda = 1.11$, showing very good agreement.

The same figure also shows calculated transmission coefficients for smaller d/λ, showing the expected change in transmission coefficient. It is evident that a variation in transmission with angle must be considered when comparing results from real defects and calibrations based on slots which are machined normal to the surface.

## 4. Near field: amplitude and frequency enhancements

An enhancement in both the amplitude and frequency content of Rayleigh waves has been observed when scanning close to a crack which is normal to the sample surface [3,6,9]. The behaviours were explored using both SLD and SLLS techniques, and using other non-contact ultrasonic techniques. We consider here the effect crack angle has on these enhancements, which has previously been ignored.

Figure 3a shows the calculated amplitude enhancement from both experimental and modelled data for the SLD technique, where the enhancement was calculated as the ratio of the highest amplitude in the windowed Rayleigh arrival time, which occurs close to the edge of the crack, to the incident Rayleigh wave amplitude. Modelled and experimental data show very good agreement, with a slightly reduced enhancement in the experimental data due to the larger detection point size [6]. Model and experimental data show an approximately exponential dependence on angle; a fit was performed using model data excluding the 10° angle crack, and is shown as a solid line. As a reference, the horizontal line indicates a ratio of 1, i.e. no enhancement. An enhancement increase is clearly visible as the crack angle decreases, with significant angle dependence for angles < 60°. Positive interference of reflected Rayleigh and mode converted waves contribute to this enhancement, as has previously been reported for cracks normal to the surface [1,6], but the very large enhancement that occurs for shallow crack angles requires further investigation, which will be reported in a later work.

Similar analysis was performed for the SLLS technique, shown in Figure 3b, where only a few data points are shown due to models taking considerably longer to run than SLD. In this case, the enhancement is due to interference effects, but also source truncation and changing boundary conditions [8,9]. Overall, like the SLD amplitude enhancement vs. crack angle, the enhancement during SLLS also has an angle dependence, but this is much less significant than that observed using SLD.

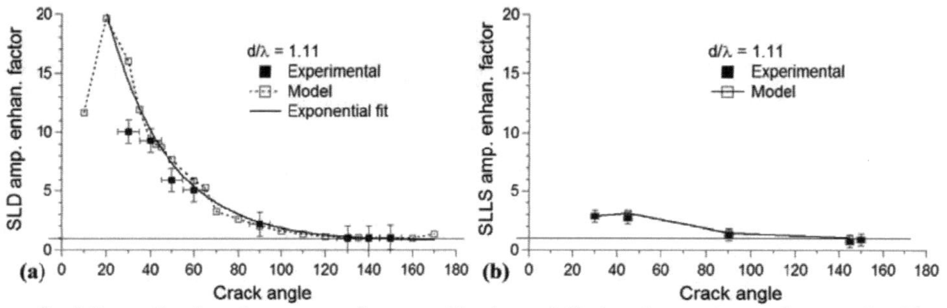


**Figure 3.** OP amplitude enhancement factors of both modelled and experimental scans for d/λ = 1.11, where (a) shows SLD, and (b) SLLS technique.

Finally we look at the frequency enhancement that occurs when scanning across a defect using the SLD technique. Two contrasting cases were selected from modelled outputs of d/λ = 1.11 for a 90° and a 20° crack angle. A magnitude fast Fourier transform (FFT) of a window around the expected Rayleigh arrival time OP displacement was performed at each position in the scan and stacked into a frequency B-Scan [9]. The corresponding scans for 90° and 20° are shown in Figures 4a and 4c, where the crack is at 45.5 mm. Frequency enhancement factors were calculated by comparing the FFT at the edge of the crack with that away from the crack (dotted horizontal lines on Figures 4a and 4c), and the corresponding relative magnitudes for a range of frequencies are shown in Figure 4b [9]. Scans for the 90° crack show a relatively small enhancement throughout the frequency range, while the 20° crack shows considerably larger enhancements. The vertical lines indicate peaks for the 20° crack, occurring at 0.3 and 2.7 MHz. High frequency enhancements due to non-linear effects have been reported for SLS [9]; enhancements at the Rayleigh central frequency, 0.81 MHz, are 23.5 and 2.1 for 20° and 90° respectively, which are similar to the enhancement in Figure 3a; enhancements occurring at frequencies lower than the Rayleigh central frequency will be discussed in a future publication.

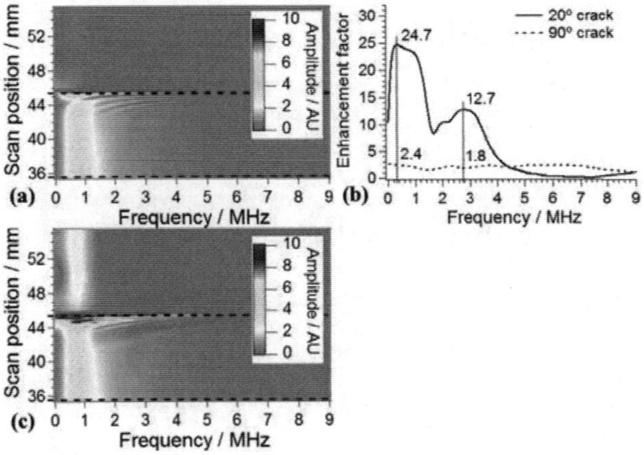


**Figure 4.** Magnitude FFT B-scan of OP displacement model output using SLD technique for (a) 90° crack and (c) 20° crack, and their corresponding frequency enhancement factors (b).

## 5. Conclusion

Modelled reflection and transmission coefficients using the SLD technique agree well with experimental data for crack angles from 30° to 150°. It is evident that variations in both reflection and transmission coefficients with angle must be considered when comparing results from real defects with calibrations based on slots which are machined normal to the surface. It was also shown for the SLD technique that very large amplitude enhancements occur at shallow crack angles, with significant angle dependence observed when comparing enhancements with those from cracks normal to the surface [6,8]. Amplitude enhancements were obtained with the SLLS technique and again show angle dependence, but at a much lower scale. For SLD, a larger frequency enhancement was found at a small angle of 20° than with the 90° crack angle for a large range of frequencies. These amplitude and frequency enhancement techniques could potentially be used to identify cracks at different angles in real samples, and used as a fingerprint of the position of a defect.

## References

[1] Jian X, Dixon S, Guo N and Edwards R 2007 *Journal of Applied Physics* **101** 7
[2] Kinra VK 1986 *Journal of the Acoustical Society of America* **79** 1688-1692
[3] Kromine AK, Fomitchov PA, Krishnaswamy S and Achenbach JD 2000 *Materials Evaluation* **58** 173-177
[4] Edwards RS, Dixon S and Jian X 2006 *Ultrasonics* **44** 93-98
[5] Vu BQ and Kinra VK 1985 *Journal of the Acoustical Society of America* **77** 1425-1430
[6] Edwards RS, Jian X, Fan Y and Dixon S 2005 *Applied Physics Letters* **87** **3**
[7] Arias I and Achenbach JD 2003 *International Journal of Solids and Structures* **40** 6917-6935
[8] Arias I and Achenbach JD 2004 Wave Motion **39** 61-75
[9] Dixon S, Cann B, Carroll DL, Fan Y and Edwards RS 2008 *Nondestructive Testing and Evaluation* **23** 25-34
[10] Aindow AM, Dewhurst RJ and Palmer SB 1982 *Optics Communications* **42** 116-120
[11] Hutchins DA 1986 *Canadian Journal of Physics* **64** 1247-1264

Journal of Physics: Conference Series **278** (2011) 012012 doi:10.1088/1742-6596/278/1/012012

# Hysteresis-like attenuation of ultrasound in severely plastically deformed nickel

**V V Kozhushko, G Paltauf and H Krenn**

Institute of Physics, Karl-Franzens-University, Universitätsplatz 5, 8010 Graz, Austria

Viktar.Kazhushko@uni-graz.at

**Abstract**. Disc-shaped specimen of nickel of about 30 mm diameter and 2.25 millimeters thick was prepared by the high pressure torsion method. The produced equivalent shear strain increased linearly from a minimum at the centre to the periphery. The cold working process multiplied the dislocations and led to the grain fragmentation. The saturation of the grain refinement was approached at the circumference of the specimen yielding a narrow distribution of the grain size with a mean value of about two hundred nanometers. The refinement of the microstructure changes the elastic properties of the material such as hardness, strength and has an influence on the attenuation and phase velocity of the ultrasound. The paper reports the measurements of the attenuation of the longitudinal wave in severely plastically deformed nickel with polycrystalline and ultrafine microstructure.

## 1. Introduction

The high pressure torsion (HPT) method produces severe plastic deformation of the specimen which simultaneously undergoes compression and shear deformation between one fixed anvil and one rotating anvil. The induced equivalent shear strain $\varepsilon$ is proportional to the radius of the specimen and the number of the rotations. The value of $\varepsilon$ of several thousand percent can be achieved by applying the necessary number of rotations. The cold working process multiplies dislocations and leads to a low angle misorientation within a coarse crystallite at the beginning of plastic deformation of originally polycrystalline (pc) material. The developed dislocations network is typical for $\varepsilon < 100\%$. Severe plastic deformation further multiplies the dislocations in pc metals that increases misorientation and induces the fragmentation. The refinement process approaches the saturation at $\varepsilon \sim 1000\%$ with a mean value of the grain size of few hundred nanometers [1]. The HPT deformed metals possess high chemical purity and increased dislocation density which is typically $> 10^{15}$ $m^{-2}$. The ultrafine (uf) microstructure demonstrates increased hardness and strength hence the application of these materials is very promising in different fields.

The estimation of the mechanical properties of these new materials was carried out by laser ultrasound [2]. The changes of the microstructure vary the phase velocity and the frequency dependence of the attenuation. The strain amplitude dependence of the ultrasound attenuation due to the nonlinear effects in polycrystals with developed dislocation network [3] or due to the increased number of grain boundary atoms in nanocrystalline materials was recently discussed [4] while there is a lack of experiments. The purpose of the work is the estimation of the longitudinal wave attenuation as a function of the pressure pulse amplitude in HPT nickel with uf microstructure and pc microstructure with a high dislocation density by means of broadband laser ultrasound spectroscopy.

## 2. Experimental

The measurements were carried out by means of immersion technique, using laser induced ultrasound pulses. A disk shaped sample with polished parallel surfaces was immersed into the water bath. Laser pulses of 5 ns duration illuminated the water metal interface. The diameter of the laser spot was about 6 mm and the laser power density did not exceed 10 MW/cm$^2$ at the surface. The absorbed laser radiation induced heat sources. The diffusion of the heat locally increased the temperature of the materials and due to the thermal expansion launched the pressure pulse. As the coefficient of thermal expansion of water is about one order of the magnitude higher in comparison with metals the contribution of the water is dominating [5]. In our experiments the illuminated surface of the metal was covered by a thin absorbing layer of black acrylic paint to absorb almost all laser radiation and to increase the efficiency of ultrasound pulse excitation by about one order of magnitude. The bandwidth of the laser induced pressure pulse covered the range from 0.1 MHz up to 150 MHz that allowed broadband ultrasound spectroscopy [2].

The ultrasound pulse was detected by a homemade piezoelectric transducer based on a 25 $\mu$m thick polyvinylidenefluoride (PVDF) foil that allowed detection of the frequencies from 5 MHz up to 90 MHz. The insert of figure 1 shows the experimental arrangement that was used to prove the frequency bandwidth within the potential dynamical range of the measurements. The primary pressure pulse was induced at the absorbing layer-water interface and arrived at about 1.35 $\mu$s. This delay is necessary to pass 2 mm of coupling water layer. The second pulse has opposite phase due to reflection from the glass-air interface and passes twice the glass plate. According to the estimation the pressure pulse with pronounced compression phase is launched. The form of the detected signals is bipolar since the measured voltage drop is proportional to the current flow and therefore to the temporal derivative of the charges induced by the pressure on the electrodes of the foil. There are small ripples on the tail of the primary pulse due to the difference of the acoustical impedance of the materials used in the construction, for more details, please, see Ref [2]. The signal was averaged 256 times with a sampling rate 1 GS/s, yielding nanosecond temporal resolution of the separate pulses. The time interval between zero crossing points gives a value of 5.52±0.02 km/s for the phase velocity in the glass plate. The spectra of the primary pulse and the first reverberation are presented in the figure 2. They show a local minimum of the sensitivity at about 100 MHz, corresponding to an acoustical wavelength in PVDF equal to the thickness of the foil. The dynamic range of the experimental setup approaches 60 dB with the estimated accuracy of the spectrum measurements of few dB.

The attenuation of 2.25±0.01 mm thick HPT nickel of 99.99wt% purity was examined at geometrical center of the specimen with pc microstructure and near the edge where the equivalent strain ε approaches 1000% and uf microstructure is observed. As the diameter of the laser beam was about of 6 mm the diffraction losses in samples were negligible for frequencies higher than 5 MHz.

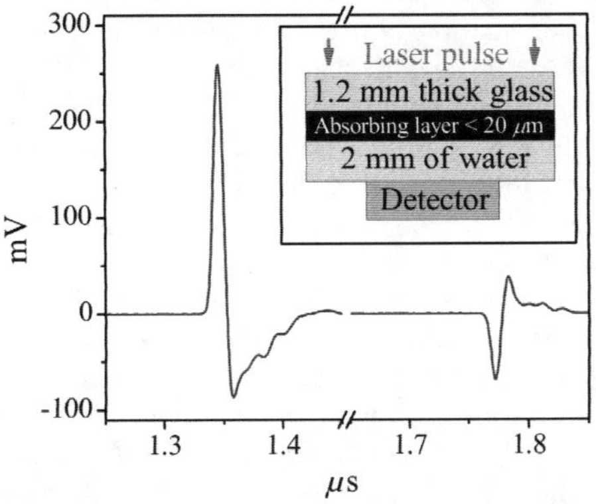


**Figure 1.** Signal detected for the estimation of the bandwidth. The insert shows the sketch of experimental arrangement.

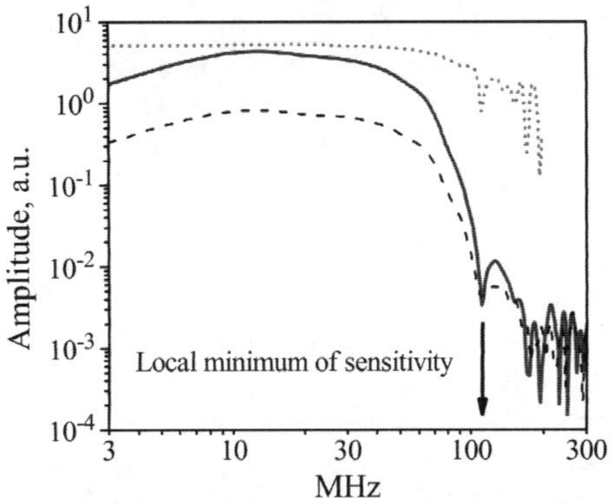


**Figure 2.** Spectra of the primary pulse (solid line) and its first reverberation (dash line); the dotted line denotes the ratio of the spectra.

The ultrasound scattering strongly influences the results of the measurements in pc metals with coarse microstructure [6]. The discrimination of the scattering and internal friction is a complicated problem. The losses due to the elastic scattering on the borders of the grains, which have different crystallographic orientation, strongly depend on the mean value of the grain size and the wave length [7]. The scattering yields small contribution in case of the attenuation estimated for HPT nickel [2], therefore, the attenuation due to the dislocation damping is dominant [8]. The attenuation was estimated as:

$$\alpha(f) = 20\lg\left(R^2 U_p(f)/U_1(f)\right)/2d, \quad (1)$$

where $f$ is the frequency, $U_p(f)$, $U_1(f)$ are the spectral amplitudes obtained from the Fourier transformation of the primary pulse and its first reverberation, respectively, $2d$ is twice the thickness, $R$ is a reflection coefficient which compensates the reduction of the pressure amplitude of the echo due to double reflection at metal-water interface. $R = (z_2 - z_1)/(z_2 + z_1) \approx -0.94$, here $z_1$ = 50 Mrayl, $z_2$ = 1.5 Mrayl are acoustical impedances of nickel and water, respectively. The detected pulses have initially pronounced compression phase while in the specimen the pulses propagating in the opposite direction away from transducer have pronounced tensile phase.

## 3. Results

In our experiment laser power density was varied to check the dependence of the attenuation on the pressure pulse amplitude. The efficiency of the laser power to pressure conversion was estimated as 1 Pa/(W/cm$^2$), giving an estimated maximum of absolute value of elastic strain of about $10^{-4}$. As the signal and its spectrum strongly depend on the local variation of the microstructure the position of the sample was fixed and only the laser power density was changed. The spectra of the primary pulses and the first echoes are presented in figure 3. The calculated attenuation spectra drastically decrease with the microstructure refinement, see figure 4.

The dislocation damping theory defines a resonance frequency as $2\pi f = \sqrt{G/\rho(1-\nu)}/L$, where $G$ is the elastic shear modulus, $\rho$ is the density of the medium, $\nu$ is Poisson's ratio and $L$ is dislocation loop length which has to be substituted by some mean or effective value for a distribution [8]. The typical loop length is about of $10^{-5}$ m in nondeformed pc metals. The estimations yield a resonance frequency in the range of $10^8$ Hz in agreement with the results obtained for non deformed pc nickel in figure 4. This phenomenological approach can explain the decrease of attenuation with grain refinement which reduces the dislocations loop length since the dislocations are pinned by the grain boundary atoms. The loop length is comparable with the linear grain size and the estimated resonance frequency is in the 100 GHz range for uf microstructure. Considering Cottrell pinning source Granato and Lücke discussed hysteretic attenuation for ultrasonic frequencies in the $10^4$ Hz range since the unloading part of the stress collapses the dislocation loops in another way than that in which they were formed. Granato-Lücke model and a recently presented revision [9] do not consider the interaction of dislocations as well as dislocations cells which are responsible for plastic deformation. The model considering the mesoscopic media predicted strong nonlinear behaviour of acoustic waves like harmonic generation and strain dependence of the attenuation because of the hysteresis [10].

In the range from 30 MHz up 55 MHz the attenuation shows small oscillations at the center of HPT nickel that proves strong nonlinear behavior of dislocation system. Due to the nanosecond pulse duration the strain rate achieves $10^4$ s$^{-1}$. The primary pressure pulse induces compression of the material with already developed network of dislocations. In the review paper [11] considering the response of the cubic pc metals on the shock three time scales, which fit to our experiment, were identified: the first one ~10 ns is assigned to dislocation slip; the second one ~100 ns is attributed to the dislocation interaction and material hardening; the third one ~1 $\mu$s is associated with the new defect source activation. The influence of dislocation motion is negligible since the Peierls energy of nickel is low [11]. It was stated that nickel with initial pc microstructure reaches the steady state conditions after ~0.5 $\mu$s by means of dislocations' cells formation [11]. Thus, the primary pulse of about 5 ns duration disturbs dislocation system which comes to a balance after at least 100 ns. A time of ~0.8 $\mu$s is required to pass twice the specimen but each reflection reverts the phase, turning pulses with pronounced compression phase into tensile pulses and vice versa. The previous strain decays in the medium with time constant defined by dislocation distribution. The variation of the attenuation for uf nickel is in the range of few dB/cm, which is about the error of the measurements but it can be more pronounced for higher frequencies. The measurements in pc nickel demonstrate a shift of the attenuation peak due to the change of pressure pulse amplitude. The connection of these attenuation peaks with a mean or effective value of length of dislocations' loops will be considered in the future.

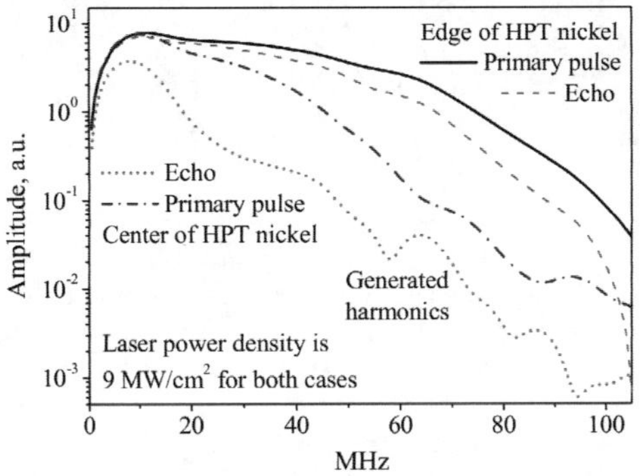


**Figure 3.** Spectra of the pulses in HPT nickel.

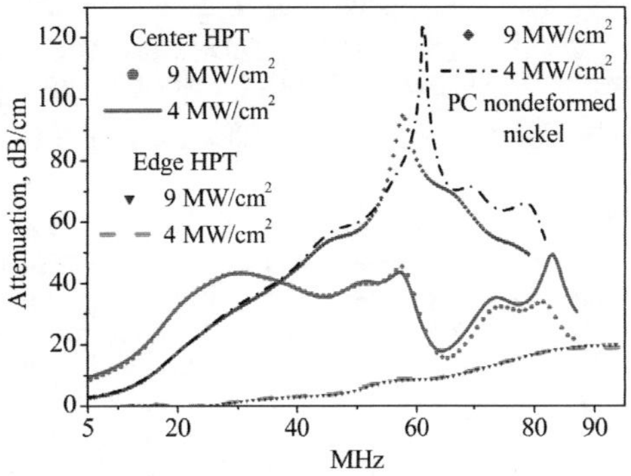


**Figure 4.** Attenuation in HPT and pc nickel.

## 4. Conclusion

The results show strong influence of the microstructural state of nickel on the ultrasound attenuation. The obtained results in HPT nickel provide a possibility to compare the variation of the ultrasound attenuation in severely deformed pc microstructure with high density of dislocations as well as in uf microstructure. The suggested experimental arrangement allows broadband detection in the frequency range 5 - 90 MHz which is comparable with the range of non-contact optical methods. The method seems to be very promising for the nondestructive evaluation of materials.

## 5. Acknowledgements

Financial support of this work by the Austrian Science Fund FWF (Projects S10407-N16) is gratefully acknowledged.

## References

[1] Pippan R, Scheriau S, Taylor A, Hafok M, Hohenwarter A and Bachmaier A 2010 *Annu. Rev. Mater. Res.* **40** 319
[2] Kozhushko V V, Paltauf G, Krenn H, Scheriau S and Pippan R 2010 *J. Appl. Phys.* **107** 094905
[3] Nazarov V E 2008 *Acoust. Phys.* **54** 283
[4] Caron A and Arnold W 2009 *Acta Mater.* **57** 4353
[5] Pelivanov I M, Kopylova D S, Podymova N B and Karabutov A A 2009 *J. Appl. Phys.* **106** 013507
[6] Kozhushko V V, Paltauf G, Krenn H, Scheriau S and Pippan R *NDT&E Int.* (submitted)
[7] Stanke F E and Kino G S 1984 *J. Acoust. Soc. Am.* **75** 665
[8] Granato A and Lücke K 1956 *J. Appl. Phys.* **27** 583
[9] Maurel A, Pagneux V, Barra F and Lund F 2005 *Phys. Rev. B.* **72** 17511
[10] Gusev V 2000 *Phys. Lett. A.* **271** 100
[11] Bourne N K, Gray III G T and Millett J C F 2009 *J Appl Phys* **106** 091301

Journal of Physics: Conference Series **278** (2011) 012013 doi:10.1088/1742-6596/278/1/012013

# Investigation of the fatigue process using nonlinear ultrasound

**R Ellwood, T Stratoudaki, S D Sharples, M Clark, M G Somekh**

Electrical Systems & Optics Research Division, Faculty of Engineering, University of Nottingham, University Park, Nottingham NG7 2RD, UK.

E-mail: `steve.sharples@nottingham.ac.uk`

**Abstract.** During normal usage components are subject to stresses that while not sufficient to cause fracture cause fatigue, which gradually weakens the component. Linear ultrasonic methods have been shown to be poor at detecting fatigue. However, there is evidence that the accumulation of damage gives the material a nonlinear elastic response that can be probed by ultrasound. By measuring the change in a material's nonlinear properties a measure of the fatigue can be obtained. Several methods of detecting material nonlinearity using acoustic waves have been proposed. The collinear mixing technique is used here. By measuring the velocity change of a probe wave due to the induced stress from a second pump wave, a measure of the nonlinearity is obtained. By generating the probe wave and detecting both waves using laser ultrasound techniques we gain the benefits of high spatial and temporal resolution. This is important when investigating the nonlinear response of a material as there is evidence that the microstructure affects the nonlinear response of a material. The change in nonlinearity over a region of a specimen (aluminium) has been monitored over several fatigue levels to investigate any relation. Early stage results are given with a discussion on the development of the technique.

## 1. Introduction

Linear ultrasound techniques are good at detecting large scale damage like gross cracks, however they are relatively poor [1][2] at detecting defects and precursors such as micro-cracks and dislocations which are associated with the initial formation of gross cracks. However nonlinear techniques show potential in detecting these damage precursors. A variety of nonlinear techniques have been developed for a range of materials to investigate this possibility. These methods include harmonic generation [3] and three phonon interaction [4]. Another group of nonlinear techniques of interest is that of collinear interaction. This group relies on a propagating wave being altered—by frequency modulation [5] or phase modulation [6]—by another wave or vibration and this method is used here. The majority of the published work in nonlinear ultrasonics is performed with bulk ultrasonics. However there is evidence [7] in some materials that micro-cracks and dislocations associated with fatigue initiate near the surface of a material. Surface acoustic waves (SAWs) are used as these are more sensitive to these defects. Material nonlinear response is not only associated to fatigue damage but also the microstructure of the material [8]. In this paper we first illustrate how this nonlinear response relates to large microstructure, then explore its relationship to fatigue.

## 2. Experimental configuration

A full description of our instrumentation can be found in a previous paper [9], the salient points are described briefly here. Figure 1 is a functional schematic of the nonlinear system.

A low frequency pump SAW acts to stress the material through which it propagates, this is excited using a standard contact transducer. A high frequency co-propagating probe SAW is affected by the stress state of the material. The probe wave is generated using a pulsed laser as this allows high frequencies and high spatial resolution, which permits greater control over interaction and simpler separation of the two waves. By varying the timing point of interaction between the two waves a different stress state is experienced by the probe wave. This causes the probe wave velocity to change due to the nonlinearity of the material. By comparing the change in phase velocity of the probe wave to the stress experienced as it propagates—see figure 2—a measure of the material nonlinearity can be obtained.

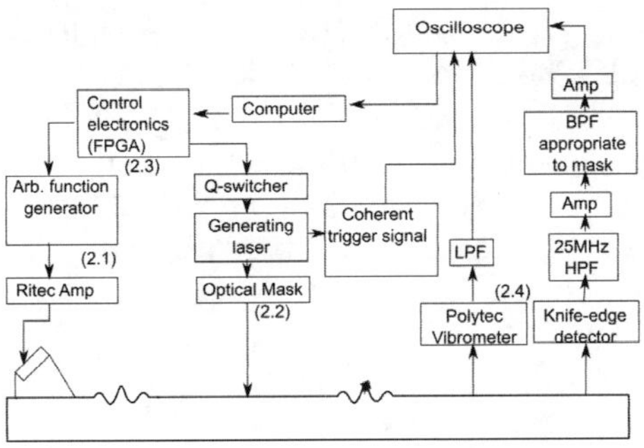


**Figure 1.** Nonlinear experiment schematic. The instrumentation is explained in the following sections in the main text: (2.1) pump; (2.2) probe; (2.3) delay control electronics; (2.4) SAW detection.

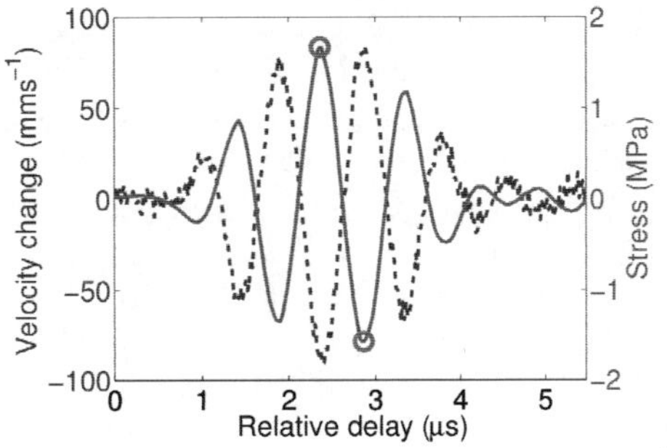


**Figure 2.** Measured velocity difference $\Delta v$ (blue - - - -) and stress $\sigma$ (green ——); maximum compressive and tensile stresses are indicated (red ∘)

### *2.1. Low frequency pump generation*

An NDT-Tech transducer with a central frequency of 1 MHz (A402S-SB) is attached to the sample, via an angle-wedge to excite the pump SAW. An arbitrary waveform generator produces a three cycle sine wave when triggered by the control electronics. This signal is then amplified and drives the transducer. Stresses of approximately 1–3 MPa are achieved.

### *2.2. High frequency probe wave generation*

The probe SAW is generated using an optical method with a Q-switched laser ($\lambda = 1064$ nm) with a 10 ns pulse width. The beam is patterned using an intensity grating mask and imaged onto the sample. A 45 $\mu$m grating is used, producing a SAW pulse packet with a central frequency of around 68 MHz.

### *2.3. Control electronics*

The stress field experienced by the probe wave is determined by where the probe wave sits on the pump wave. The interaction between these two waves is controlled by precisely timing the points at which they are generated. An FPGA (field programmable gate array) controls the triggering of the pump and probe wave by a computer-set variable delay. The velocity changes due to the imposed stress are very small, of the order of 0.01 $ms^{-1}$. This requires a large number of averages—taking tens of seconds—to ensure a sufficiently high SNR (signal to noise ratio). The velocity measurement could easily be overwhelmed by gradually changing environmental factors such as temperature, as the coefficient relating SAW velocity to temperature is approximately 1–4 $ms^{-1}\,K^{-1}$. In addition to some basic temperature control, we implemented a differential interleaving process where by the probe wave is switched rapidly (30 Hz) between interacting with a stress state (target state) and a zero stress state (reference state). The reference wave will be subject to the same environmental conditions as it is interleaved with the target wave. By subtracting the velocity change of the reference state from that of the target state it is possible to eliminate environmental factors. With a sufficiently high number of averages (16k), the standard deviation on the measured velocity change is less than 6 $mm\,s^{-1}$.

### *2.4. SAW detection and data processing*

The pump wave is detected using a Polytec vibrometer (30 kHz–24 MHz), this provides a calibrated displacement that can be converted to stress. The probe wave is detected using a split knife-edge (beam deflection) detector with a broadband response (400 kHz–450 MHz).

The time shift between the detected target and reference probe waveforms is determined using a full spectrum method: cross correlation is used to determine the approximate delay, then the slope of the phase of all frequencies—weighted by the energy at each discrete frequency—is determined using a least mean squares fit. This time shift can then be converted into a velocity shift. The probe wave has a significantly smaller wavelength than that of the pump wave, this means that the probe wave energy is confined to the top region of the pump wave where the stress is essentially longitudinal [10]. The probe velocity change $\Delta v$ and the pump stress $\sigma$ are combined to give the nonlinearity coefficient $\Delta v/\sigma$ in $mm\,s^{-1}/MPa$.

## 3. Experimental results

By reducing the number of interaction points between the pump and probe waves to two—the peak and trough of the pump wave, see figure 2—the measurement time is reduced to 40 seconds per point. This allows imaging of nonlinearity over the material surface within a sensible time frame. The lateral resolution is determined by the pump/probe interaction distance. On the one hand, longer distances mean bigger time shifts for a given change in velocity, improving SNR; on the other, attenuation and acoustic aberration reduce the SNR, especially for higher frequencies. The optimal distance for our setup is 3 mm.

Experiments were conducted on an aluminium 6061 sample with grains 3–4 mm in size. The linear SAW velocity was imaged using spatially resolved acoustic spectroscopy (SRAS) [11], revealing the microstructure of the sample, see figure 3(A). The area of the nonlinear scan is outlined in the dashed box, taking into account the 3 mm interaction distance (the box is offset by 1.5 mm). This scan area was selected because it covers an area containing part of a very large grain. The region on the right of figure 3(B) indicates nonlinearity of approximately $-60$ $mm\,s^{-1}/MPa$, in the rest of the scanned area the value is closer to $-20$ $mm\,s^{-1}/MPa$. The boundary between these two areas coincides with the edge of a large grain, indicating that there is a relationship between the measured material nonlinearity and the microstructure.

Preliminary measurements have been taken to determine the effects of fatigue on nonlinearity. A $2\times2$ mm area of an aluminium 2024-T351 sample was imaged at 0 %, 60 % and 80 % of its fatigue life (figure 4A–C). There was a overall increase in the mean nonlinear value from $-56.6$ to $-61.4$ $mm\,s^{-1}/MPa$. The most striking trend is that the standard deviation changed from 4.4 to 2.8 $mm\,s^{-1}/MPa$. To put this in context, the standard deviation measurement is repeatable to $< 0.1$ $mm\,s^{-1}/MPa$.

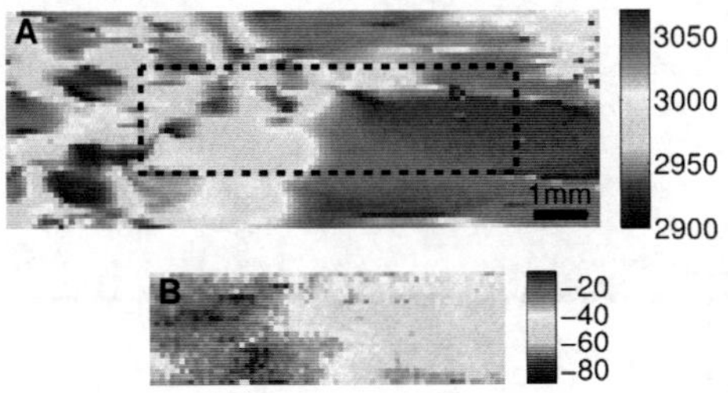


**Figure 3.** (A) Linear SAW velocity ($ms^{-1}$), revealing the microstructure of the area around the nonlinearity scan below, which is indicated by a dashed box. (B) Nonlinearity ($mm\,s^{-1}$/MPa) for the area in the dashed box above, taking into account the 3 mm pump/probe interaction distance.

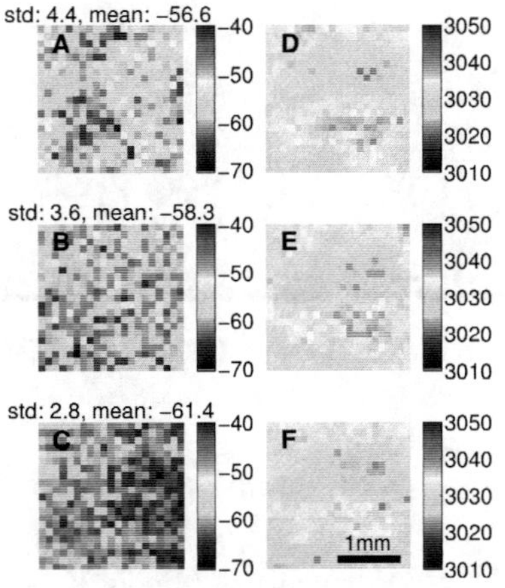


**Figure 4.** Left: nonlinearity ($mm\,s^{-1}$/MPa) at 0 % (A), 60 % (B), 80 % (C) of fatigue life. Right: linear velocity ($ms^{-1}$) measured in the same area at the same time (D–F).

## 4. Discussion and further work

In this paper we demonstrate a method to image the nonlinear response of aluminium based on the collinear mixing technique. Scans of the nonlinear response of an unfatigued large grained sample show a relationship between the nonlinear response of the material and the microstructure. A scan shows wide spatial variation of nonlinearity between −20 and −60 $mm\,s^{-1}$/MPa. This implies that either careful registration, or averaging over a large region, is required for experiments monitoring fatigue. Initial results for a fatigued sample have shown a small increase in the mean nonlinear response across the sample microstructure. The reduction in the variability in the material nonlinearity is interesting, and more experimental evidence is required to determine whether this is a viable metric for fatigue life or not, and if so, then what physical mechanism is responsible. Further proposed work includes investigation into the spatial variation of material nonlinearity with fatigue. Samples that have microstructure considerably smaller—and larger—than the co-propagation distance will also be investigated. This work is supported by the UK RCNDE (Research Centre in Nondestructive Evaluation), and the UK TSB (Technology Strategy Board).

## References

[1] Nagy P, Blaho G and Adler L, 1993 *Rev. Prog. in QNDE* **13**, 1987-94
[2] Frouin J, Sathish S, Matikas T and NA J 1999 *Journal of Materials Research* **14** 1295-98
[3] Dace G, Thompson R and Buck O 1991 *Rev. Prog. in QNDE* **11**, 1685-92
[4] Rollins F Jr, Taylor L H and Todd P H Jr, 1964 *Physical Review* **136 3A** A591 -601
[5] Li P, Winfree W, Yost W and Cantrell J *Proceedings of the 1983 IEEE Ultrasonics Symposium*,(Atlanta, 1983), 1152-56
[6] Jacob X, Barriere C and Royer D 2003 *Appl. Phys. letters* **82**, 886-8
[7] Buck O, Morris W L and Richardson J M 1978 *Appl. Phys. letters* **33(5)** 371-3.
[8] *Nonlinear Surface Acoustic Waves in Cubic Crystals* Kumon R, The University of Texas at Austin (December 1999).
[9] Sharples S, Stratoudaki T, Ellwood R, Collison I, Clark M and Somekh M 2009 *Rev. Prog. in QNDE* **29A**, 287-94
[10] Viktorov I 1967 *Rayleigh and Lamb Waves: Physical Theory and Applications* (New York: Plenum Press)
[11] Sharples S, Clark M and Somekh M G, 2005 *IEEE Ultrasonics Symposium* 886-9

Journal of Physics: Conference Series **278** (2011) 012014 doi:10.1088/1742-6596/278/1/012014

# Laser ultrasonics for civil engineering : some applications in development for concrete non destructive testing

**O Abraham[1], JS Popovics[2], L-M Cottineau[1], O Durand[1]**

[1] LCPC, MACS Dept., Route de Bouaye, BP4129, 44341 Bouguenais cedex
[2] University of Illinois 205 N. Mathews Ave. MC 250 Urbana IL 61801 USA

odile.abraham@lcpc.fr

**Abstract**. Non destructive testing of civil engineering infrastructures is becoming of primary importance for their diagnosis, residual time life estimation and/or structural health monitoring. A particularity of civil engineering application is the large size of the survey zones and the expected low cost of inspection. In this context non contact ultrasonics may offer the possibility to built robots that can automatically scan large areas (or eventually be integrated in moving vehicles) to recover mechanical properties of material or to perform imagery for geometrical information recovery. In this paper we present two possible applications of *in situ* laser ultrasonics : one is the detection of voids in tendon duct with the impact echo method, the other is the use of surface waves to recover mechanical properties of the first centimetres of concrete structures (here after called cover concrete).

## 1. Introduction

The use of non contact ultrasonic measurements is becoming more and more popular in the field of civil engineering as the surfaces to be investigated are very large and speed is a key factor to come up with economically viable non destructive testing methods. Alongside with air coupled sensors some developments are underway at LCPC with laser interferometers. For on-site measurements a Polythec PI (OFV-505 sensor and OFV-5000 controller) with a VD-02 demodulator is currently used at LCPC. Two applications are described below: one, called impact echo, in the frequency range 1kHz – 60 kHz for the detection of voids of decimeter size in concrete wall, the other , surface waves analysis, in the frequency range 50 kHz – 200 kHz for the mechanical characterization of cover concrete.

Impact echo is a resonance method that has been developed more than 20 years in concrete slabs to measure thickness and to detected interface between layers. Recent works have clearly established than the underlying physics is a resonance phenomenon associated to the Zero Group Velocity frequency of the first symmetric Lamb mode [2,3]. Here we investigate the modification of the resonance response (measure with laser interferometer mounted on a scanning robot) of concrete wall where pipes of various filling have been embedded.

For material characterization of cover concrete we have developed a robot that records seismograms with a laser interferometer. We focus here on surface waves that offers from measurements at the surface the possibility to reconstructed shear wave velocity profiles as a function of depth. The dispersion curves obtained on 8 concrete mixes of various porosity at water saturation ranging from 0 to 100% are commented.

## 2. Impact echo for the detection of voids in tendon ducts

The detection of voids in tendon ducts with the impact echo method is a real challenge as it could result in the diminution of gammagraphy need. Two major phenomena are indicative of the presence of a void in the impact echo method: a decrease of the thickness resonance frequency $f_e$ [1] and the apparition of a higher frequency in the literature named $f_{void}$ . The first observable alone might be in a number of case misleading as a frequency shift has been sometime observed above tendon that are diagnosed as fully filled by gammagraphy. In this paper we follow the thickness frequency shift above various tendon filling showing that this single phenomenon is not sufficient to diagnose of void.

### *2.1. Experimental set-up*

In order to carry out reproducible and reliable impact echo diagnosis imagery, using B-SCAN or C-SCAN, is recommended. This is why a robot we designed to conduct measurements over a surface of almost 4 m$^2$ with a precision of 5 mm in both vertical and horizontal direction (figure 1). The source system is a steel ball impacting the concrete surface driven by an electromagnet. The source is maintained at a fixed distance from the wall with a wheel so that the impact is reproducible. The sensitivity of the laser interferometer is set at 5 mm.s$^{-1}$.V$^{-1}$ with a corresponding bandwidth from 0 to 250 kHz. One originality of the robot is the possible movements of the source and the receiver independently one from another in the horizontal direction. Indeed, our objective in a near future is to take advantage of this degree of freedom to perform spatial stacking to improve signal to noise ratio and to remove the Lamb fundamental antisymmetric mode perturbation.

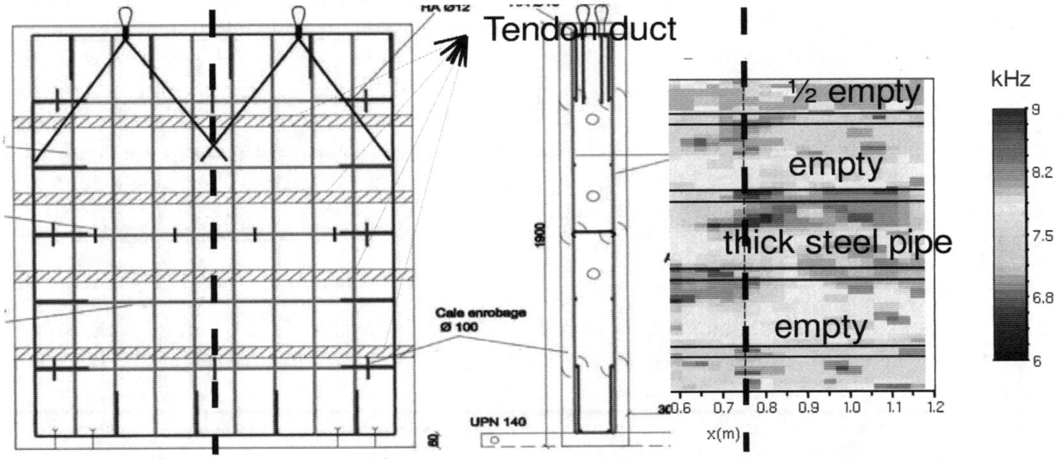


**Figure 1.** Left : plan of the wall Right: frequency of the maximum of the FFT (C-SCAN)

### *2.2. Results*

A wall 0.25 m thick including 8 tendon ducts with various filling has been built at LCPC (figure 1 left). The source used here has a diameter of 19 mm. No wall preparation is required for the measurement as the energy level is sufficient to record signal with a good signal to noise ratio. Figure 1 (right) shows a C-SCAN of the frequency of the maximum of the FFT of each signal. As expected above empty tendon duct we see a decrease of the thickness resonance frequency. In the case of a filled pipe with epoxy having a rigidity lower than normal cement grout the shift is also visible. Finally this shift is not present in the case of a empty thick steel pipe. The resonance frequency is linked to the global local stiffness of the slab.

## 3. Surface waves for concrete characterization

Within the scope of the French National Agency project SENSO several non destructive methods have been studied with the aim of characterizing concrete mixes of various compositions including variation of aggregate type and size, porosity, water content, chloride content and carbonation depth.

Surface Waves (SW) are a good candidate to evaluate non destructively the mechanical properties of cover concrete and thus to tackle the problem of assessing *in situ*, non destructively, some durability indicators. SW can be easily generated and represent the more energetic wave train recorded at the surface. Surface wave investigation depth is approximately equal to half their wavelength, so that, their velocity varies with frequency as soon as the material presents some mechanical characteristics varying with depth. The phase velocity dispersion curve, namely the phase velocity as a function of frequency, becomes the input to an inverse problem which aim is to recover as a function of depth the shear wave velocity $V_S$ .

In the SENSO project, 9 concrete mixes have been studied. For each concrete mix 8 slabs, of 0.5#m x 0.25 m x 0.12 m size, are available to perform the parametric study. For dry and fully saturated concrete the 8 slabs are available. For subsequent water content (20 %, 40 %, 80 %), carbonation depth and chloride content, this number is reduced. In order to perform the requested large number of measurements a robot that uses a non contact receiver has been designed.

### *3.1. Experimental set-up*

SW are generated with a sensor in contact. This source has been designed especially for this purpose by the French society IMASONIC. It is made of a matrix of piezoelectric components. Its is a large band transducer with a central frequency at 110 kHz [50 kHz – 300 kHz at -3 dB]. A wedge adapted for a Rayleigh wave speed around 2100 m/s is used to favor the generation of SW. The source is amplified by a Ritec RAM 500 amplifier. The receiver is a laser interferometer from Polythec PI (OFV-505 sensor and OFV-5000 controller) with a VD-02 demodulator that has a sensitivity of 25 $mm.s^{-1}.V^{-1}$ and a bandwidth from 0 to 1.5 MHz. The measurements points are aligned with the source. An aluminum tape is glued onto the line to improve the reflectivity of the surface. The laser interferometer is moved automatically every 0.005 m, from 0.01 m to the source up to 0.43 m, the largest offset.

### *3.2. Results*

Figure 2 shows one recorded seismogram and a zoom on one signal situated at 0.2 m from the source. All signals are windowed automatically based on an automatic procedure that picks the maximum amplitude of each trace to center the window.

We have decided to follow within the SENSO project the phase wave velocity $V\varphi$ at given wavelengths. Our aim was to always investigate the slabs at given depths range: six values $V\varphi$/#called observables, corresponding to wavelengths from 0.01 m to 0.06 m, were extracted from the dispersion curves. The observable that has proven to be the more relevant, on the basis of statistical studies carried out by other partners of the project [6] was $V\varphi$#having a wavelength equal to 0.03 m. Obviously smaller wavelengths are more sensitive to the heterogeneousness of concrete, while larger wavelengths could be perturbed by the thickness of the slab here equal to 0.12 m. Finally this wavelength was in the frequency region where our source was the more energetic so that signal to noise ratio was optimal.

The phase velocity dispersion curves are computed with the slant stack method in the frequency domain [4].

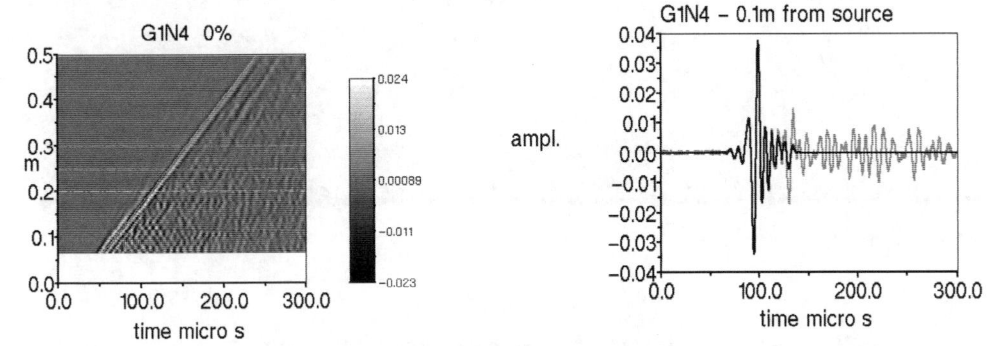


**Figure 2.** Left : one seismogram. Right : one signal at 0.2 m from the source.

Figure 3 shows examples of phase velocity dispersion curves obtained for the more (G8) and less (G1) porous concrete in the fully saturated (100%) and dry (0%) cases. The water to cement ratio respectively equals to 0.9 for G8 and 0.3 for G1. The inversion of the surface wave dispersion curves inform on the homogeneity of the water content with depth.

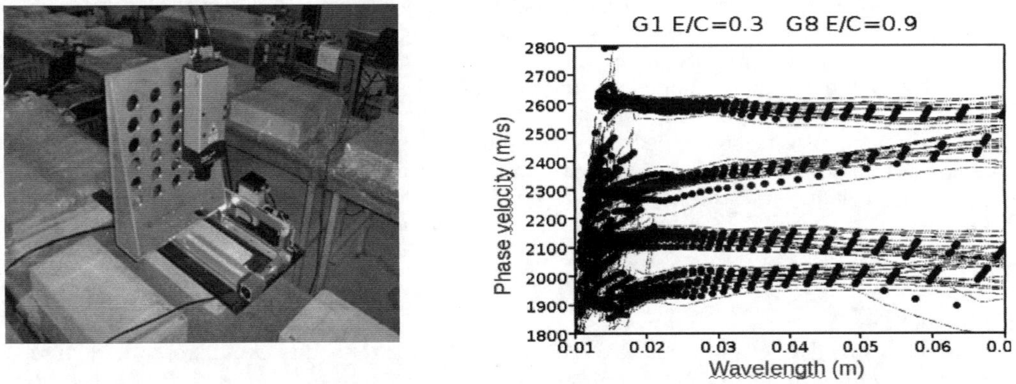


**Figure 3.** Left : robot Right : *Phase velocity dispersion curves for G8 (W/C=0.9) and G1 (W/C=0.3)] for the [saturated case – 100%] and [dry case – 0%]*

**Acknoledgement**

We would like to thank the French National Agency for Research (ANR) and Electricité De France (EDF) for funding the research projects ACTENA and SENSO. We would like to thank Fabrice Blaineau (LCPC) for the design of the robot and Alain Grosseau (LCPC) for its realization. This work received also help from Michel Valade from Lyon LRPC, France, for in situ impact echo measurement and from Safia Bedaoui and Pierre Argoul from LAMI, ENPC, France, for signal processing.

**References**

[1] Abraham O. and Côte Ph. 2002 *ACI Struct. J.* **99**(3) 239-247
[2] Gibson A. and Popovics J S. 2005 ASCE *J. of Eng. Mech.* **131**(4) 438–443
[3] Prada C., Clorennec D. and Royer D. 2008 *Wave Motion* **45**(6) 723–728
[4] Mokhtar T. A., Herrmann R. B. and Russel D.R. 1988 *Geophysics* **53**(11) 1379-1387

Journal of Physics: Conference Series **278** (2011) 012015 doi:10.1088/1742-6596/278/1/012015

# Computer simulation of ultrasonic testing for aerospace vehicle.

**H Yamawaki[1], S Moriya[2], T Masuoka[2] and J Takatsubo[3]**
[1]National Institute for Materials Science, 1-2-1, Sengen, 305-0047 Tsukuba, Japan
[2]Japan Aerospace Exploration Agency, 1 Koganesawa, Kimigawa, 981-1525 Kakuda, Japan
[3]Advanced Industrial Science and Technology, AIST Tsukuba Central 2, 1-1-1 Umezono, 305-8568 Tsukuba, Japan

E-mail: [1]yamawaki.hisashi@nims.go.jp

**Abstract.** Non-destructive testing techniques are developed to secure reliability of aerospace vehicles used repetitively. In the case of cracks caused by thermal stress on walls in combustion chambers of liquid-fuel rockets, it is examined by ultrasonic waves visualization technique developed in AIST. The technique is composed with non-contact ultrasonic generation by pulsed-laser scanning, piezoelectric transducer for the ultrasonic detection, and image reconstruction processing. It enables detection of defects by visualization of ultrasonic waves scattered by the defects. In NIMS, the condition of the detection by the visualization is investigated using computer simulation for ultrasonic propagation that has capability of fast 3-D calculation. The simulation technique is based on finite-difference method and two-step elastic wave equations. It is reported about the investigation by the calculation, and shows availability of the simulation for the ultrasonic testing technique of the wall cracks.

## 1. Introduction

Various nondestructive testing methods are examined for the reliability evaluation of structures for the space. Development of health evaluation method before and after the rocket use is a current target for the liquid-fuel rocket repetitively used. One of the developments is about non-destructive inspection for the combustion chamber of the liquid fuel engine. Inner wall of the chamber is affected by severe stress/thermal environment, and it may causes thinning and cracking of the wall[1,2]. As shown in figure 1, the wall has small channels where liquid hydrogen flows, and is made with Cu-alloy. The thinning occurs at the thin wall between the inner surface and the channels. Following the thinning, crack initiation and its growth occurs. Small defects of the combustion chamber may lead to severe accident of the rocket, so that precise investigation of the wall is required.

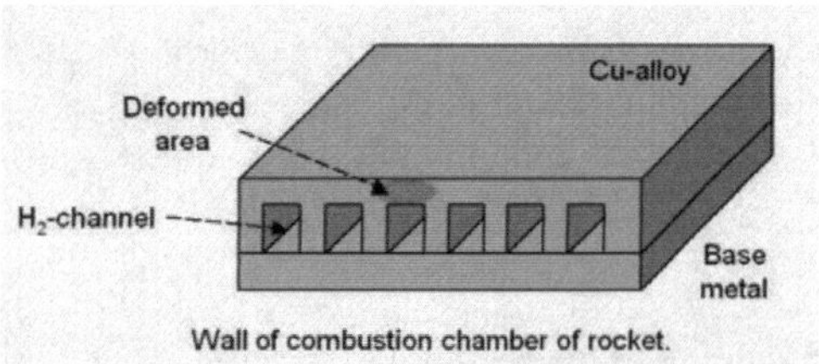


**Figure 1.** Schematic of wall of the combustion chamber. Upside is the inner surface. Repetition of combustion gas pressure and thermal stress causes deformation of the indicated area. Crack occurs from inside surface of H2-channel. Thickness of the area is around 1mm.

For the purpose, non-contact visualisation technique of ultrasonic propagation developed in AIST[3,4] is examined. The technique is composed with non-contact ultrasonic generation by scanning pulsed-laser with weak power, a position-fixed piezoelectric transducer for the ultrasonic reception, and image reconstruction process based on reciprocity theorem. The image reconstruction by reciprocity theorem gives the animation as if the ultrasonic surface waves and/or Lamb waves spread from the transducer along specimen surface. Existence of defects located near surface can be recognized as scattered surface/Lamb waves in the animation.

In parallel with the experiment of the imaging, the computer simulation of the visualisation was examined. Computer simulation of ultrasonic propagation is expected as tool for analysis and investigation of experimental condition. In this report, the computer simulation for the NDT of inner wall of the combustion chamber is described.

## 2. Computer simulation method

The calculation method[5] for ultrasonic propagation is based on the finite-difference method. The method uses 2-stages elastic wave equations composed with two independent variables of $v$: particle velocity and $T$: stress, as shown in eq.(1)[9], where ρ: density, and $C$: stiffness matrix. Originally, the 2-stages equations are used in FDTD[7] and EFIT[8] methods. Difference between the method and FDTD/EFIT exists on discretized elastic material models. Figure 2 shows the model used in the previous technique of NIMS[6], and is also used in the new technique used here. This model gives easy interpretation of calculated results, and easy applicability for combined materials. Benefits of employment of eq.(1) were clarified by actual calculations as improvement of calculation speed, calculation stability at boundary, and applicability for air-coupled ultrasound.

Calculation model of the wall in figure 1 is constructed using combination of the unit cubes shown in figure 2. Each unit cube owns density and stiffness matrix of Cu-alloy. Crack is constructed as vacancy of unit cubes. Size of unit cube (spatial step) and duration of time step are determined by the same way as the finite-difference method.

$$\rho\frac{\partial v}{\partial t} = \nabla \cdot T \qquad \frac{\partial T}{\partial t} = C\nabla_s v \qquad (1)$$

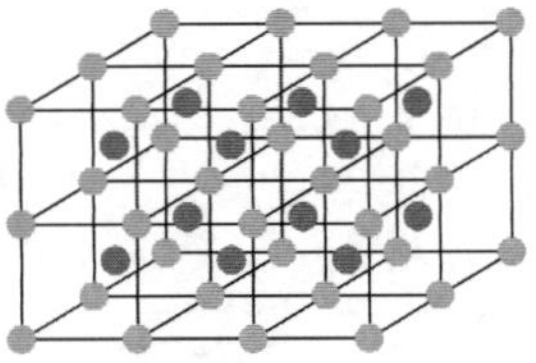

**Figure 2.** Discretized 3-D elastic material model for calculation. The corner points of each cube represents particle velocities, and the center point of each cube represents stress.

## 3. Two-Dimensional calculation for crack detection using surface wave

The surface wave (Rayleigh wave) propagates in surface layer of solid, and penetration depth of surface waves depends on the wavelength, so that it is suitable for detection of defects located near surface. By selecting frequency of surface waves, subsurface defects could be detected without disturbance of subsurface structure such as H2-channel grooves in figure 1. Figure 3 shows 2-D calculation model to verify reflection of surface wave from a subsurface crack (slit). Only 3 subsurface grooves structure in wall of combustion chamber is modeled for the simplification of the calculation. The wall thickness is equal to 2 wavelengths of surface wave. The surface wave is generated selectively by applying sinusoidal stress of three cycles in the manor of the phased array. The slit with one spatial step in width is installed from the groove side in center part of the second groove. The relation between intensity of reflected wave from the slit and depth of the slit was examined.

In figure 3, reflected surface wave by the subsurface slit is observed clearly. In figure 4, waveform observed near the slit (point B) shows waves of various modes generated on the slit, however only the wave of mode of surface wave propagates stably along the surface (point A). The graph in figure 4 shows that deeper (longer in depth) slit or shorter residual thickness at slit position gives stronger reflection of surface waves. Especially the shorter residual thickness than one wavelength of surface wave (>50% in the graph) shows larger increasing rate of the reflected surface waves. From these results, by selecting the appropriate wavelength of surface wave, even with complex subsurface structure such as the grooves, it is expected that the reflected surface wave from the crack can be distinguished, and there is a possibility of estimation of the depth of the subsurface cracks.

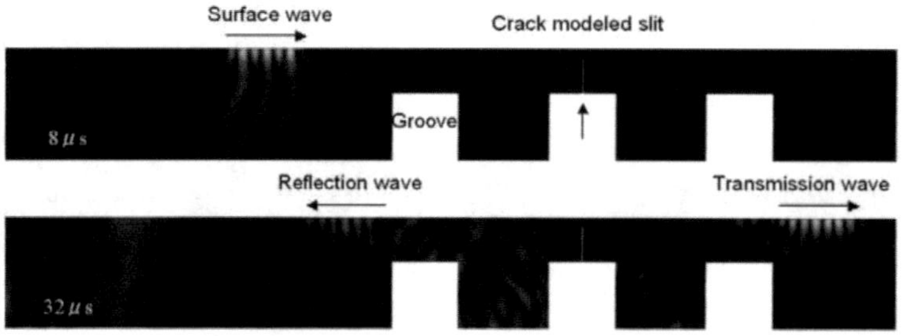


**Figure 3.** 2-D Calculation model of the wall with crack (slit), and visualization of ultrasonic waves (shown in displacement). Depth of slit is 80% of wall thickness. Wavelength of Surface wave is 50% of the wall thickness at the grooves. Model size is 1600 x 200 spatial steps.

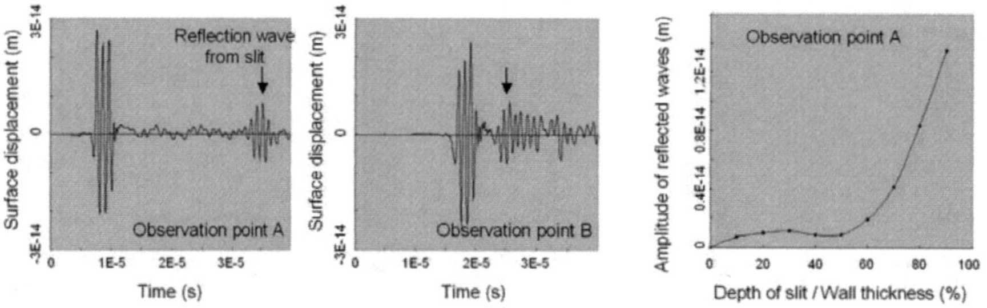


**Figure 4.** Waveforms of incident surface wave and reflected waves calculated by the simulation (case of slit depth in figure 3), and relation between depth of slit and strength of reflected wave from the slit. Point A is on the surface of the generation area, and point B is on the middle surface between first and second grooves, in figure 3.

## 4. Three-dimensional calculation for visualisation of surface waves with cracks

Currently, transfer function of the receiving transducer is not taken into account in the calculation, so results obtained by the calculation are not perfectly corresponding to the experimental results. However, it is obvious that the ultrasonic visualisation technique of AIST and the computer simulation have high adaptability. Figure 5 shows 3-D calculation model for the wall. The new simulation method enables efficient 3-D calculation of ultrasound using personal computers.

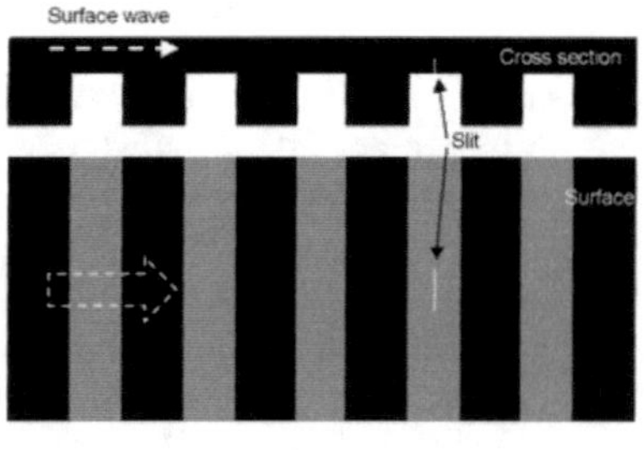


**Figure 5.** 3-D Calculation model for the wall with grooves and slit. Plane surface wave propagates toward to area with a subsurface slit. Model size is 720 x 300 x 100 spatial steps

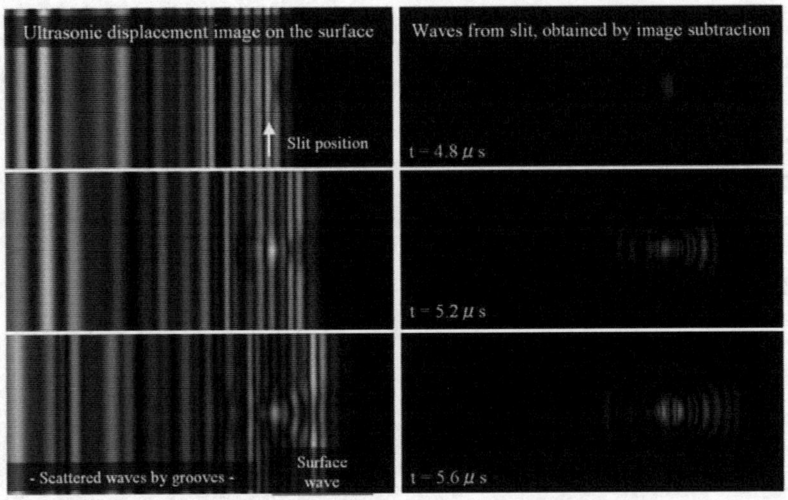


**Figure 6.** Snapshots of ultrasonic displacement at the surface. Three cycles sinusoidal surface wave propagates. Wavelength of the surface wave is equal to the wall thickness (0.8mm), and depth of the slit is 50% (0.4mm) of the thickness. After the surface waves, various scattered waves by grooves are observed in left images. Right images are obtained by image subtraction between images with the slit and without it. Scattered waves by the slit are selectively observed by the subtraction.

In figure 6, existence of slit is recognized clearly by the deformed wave front. In this case, wavelength of the surface wave is large in comparison with wall thickness, so that scattered waves by the grooves are remarkable. In the actual experiments, signal processing to eliminate progressive wave is performed. By the processing, effect similar to the image subtraction in figure 6 is performed.

## 5. Conclusions

By the computer simulation for ultrasonic propagation, it was clarified that the non-contact visualisation technique of ultrasonic propagation is effective for the defect detection of combustion chamber of liquid-fuel rocket. Also enhancement of detectability of defects by image processing was demonstrated in the simulation. It was confirmed that the simulation was effective to secure reliability of the rocket. It is expected wider application of the computer simulation for ultrasonic non-destructive testing.

## References

[1] Cook R T, Fryk E E and Newell J F 1983 SSME Main Combustion Chamber Life Prediction *NASA report* CR-168215.
[2] Popp M 2003 Cooling and Life Issues for Long Life Combustion Chambers *Proc. 5th Int. Symposium on Liquid Space Propulsion* (Chattanooga, Tennessee, October 27-30 2003)
[3] Takatsubo J, Wang B, Tsuda H and Toyama N 2007 *J. Solid Mechanics and Materials Engineering* **1** 1405
[4] Lee J, Takatsubo J, Toyama N and Kang D 2007 *Measurement Science & Technology.* **18** 3816
[5] Yamawaki H 2010 *J. of the Japanese Society for Non-Destructive Inspection* **59** 624 (in Japanese)
[6] Yamawaki H and Saito T 2000 *NDT&E International* **33** 489
[7] Sato M 1995 Jpn. J. Appl. Phys. **34** 2808
[8] Langenberg K J, Hannemann R, Kaczorowski T, Marklein R, Koehler B, Schurig C and Walte F 2000 *NDT&E International* **33** 465
[9] Auld B A 1990 *Acoustic Fields and Waves in Solids,* Vol.I (New York: Krieger) 164

Journal of Physics: Conference Series **278** (2011) 012016 doi:10.1088/1742-6596/278/1/012016

# Interaction between femtosecond radiation and sound in a light dispersive delay lines using effect of strong elastic anisotropy

**V Ya Molchanov, S I Chizhikov, O Yu Makarov**

Acoustooptical Research Center, National University of Science and Technology "MISIS", Leninsky prospect, 4, 119049 Moscow, Russia

E-mail: v.ya.molchanov@gmail.com

**Abstract.** Femtosecond optical amplifiers typically have high nonlinear spectral dispersion breaking relationships of the spectral phases of femtosecond pulse. It is necessary to use some additional dispersive device for controlling of spectral amplitude and phases of amplified laser pulses. We discuss the investigation of light dispersive delay lines based on the light-sound interaction in crystals. The light dispersive delay lines geometry uses the effect of strong elastic, photoelastic and optic anisotropy in the $TeO_2$ single crystal. Different experimental delay lines were designed and tested. The problem of femtosecond laser beam angular chirp compensation was solved also. The experiments were performed at the OPCPA femtosecond laser system.

## 1. Introduction

The creation of superpowerful femtosecond laser systems has been one of the most striking breakthroughs of physics at the turn of the $20^{th}$-$21^{st}$ century. The most recent decade is characterized by the conquest of new frontiers in energetic and temporal characteristics of femtosecond laser radiation. Previously reached record power, and impulse duration values have been exceeded by several orders of magnitude.

Today, peak laser impulse power achieved in just several of the world's labs amounts to 1 petawatt, a hundred times higher than the sum total of all energy sources on our planet. In the nearest future, such laser sources will provide the opportunity to conduct fundamental research in absolutely new areas. For instance, it will be possible to study the properties of vacuum in the presence of superstrong fields, to model in laboratory conditions the processes at the core of stars and planets, or to perform quantum control of intromolecular and interatomic processes. The newest line of research is the creation of laser-powered charged-particle accelerators based on the impact of petawatt optical impulses on gas and solid state targets, creating new possibilities for low-dose x-ray phase contrast tomography, hadron therapy for cancer, protonography, etc.

In most applications, the quality of compressed pulses (pulse duration, lack of prepulses, and high contrast ratio) are critical. The pulse quality is determined by a noncompensated high-order dispersions and spectral deformation in the amplifiers. An adaptive device that compensates for the induced high-order dispersion and spectrum distortion in the system by modifying the spectral amplitude and phase of the seed femtosecond pulse can be used to decrease the duration of the compressed pulse and suppress the prepulses. This conceptual solution helps increase femtosecond radiation power not by increasing the energy of the laser system, but by decreasing the duration of the pulse due to the correctly arranged spectral phase. Independent control of the spectral amplitude helps produce pulses of required shape, oriented at optimal conditions for amplifying the subsequent cascades.

Currently, three principal methods for controlling the spectral phase are known: chirped mirrors, a line of liquid crystal-based phase-modulated transparencies, and wave-electronics devices: adaptive time-delay

Journal of Physics: Conference Series **278** (2011) 012016 doi:10.1088/1742-6596/278/1/012016

lines based on the acoustooptical effect. The experience of working with powerful laser systems shows that the third method has a number of incontestable advantages, such as linear geometry that doesn't require to be rigidly coordinated with stretcher geometry; continuous phase control; the possibility of independently controlling not only spectral phase, but also amplitude; high-speed adaptation facilities.

Thanks to these advantages, acoustooptical time-delay lines have proven themselves to be the best among such devices.

## 2. Light spectra processing by acoustooptical light dispersive delay lines

A collinear acoustooptic filter for compression of ultrashort laser pulses was first proposed in 1986 by Pustovoit and Pozhar [1, 2]. According to their concept the anisotropic acoustooptical interaction converts the fast (slow) input spectral component into a slow (fast) diffracted component. The amplitude and the phase of each spectral component is determined by the sound spectral amplitudes and the frequencies. The control of sound waves provides laser pulse shaping both in the spectral phase delay and the amplitude; a concept that proved to be practical. Tournois [3, 4] proposed in 1997 the same technical solution for femtosecond pulse shaping termed an acoustooptic programmable dispersive filter (AOPDF). Now this system, the Dazzler, manufactured by Fastlite, France, is widely used in femtosecond laser equipment.

Acoustooptical light dispersive delay lines (LDDL) remain a virtually unstudied new generation of photonics devices. Consumer feedback shows that the Dazzler has a number of typical weaknesses: low diffraction efficiency, low spectral resolution, limited optical delay and low adaptive ability [5, 6]. The aim of this paper is the further development of LDDL based on $TeO_2$ crystal for optimal performance.

We would like to point attention on the features of LDDL. Acoustooptic interaction always performs consequent spectrum analysis. LDDL is the only type among all other acoustooptical devices providing parallel but not consistent light spectrum processing. For example, acoustooptical modulators and deflectors control monochromatic laser radiation by multifrequency sound. Acoustooptical filter provides consistent light spectrum analysis by tunable monochromatic sound. Factually, LDDL can be considered as tunable acoustooptical filter, having adaptive spectral transmission function (spectral window) synthesized by multifrequency sound injected into the crystal according to the given rules [7, 8].

## 3. Theoretical consideration

The analysis has been carried out for plane (1-10) of paratellurite crystal according [9]. The well-known formalism of wave-vector diagrams for the uniaxial crystal was used. This formalism defines the relationship between the frequency of Bragg synchronism and the angles of propagation of light and sound relative to crystallographic axes with the understanding that waves are considered to be plane. The efficiency of interaction of light with sound is determined by the effective photoelastic constant. From the definition of the quasicollinear acoustooptical interaction follows that wave vector $\boldsymbol{k}_i$ of incident light propagating at angle $\varphi$ should be collinear to acoustic group velocity $\boldsymbol{V}_{g1}$ defined in the same coordinate system (figure 1).

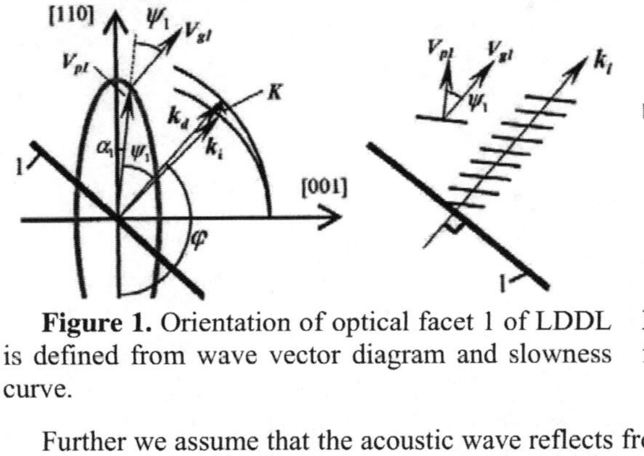


**Figure 1.** Orientation of optical facet 1 of LDDL is defined from wave vector diagram and slowness curve.

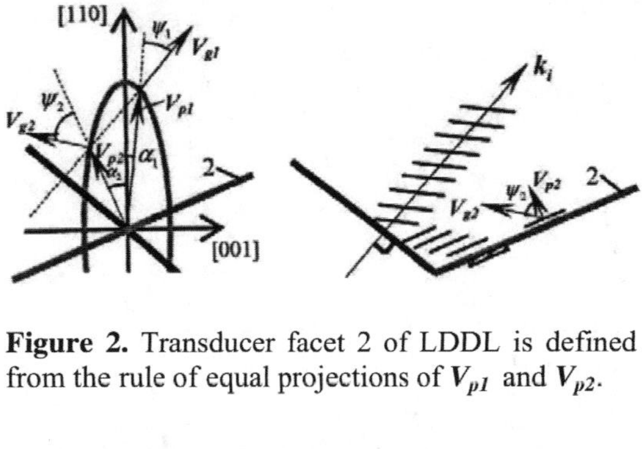


**Figure 2.** Transducer facet 2 of LDDL is defined from the rule of equal projections of $\boldsymbol{V}_{p1}$ and $\boldsymbol{V}_{p2}$.

Further we assume that the acoustic wave reflects from the input facet of crystal. Here: $\boldsymbol{k}_d$ and $\boldsymbol{K}$ – wave vectors of diffracted light and sound, $n_o$ and $n_e$ – ordinary and extraordinary indexes of refraction; $\psi_1$ - walk-off angle between phase and group velocities of sound. We introduce the main parameter of filter $\alpha_1$

as an angle between phase velocity $V_{p1}$ of the acoustic wave reflected from the input facet of the crystal and the axis of the crystal [110]. We also define: $\alpha_2$ as an angle between phase velocity $V_{p2}$ of the acoustic wave generated by piezotransducer, and [110] axis, $\psi_2$ as a walk-off angle between phase $V_{p2}$ and group $V_{g1}$ velocities of the acoustic wave (figure 2).

The orientation of the input optical facet is defined by angle $\pi/2-(\alpha_2+\psi_1)$ between the facet plane and axis [110] of the crystal (figure 1). The orientation of the transducer facet is defined by angle $\pi/2-$ $-\alpha_2$ between the facet plane and axis [110] of the crystal (figure 2). The value of angle $\alpha_2$ is determined as a function of $\alpha_1$ [6, 9]. When $\alpha_1=1.78^{\circ}$ and $\alpha_1=13.31^{\circ}$ the optical and acoustical facets are orthogonal. Acoustic wave vectors of incident and reflected sound columns have to satisfy the requirement of equal projection in the crystal.

Different LDDL were designed using the mathematical approach described above. Operating points of LDDL are represented by the surface shown of figure 3. When $\alpha_1=1.78^{\circ}$ the operating point is point $A$.

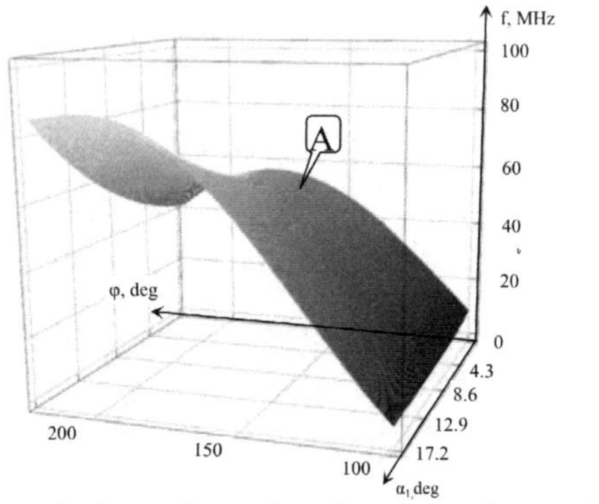


**Figure 3.** Operating point of LDDL at $\lambda$ = 1550 nm. Point $A$ corresponds to $\alpha_1 = 1.78^{\circ}$, $f$ = 52.2 MHz, $\varphi$ = 160.1 deg.

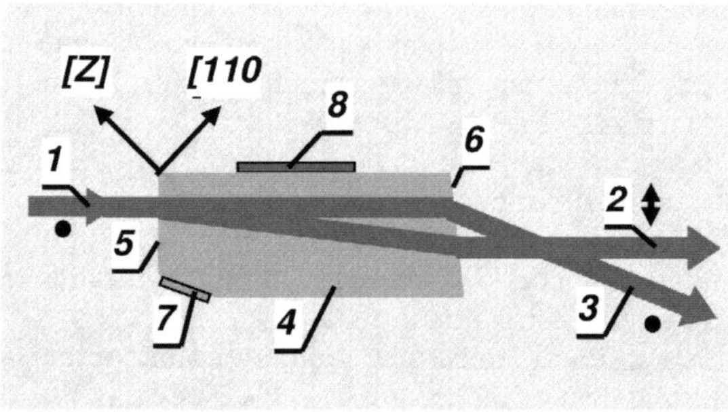


**Figure 4.** Optical scheme of LDDL ($1.78^{\circ}<\alpha_1<13.31^{\circ}$). 1. incoming beam; 2. diffracted beam; 3. non-diffracted beam; 4. $TeO_2$ crystal; 5. input optical facet; 6. output facet; 7. transducer; 8. acoustic absorber.

## 4. Features of adaptive optical delay line

Optical scheme of LDDL is presented in figure 4. The input facet (5) of the crystal (4) is orthogonal to the wave vector (1) of the incoming laser radiation. Polarization of laser radiation is shown schematically in figure 4. The acoustic wave generating by transducer (7) reflects from the input facet (5) of crystal. The quasicollinear interaction takes place within the sound column reflected from the input optical facet (5) of the LDDL. The sound column is dissipated by acoustic absorber (8). Experimental devices were fabricated by bonding method of our own utilizing the vacuum interdiffusion of atoms in In-Au nanostructures.

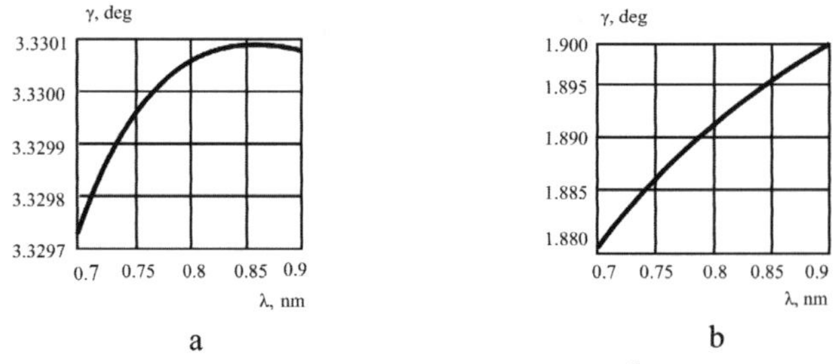


**Figure 5.** The chirp angle $\gamma$ inside crystal versus $\lambda$. $\alpha_1=1.78^{\circ}$. a) Full compensation; b) Particularly compensation.

The angular dispersion in LDDL leading to an angular chirp of femtosecond laser beam originates from the angular spectral dependence of Bragg synchronism. The reduction of angular chirp (figure 5a) can be done by adding a compensating prism with inverse dispersion to the output facet (6) of the crystal as it is shown in figure 4. In practice it is convenient to employ LDDL when the diffracted beam (2) propagates collinear to the input beam (1). In this case the angular chirp compensation will be not so perfect (figure 5b). The angular chirp will be nevertheless less, than a typical divergence of femtosecond laser beam.

The spectral transmission function bandwidth of one of LDDL utilizing the angle $\alpha_l$=1.78° measured in a single frequency regime by optical spectrum analyzer at -3 dB level is about 0.24 nm at 1550 nm. The crystal length is 67 mm, the optical aperture is 4x4 mm. Other specifications are as follows: central laser wavelength – 800 nm, maximum optical delay exceeds 20 psec. The frequency of Bragg synchronism at 800 nm is 105 MHz. Owing to our origin transducer's bonding technology the driving RF power in a single frequency regime was extremely low (60 mW at 90% efficiency) that permitted to control the whole femtosecond spectrum up to the 200 nm. The general view of experimental device is presented in figure 6.

**Figure 6.** High-performance LDDL ($\alpha_l$=1.78°).

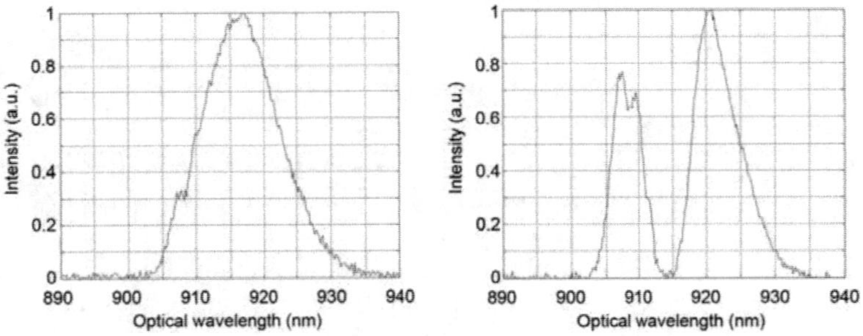


**Figure 7.** Signal pulse spectrum before (left) and after (right) LDDL's transformation ($\alpha_l$=3.8°) in OPCPA laser system [6].

The experiments with another LDDL ($\alpha_1$=3.8°) has been carried out at subpetawatt OPCPA laser system created by the Institute of Applied Physics of the Russian Academy of Sciences [6, 10]. An LDDL was placed between a femtosecond oscillator and a stretcher. The femtosecond master oscillator was a Cr:forsterite laser generates 40 fs pulses at a central wavelength of 1250 nm. The measured efficiency of LDDL over the whole femtosecond pulse spectrum bandwidth 120nm was 70%. The spectrum bandwidth of the signal radiation in the OPA was approximately 12 nm. Shaping of the sound amplitude distribution that provides a Gaussian valley in transforming function produced corresponding modifications in the signal spectrum (figure 7).

## 5. Conclusion

Femtosecond light delay lines are the only class among all other types of acoustooptical device because they provide parallel but not consistent optical spectral processing. The designed LDDL systems have revealed high efficiency, high spectral resolution, flexibility of pulse control and simplicity in operation.

## Acknowledgements

This work was supported by RFBR grants 09-02-12400 and 09-07-00365. The authors are grateful to N. Solodovnikov from Acoustooptical Research Center of the National University of Science and Technology "MISIS" and E. Khazanov, V. Ginzburg, I. Yakovlev, V. Lozhkarev from the Institute of Applied Physics of the Russian Academy of Sciences for experiments performance and helpful discussions.

## References

[1] Pozhar V and Pustovoit V 1987 *J. Sov. Quant. Electronics* **14** (4) 811-813.
[2] Pustovoit V and Pozhar V 1994 *Photonics and optoelectronics* **2** (2) 53-69.
[3] Tournois P 1997 *J. Optics Communications* **140** 245-249.
[4] Verluise F, Laude V, Huignard J and Tournois P 2000 *J. Opt. Soc. Am.* B **17** (1) 138-145.
[5] Danailov M, Demidovich A and Ivanov R 2006 *Proc. of FEL, BESSY, Berlin, Germany* 617-620.
[6] Molchanov V, Chizhikov S, Makarov O at al. 2009 *Appl.Optics* **48** (7) 118-124.
[7] Magdich L 1980 *J. Optics and Spectroscopy* **49** 387-390.
[8] Magdich L, Molchanov V and Ponomareva I 1984 *J. Optics and Spectroscopy* **56** 736-739.
[9] Molchanov V, Voloshinov V and Makarov O 2009 *Quantum Electronics* **39** (4) 353-360.
[10] Molchanov V, Chizhikov S, Makarov O at al. 2009 *Acta Physica Polonica A* **116** (3) 355-358.

Journal of Physics: Conference Series **278** (2011) 012017 doi:10.1088/1742-6596/278/1/012017

# Laser ultrasonic measurements in a diamond anvil cell on Fe and the KBr pressure medium

**N Chigarev[a1], P Zinin[b], D Mounier[c], A Bulou[c], A Zerr[d], L C Ming[b] and V Gusev[c]**

[a]LAUM, CNRS, PRES UNAM, Université du Maine, Av. O. Messiaen, 72085 Le Mans, France
[b]HIPG, University of Hawaii, Honolulu, USA
[c]LPEC, CNRS, PRES UNAM, Université du Maine, Av. O. Messiaen, 72085 Le Mans, France
[d]LPMTM, CNRS, Université Paris 13, av. J. B. Clement, 93430 Villetaneuse, France

E-mail: nikolay.chigarev@univ-lemans.fr

**Abstract.** Laser ultrasonic technique has been successfully applied to a sample embedded in a transparent pressure transmitting medium to measure sound velocity at high pressures in a diamond anvil cell. Shear and longitudinal acoustic wave velocities are measured both for an opaque sample, Fe, and a transparent pressure transmitting medium, KBr, at pressures ranging from 0.3-24.8 GPa. The small sample size used in this experiment for sound velocity measurements suggests great potential for using this technique to evaluate the elastic properties of solids at pressures approaching 100 GPa and for other samples embedded in soft, transparent pressure transmitting media such as KBr or argon.

## 1. Introduction

Direct measurements of velocities and absorption of acoustic waves in minerals at elevated pressures and temperatures are key to understanding seismic information [1]. For samples compressed in a diamond anvil cell (DAC), the laser ultrasonics (LU) was first used to generate and detect surface acoustic waves [2] and later to measure the longitudinal sound velocities [3,4]. Recently it has been demonstrated that it is possible to detect both longitudinal and shear waves as well as measure their velocities for a non-transparent iron layer compressed in a DAC to pressures approaching 23 GPa by using LU in a point-source–point-receiver [5,6] or line-source–point-receiver [7] configurations. These experiments were conducted without a pressure transmitting medium, which led to non-hydrostatic pressure conditions in the sample. Soft pressure transmitting media (PTM) are used in high pressure (HP) experiments in a DAC to eliminate or reduce pressure gradients in the sample because they could bias experimental data. The purpose of the present study is to examine the effect of a pressure transmitting medium (KBr) on the monitoring of acoustic waves in experiments with the LU-DAC.

[1]To whom any correspondence should be addressed.

## 2. Method

A set of LU-DAC experiments in line-source–point-receiver configuration was conducted using KBr powder as a PTM (figure 1). We used KBr powder because it is very soft [8] and is considered one of the best pressure transmitting media [9]. Schematics of the sample assemblage and of the sound wave trajectories in our LU experiments in a DAC are shown in figure 1.

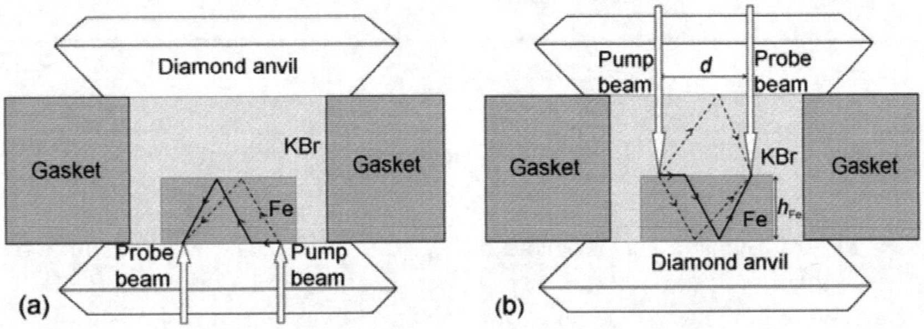


**Figure 1.** Sketch of the sound wave propagation in a flat iron specimen embedded in the KBr pressure medium for the experimental configurations D/Fe (a) and KBr/Fe (b). Dashed line - bulk waves; solid line - a skimming bulk wave emitting a head wave; $d$ - distance between the points of the sound generation and detection; $h_{Fe}$ - thickness of the Fe-specimen.

A flat iron specimen of about 150 $\mu$m in width is pressed to the lower diamond anvil and separated from the upper one by a KBr layer (figure 1). The way the sample is loaded in DAC is similar to that used in impulsive stimulated light scattering [2] and gigahertz ultrasonic interferometry [10,11] in the DAC. The initial thickness of the iron specimen, $h_{Fe}$, was estimated to be about 15 $\mu$m. For the sample geometry presented in figure 1, the propagation time of the acoustic waves inside the Fe-specimen can be measured in two configurations: either when the pump and the probe laser beams are focused on the diamond/Fe interface (D/Fe configuration, figure 1a) or when the pump and the probe lasers are focused at the KBr/Fe interface (KBr/Fe configuration, figure 1b).

## 3. Results

The photoacoustic signals measured at 20.8 GPa are presented in figure 2. Following the notations introduced in [5-7], the peaks corresponding to the waves both excited and reflected from one of the interfaces as longitudinal (L) or transversal (T) are marked as $LL_i$ or $TT_i$, respectively, where index "i" stands for the medium where the wave propagates (Fe, KBr, or D for diamond). The peaks marked as $SL_i$ or $ST_i$ correspond to the waves propagating as skimming L or T, respectively [12]. As mentioned in [7], skimming waves SL and ST could combine, under certain conditions, with $LL_i$ or $TT_i$ waves leading to the so-called "head wave arrivals" (see $SL_D$-$LL_{Fe}$, $SL_{Fe}$-$TT_{Fe}$ and $SL_{KBr}$-$LL_{Fe}$ peaks in figure 2).

In figures 3a and 3b the propagation times ($\tau$) of the acoustic waves, measured for different distances $d$ between the generation and detection points, are presented for the pressures at 9 and 20.8 GPa, respectively. A method of analysis of $TT_{Fe}$ echoes in the Fe-specimen and $LL_{KBr}$ and $TT_{KBr}$ echoes in KBr can be found elsewhere [5-7]. Least square fits of two-parameters (sound velocity and sample thickness) to the experimental data $\tau(d)$ allowed us to determine the T-wave velocities in iron, $C_{TFe}$, as well as the L- and T-wave velocities in KBr ($C_{LKBr}$ and $C_{TKBr}$) (figure 3c).

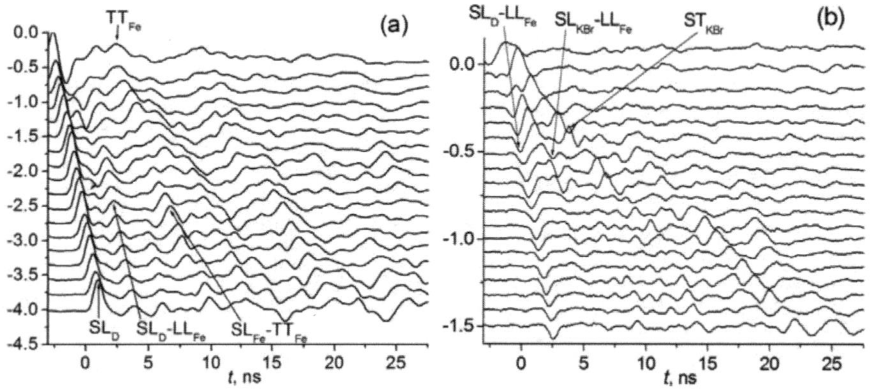


**Figure 2.** Photoacoustic signals measured at 20.8 GPa for $d$ varying from 7.4 to 70.5 $\mu$m and from 4.0 to 70.6 $\mu$m in the configurations D/Fe (a) and KBr/Fe (b), respectively. In both measurements $d$ was changed in increments of 3.7 $\mu$m.

Due to overlapping of the $LL_{Fe}$ and $SL_D$-$LL_{Fe}$ signals (figure 3a), we could not determine the longitudinal wave velocity in Fe, $C_{LFe}$, using the analysis method discussed in [5,6]. Nevertheless, we found that the velocity $C_{LFe}$ could be obtained from $\tau$ -values measured for the skimming longitudinal wave $SL_{Fe}$ propagating in iron parallel to the interface with the diamond. These waves were detectable, however, only at pressures between 9 and 12 GPa (figure 3c).

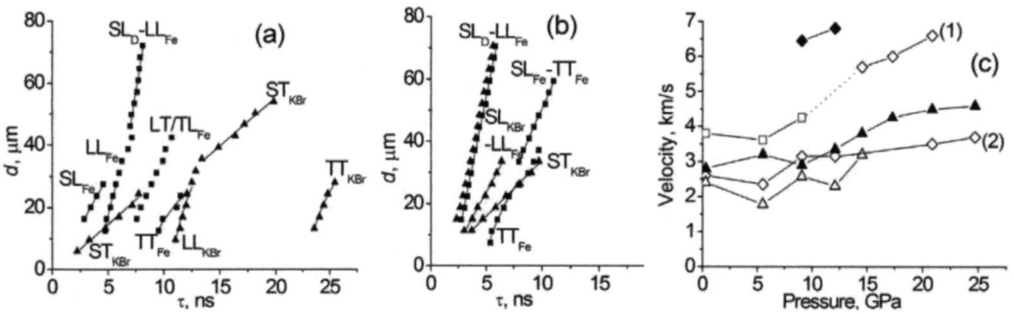


**Figure 3.** Propagation time $\tau$ of the acoustic signals measured for different generator-detector distances, $d$, at 9 GPa (a) and 20.8 GPa (b). Squares and triangles represent the data for the configuration D/Fe and KBr/Fe, respectively. Solid curves represent the fit results.
Pressure dependences of the measured velocities (c): solid rhombi - $C_{LFe}$, obtained from the $SL_{Fe}$ wave; solid triangles - $C_{TFe}$ from the $TT_{Fe}$ wave; open rhombi marked (1) - $C_{LKBr}$ from the $SL_{KBr}$-$LL_{Fe}$ wave; open squares - $C_{LKBr}$ from the $LL_{KBr}$ wave; open rhombi marked (2) - $C_{TKBr}$ from the $ST_{KBr}$ wave; open triangles - $C_{TKBr}$ from the $TT_{KBr}$ wave.

Because the arrivals of the longitudinal $LL_{KBr}$ and transversal $TT_{KBr}$ waves could not be recognized at 20.8 GPa (figure 2b), the values of $C_{TKBr}$ and $C_{LKBr}$ were obtained from the skimming $ST_{KBr}$ and mixed $SL_{KBr}$-$LL_{Fe}$ waves, respectively. As shown in figure 3b, the signal $SL_D$-$LL_{Fe}$ has been observed in both D/Fe and KBr/Fe experimental configurations.

Sound velocities of the acoustic waves measured in the present experiment at pressures up to 24.8 GPa are summarized in figure 3c. Both the $C_{LFe}$ and $C_{TFe}$ values of iron, determined in this work from the $SL_{Fe}$ - and $TT_{Fe}$-signals, respectively, are in a good agreement with the results obtained earlier using the LU-DAC technique without a PTM [5-7].

## 4. Conclusion

The experimental results presented in this article demonstrate that the LU-DAC technique can be applied for sound velocity measurements in small specimens compressed in soft pressure transmitting media to very high pressures. Moreover, our results show that this technique allows the determination of acoustical properties of both the specimen and the pressure medium.

## 5. Acknowledgement

This work was supported by U.S. Department of Energy Grant NO. DE-FG02-07ER46408.

## References

[1] Bass J D 2008 *Phys. Earth Planet. Inter.* **170** 207
[2] Crowhurst J C, Abramson E H, Slutsky L J, Brown J M, Zaug J M and Harrell M D 2001 *Phys. Rev. B* **64** 100103
[3] Villagran-Muniz M, Navarrete M and Mejia-Uriarte E V 2003 *Rev. Sci. Instrum.* **74** 732
[4] Decremps F, Belliard L, Perrin B, and Gauthier M 2008 *Phys. Rev. Lett.* **100** 035502
[5] Chigarev N, Zinin P, Ming L C, Amulele G, Bulou A and Gusev V 2008 *Appl. Phys. Lett.* **93** 181905
[6] Chigarev N, Zinin P, Mounier D, Bulou A, Ming L C, Acosta T and Gusev V 2010 *High Pressure Research* **30** 78
[7] Zinin P, Chigarev N, Mounier D, Bulou A, Ming L C, Acosta T and Gusev V 2010 *J. Phys. Conf. Ser.* **215** 012053
[8] Singh S, Singh R K, Singh B P, Singhal S K and Chopra R 2000 *Physica Status Solidi a-Applied Research* **180** 459
[9] Shen A H, Reichmann H J, Ghen G, Angel R J, Bassett W A and Spetzler H 1998 *Geophys. Monogr. Ser.* **101** 71
[10] Jacobsen S D, Spetzler H A, Reichmann H J, Smyth J R, Mackwell S J, Angel R J and Bassett W A 2002 *J. Phys.-Condes. Matter* **14** 11525
[11] Bassett W A, Reichmann H J, Angel R J, Spetzler H and Smyth J R 2000 *Am. Miner.* **85** 283
[12] Beghi M G, Every A G, and Zinin P V 2004 *Ultrasonic Nondestructive Evaluation: Engineering and Biological Material Characterization*, ed T Kundu (Boca Raton: CRC Press)

Journal of Physics: Conference Series **278** (2011) 012018 doi:10.1088/1742-6596/278/1/012018

# Time-resolved X-ray diffraction studies of laser-induced acoustic pulse generation in semiconductors using synchrotron radiation

**Y Tanaka[1-3], Y Uozaki[2], K Nozaki[2], K Ito[1], K Yamasaki[2], H Terauchi[2], I Takahashi[2], K Tahara[1], T Ishikawa[1]**

[1]RIKEN/SPring-8 Center, 1-1-1, Kouto, Sayo-cho, Sayo-gun, Hyogo 679-5148 Japan
[2]School of Science and Technology, Kwansei-Gakuin University, 2-1Gakuen, Sanda, Hyogo 669-1337 Japan,
[3]E-mail: yotanaka@riken.jp

**Abstract**. Laser-induced lattice deformation generating acoustic pulses has been studied by time-resolved X-ray diffraction method using a brilliant synchrotron radiation source. We constructed a time-resolved X-ray diffraction system equipped with a synchronized femtosecond pulsed laser at an undulator beamline of SPring-8 facility, and observed laser-induced acoustic pulse in a semiconductor wafer of GaAs, where the laser irradiation causes initial strain of expansion with a response time of around 200 ps. High resolution X-ray diffractometry in asymmetric configuration combined with laser-pump X-ray probe method revealed that the strain is formed with coherent longitudinal acoustic phonons and lattice expansion along the surface normal. The laser power dependence shows the saturation of the lattice expansion ratio with the lengthened relaxation time for high power excitation.

## 1. Introduction

When the surface of a semiconductor plate is irradiated by a femtosecond pulsed laser, transient lattice deformation is generated around the surface and induces acoustic pulse propagating toward the rear surface. The pulse is then reflected at the rear surface and the deformation is again observed at the front surface at a certain delayed time. The repetitive reflection is known as acoustic pulse echoes. This phenomenon has been investigated by time-resolved optical monitoring method [1] which reveals the surface displacement and the shape. On the other hand, X-ray diffraction (XRD) is powerful method to investigate the lattice deformation. At the same time, a brilliant X-ray source is required when the XRD is combined with time-resolved technique to watch the transient lattice deformation in acoustic pulse echoes. Significant progress in recent development of synchrotron radiation (SR) X-ray sources has offered the opportunity to observe the laser induced acoustic pulse generation and the echoes by time-resolved (TR) XRD method [2,3]. The high quality of SR X-ray beam enabled to combine time-resolved technique and XRD method with high momentum resolution such as triple crystal diffractometry (TCD). By the TCD, where three crystals are the monochromator to generate a collimated and monochromatic incident X-ray beam, a sample, and an angle analyzer for diffracted beam, we can identify the type of lattice strain and know the direction and degree of lattice deformation [3]. We have been investigating the laser-induced acoustic pulse generation and the echoes in semiconductor wafers of gallium arsenide (GaAs) and silicon (Si) by TR-TCD at a SPring-8 undulator beamline, equipped with a femtosecond pulsed laser system. Initial strains are expansion and shrinkage for GaAs(100) crystal and Si(111) crystal, respectively, and their echo pulses with several-tens repetition have the corresponding phase [2]. Since the lattice of GaAs expands with fast response time of about 200 ps, we conducted pump-probe method where the time-resolution is determined by the SR pulse duration of 40 ps (FWHM).

In this paper, we show the TR-TCD method with asymmetric configuration to investigate the initial strain generating acoustic echoes in a single crystal of GaAs, and report the deformation direction, phases of acoustic waves, and the laser power dependence of deformation.

## 2. Experimental

The TR-TCD measurements were conducted at an undulator beamline of BL19LXU of SPring-8 SR facility. Figure 1(a) shows the experimental setup. A 0.6 mm-thick, 76 mm-diameter, non-doped GaAs(100) wafer was aligned using a high precision diffractometer in the experimental hutch. The incident X-rays were monochromatized to be 21.3 keV, so that the X-rays are diffracted by the 620 plane of GaAs with a grazing exit angle of about 0.5°. In order to get a smaller energy bandwidth of incident X-rays, a Si 440 channel-cut crystal was used, in addition to a symmetric Si 111 double-crystal monochromator. The diffraction intensity from the sample is detected through a Si 111 analyzer crystal. The energy bandwidth and angle resolution were about $\Delta E/E = 9\times 10^{-6}$ and $\Delta\theta$=30 μrad, respectively. These correspond to a resolution window of $3.2\times 10^{-4}$ Å$^{-1}$ (in the direction of $\Delta\theta$) by $6.3\times 10^{-5}$ Å$^{-1}$ ($\Delta G$). A diffraction intensity map in reciprocal space can be obtained by scanning the angles of diffractometers of D1 and D2.

A mode-locked Ti:sapphire laser with a regenerative amplifier, which has a pulse duration of 130 fs, wavelength of 800 nm, a pulse energy of 130 μJ, and repetition rate of 1 kHz, was used for irradiation of the GaAs. The laser beam diameter at the sample surface was about 1.8 mm, covering the X-ray diffracting area (100 μm by 100 μm). The penetration depth of the laser, 0.7 μm, is about 2 times larger than the extinction depth of X-rays. The timing of the laser shot was controlled by the laser trigger applied through an electronic circuits composed of RF phase shifter from the radio frequency (RF) signal of the master oscillator of the SR storage ring determining the SR X-ray pulse timing. The stability of the timing between the laser and SR pulses has been evaluated to be less than 5 ps [4], which is much better than the X-ray pulse duration of 40 ps (FWHM). The X-ray photons diffracted from GaAs 620 were detected by an avalanche photodiode. The counts only from the target SR pulse are picked up by using an electronic gate synchronized with the RF master oscillator of the ring. The data were accumulated for 10 s for each point in a reciprocal space map.

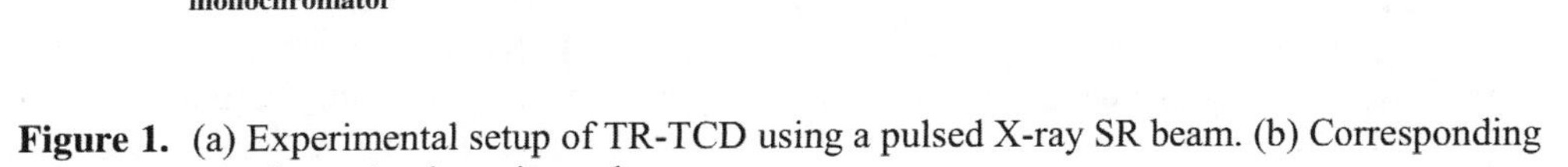


**Figure 1.** (a) Experimental setup of TR-TCD using a pulsed X-ray SR beam. (b) Corresponding configuration in reciprocal space.

## 3. Results and discussion

Figure 2(a), (b) and (c) show the X-ray diffraction snapshots of GaAs 620 in reciprocal space, obtained for no laser irradiation, Δt=180 ps, and Δt=1 ms after the irradiation by the laser with a pulse energy of 5 mJ/cm$^2$, respectively. Asymmetric distribution along the [100]-axis is seen in figure 2(b), which is due to acoustic phonon generation. Figure 3 shows the plot of the peak position of the diffraction intensity distribution for Δt= (i)1 ms, (ii)180 ps, (iii) 300 ps, and (iv) 800 ps. It is found that the transient expansion occurs along a dashed vertical line parallel to the surface normal direction, and that the diffraction peak for about 1 ms after the irradiation is shifted towards an inclined dashed line showing the direction of $\mathbf{G}_{620}$ with respect to the position without laser irradiation **O'**. The isotropic lattice expansion, $|\Delta\ d_{620}|/d_{620}$, is estimated from figure 3, to be 1.4× 10$^{-5}$, which corresponds to the heating by 2.4 K. The maximum anisotropic transient expansion in the direction of the surface normal, is also evaluated to be$|\Delta\ d_{100}|/d_{100}$ =2.2× 10$^{-5}$ at a delay time of 300 ps.

The fast lattice expansion is, then, observed by measuring the diffraction intensity as a function of the angle of analyzer crystal because Δθ direction is almost parallel to the [100]-direction as shown in figure 1(b). Figure 4 shows time dependence of the diffraction intensity profile in the Δθ direction, as obtained by changing an analyzer angle with a fixed sample angle. In order to discuss the lattice expansion and phonon oscillation separately, the shift, $\theta_p$ of principal peak in figure 4 is plotted in figure 5(a). The peak shift of the profile to minus angle direction indicates the lattice expansion whose maximal spacing is achieved at around 300 ps. The time dependence of the diffraction intensity profile with an axis of $\Delta\theta$–$\theta_p$ (=Δθ') was drawn in order to pick up the acoustic phonon oscillation. The time-dependence of diffraction intensity at Δθ' is shown in figure 5(b). Acousto-optic effect produced by laser-induced longitudinal acoustic (LA) phonon and incident X-rays is shown as out of phase oscillation with a frequency of GHz region at side bands around the principal peak shifted due to lattice expansion. Figure 5(b) shows the diffraction intensity modulation depending on the off-set angle, which should have the relation of $\Omega = v|\mathbf{k}||\Delta\theta'|$, where Ω is the angular frequency of the intensity modulation, v, the speed of sound, and **k**, the wavevector of X-rays [5].

From the relation between the off-set momentum and the oscillation frequency as shown in figure 5(c), the speed of acoustic phonon is roughly estimated using the above relation, to be v=3400 m/s, which is comparable with the speed of sound of GaAs in bulk.

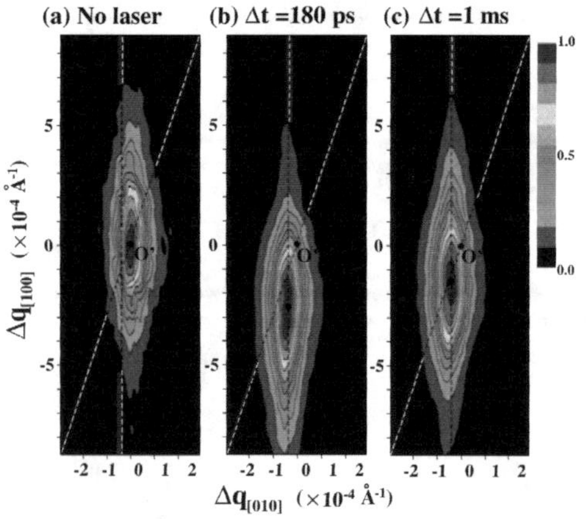


**Figure 2.** The snapshots of GaAs 620-Bragg diffraction intensity distribution for (b) Δt=180 ps and (c) Δt=1 ms. (a) is the Bragg spot without laser irradiation.

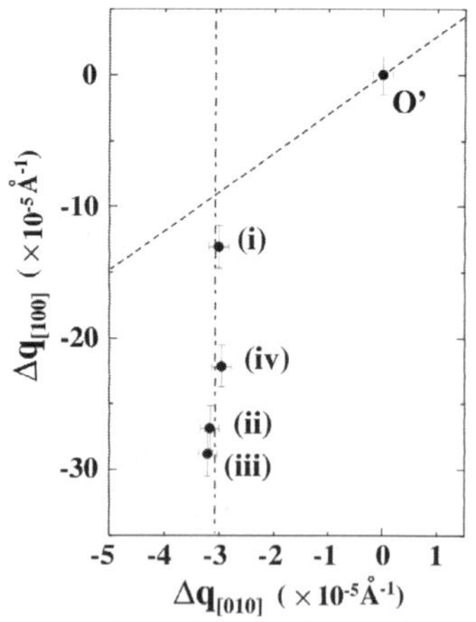


**Figure 3.** The peak position of the diffraction intensity distribution moving with Δt of (i)1 ms, (ii)180 ps, (iii) 300 ps, and (iv) 800 ps. An inclined dashed line indicates the direction of $\mathbf{G}_{620}$.

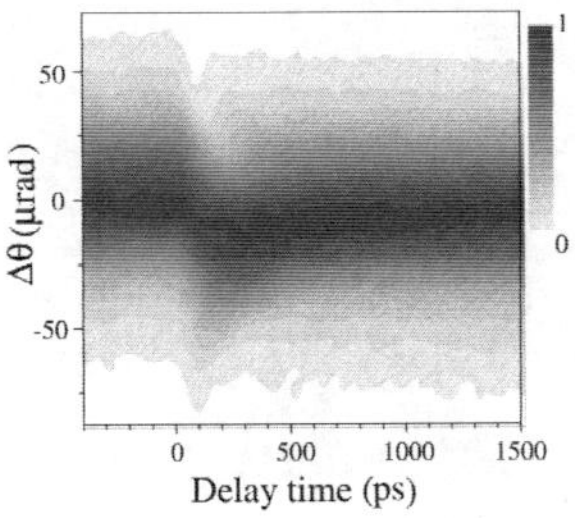


**Figure 4.** Time-dependent angular distribution of X-ray diffraction intensity for GaAs 620.

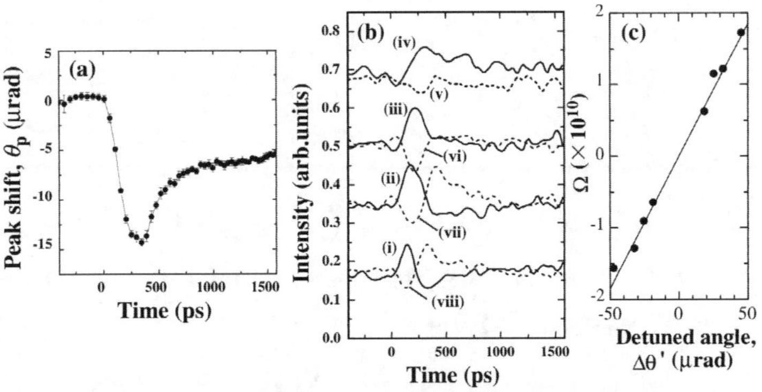


**Figure 5.** (a)The peak shift, $\theta_p$, and (b) oscillatory part at detuned angles $\Delta\theta' = \Delta\theta - \theta_p$, obtained from figure 4. In (b), the detuned angles are, $\Delta\theta'$= (i) 46, (ii) 33, (iii) 26, (iv) 19, (v) -18, (vi) -25, (vii) -32, and (viii) -47 μrad. Figure 5(c) shows the $\Delta\theta'$-dependence of the frequency of diffraction intensity oscillation, Ω.

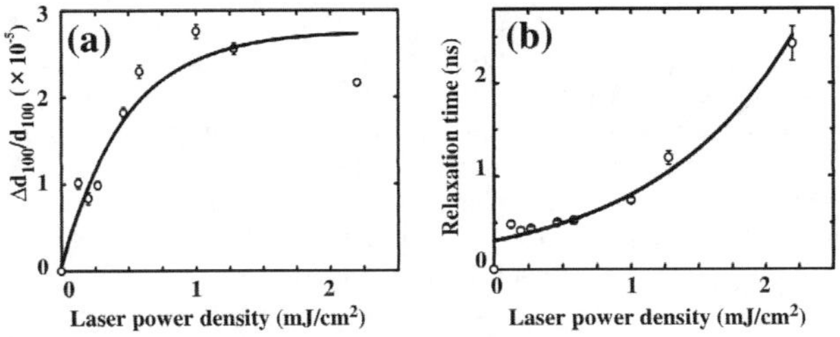


**Figure 6.** The laser power dependence of the lattice expansion measured for 400-GaAs diffraction.
(a) The maximum expansion ratio and (b) relaxation time.

The laser power dependence of the lattice expansion obtained by the peak shift was also measured for 400-Bragg diffraction from a GaAs wafer. Figure 6 (a) and (b) show the laser power dependences of maximum expansion ratio and relaxation time, respectively. The maximum expansion ratio is almost proportional to the laser power for relatively small excitation; for the high power excitation, the ratio is saturated as the relaxation time becomes longer.

## 4. Conclusions

Laser-induced lattice deformation generating acoustic pulses in a semiconductor wafer of GaAs has been studied by TR-TCD using the SPring-8 SR source. The time-evolution of diffraction profile shows that the strain is formed with coherent LA phonon generation and lattice expansion along the surface normal. The maximum lattice expansion ratio is saturated for high power laser excitation.

## References

[1] Wright O B and Kawashima K 1992 *Phys. Rev. Lett.* **69** 1668
[2] Hayashi Y, Tanaka Y, Kirimura T, Tsukuda N, Kuramoto E and Ishikawa T 2006 *Phys. Rev. Lett.* **96** 115505
[3] Hayashi Y, Tsukuda N, Kuramoto E, Tanaka Y and Ishikawa T 2005 *J. Synchrotron Rad.* **12** 685
[4] Tanaka Y, Hara T, Kitamura H and Ishikawa T 2000 *Rev. Sci. Instrum.***71** 1268
[5] Larsson J, Allen A, Bucksbaum P H, Falcone R W, Lindenberg A, Naylor G, Missalla T, Reis D A, Scheidt K, Sjogren A, Sondhauss P, Wulff M and Wark J S 2002 *Appl. Phys.* **A75** 467

Journal of Physics: Conference Series **278** (2011) 012019 doi:10.1088/1742-6596/278/1/012019

# Graphitized layer buried in a diamond: photothermal properties and hypersound generation under picosecond optical excitation

**A Klokov, M Kochiev, A Sharkov, V Tsvetkov and R Khmelnitskiy**

P. N. Lebedev Physical Institute RAS, Leninskii pr. 53, 119991 Moscow, Russian Federation

klokov@lebedev.ru

**Abstract**. We assume that a coherent phonon generation was found out in structures with a single graphitized layer under picosecond laser excitation by means of the Pump/Probe method. The thermal component of the response was calculated that allowed us to define the photothermal coefficient of the graphitized layer as well as its thermal conductivity from the experimental results. The thermal-conductivity coefficient of the graphitized layer appears to grow with the thickness decrease. This makes it possible to assume that the microstructure of the graphitized layer produced during annealing depends on its thickness.

## 1. Introduction

A unique combination of electrical, optical, structural and thermal properties of diamond makes it an attractive material for production of MEMS as well acoustoelectronic devices in the gigahertz range (filters, surface and bulk waves cavities) [1]. Ion implantation provides a possibility of formation of embedded layers (amorphized and graphitized) built-in the diamond matrix [2]. The thicknesses of the layers produced are typically of about tens and even more nanometers being conductive and perfectly absorbing optical radiation. Optical and thermal properties of such layers are studied in a number of publications [3,4]. However, the structure and acoustic properties excluding [5], which is devoted to the generation of noncoherent acoustic phonons, virtually are not studied. One can obtain information on the sound velocity, attenuation and acoustic mismatch of the implanted layer and a diamond matrix, as well as thermal properties, with the help of picosecond acoustic methods [6]. Such data are of interest both from fundamental – investigation of phase transitions in radiation-damaged materials, and practical points of view – possible using of such layers to form various acoustic wave-guiding systems and phonon crystals. The purpose of this work is to investigate the possibilities to excite coherent phonons by exciting single graphitized layers buried in diamond with a picosecond laser.

## 2. Experiment

The buried graphitized layers (see table 1) were created by implanting C+ ions with the energy of 350 keV into a IIA-type ~300 μm thick natural diamond wafer. The diamond plane (100) was misoriented by 3.12° against the wafer surface. The implantation was followed by one-hour vacuum annealing at ~1600°C. The depth of location and thickness of the layers were determined by optical methods similarly to [7] and are shown in table 1.

**Table 1.** Condition of diamond implantation and parameters of graphitized layers. ($d_1$ is the depth, $d_2$ is the layer thickness, $\kappa_{Gr}$ is the layer thermal-conductivity coefficient)

| Structure no. | Dose, $\times 10^{15}$ cm$^{-2}$ | $d_1$, nm | $d_2$, nm | $\kappa_{Gr}$, W·cm$^{-1}$·K$^{-1}$ |
|---|---|---|---|---|
| 1 | 12 | 235 | 283 | 3.4 |
| 2 | 8 | 250 | 249 | 8.5 |
| 3 | 6 | 295 | 160 | 17 |

The complex refractive index of the layer was $n_{Gr}$=2.2+j0.88. The reflected light color of those areas with different implantation doses changed from green to dark blue, which was caused by the interference in optical wavelength range. This allowed us to assume the optical perfection of the diamond/graphitized layer interface [8].

The conventional Pump-Probe method was used to investigate the coherent phonon excitation. The experiment setup is shown in figure1.

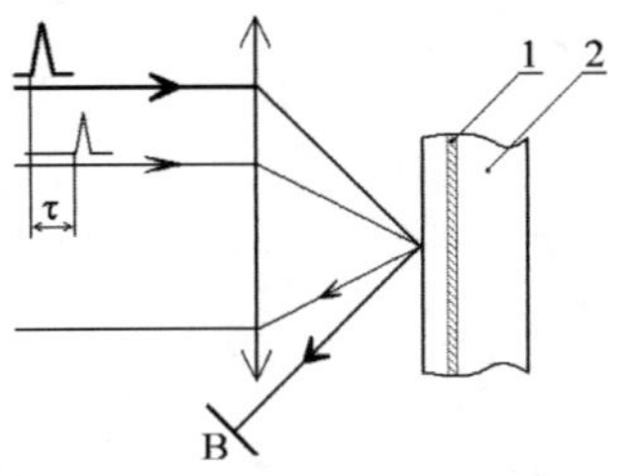


**Figure 1.** Schematic experimental setup: thick line – pump beam, thin line – probe beam. B – beam stopper. 1 – is a buried graphitized layer and 2 – diamond matrix. Laser irradiation falls on the implanted side of the sample.

Coherent phonons were excited as a result of the thermoelastic effect arising in the buried graphitized layer (layer 1 in figure1) being irradiated by picosecond pulses of the MIRA-900P titanium-sapphire laser (wavelength 797 nm, pulse duration 1.5-2.5 ps, repetition frequency 76 MHz). As the heat relaxation time in structures similar to that one under study was ~100 ns, a pulse picker was used to make repetition frequency ~4 MHz. The laser pulse was split into exciting and probing beams with crossed polarizations. The probing beam could be delayed against the exciting beam by means of the optical delay line based on the translation stage. Beams were focused both at one point by means of the microscope long-focus objective. The size of the laser radiation spot on the sample was ~12 μm. Polarization of the probing beam was perpendicular to the incidence plane with the incidence angle being ~4°. The maximum incident fluence of the pump beam was 8 J·m$^{-2}$. The exciting beam was modulated by means of the Pockels cell in order to implement the photodiode output synchronous detection method that allowed registering the probing beam reflected from the sample.

## 3. Discussion

Figure 2 shows responses of the sample with buried graphitized layers. The responses of structures 1 and 2 had a short ~3 ps spike changing into an abrupt dip followed by a slow increase. The maximum of response (depending on depth of location and thickness of the layer) was found 1÷2 ns after the excitation pulse. The response of structure 3 had an abrupt front and a subsequent decay close to exponential. All responses had evident fluctuations with a maximum value reaching $4 \cdot 10^{-5}$. A short spike at the beginning of the response arose due to the generation of a large number of hot carriers in the graphitized layer, caused by the laser pulse absorbing. Within a few picoseconds the carriers were thermalized and transferred energy to the phonon system causing the layer temperature to increase by several degrees. The stress induced by this process resulted in exciting elastic waves (coherent phonons). The probing beam reflection coefficient varied due to the disturbance of optical and geometrical parameters of the structure under study, caused by both nonstationary heating and elastic waves [9]. The location depth of the graphitized layer essentially affected the form of the response as it determines the optical path difference of the beams reflected from the sample surface and the interface between graphitized layer and diamond.

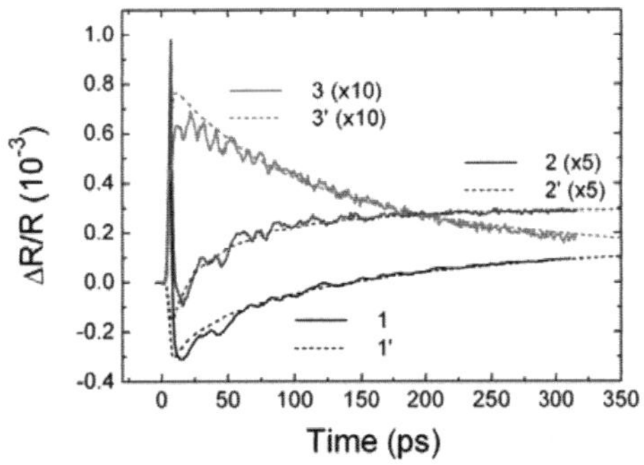


**Figure 2.** Responses of the sample with buried graphitized layers obtained under picosecond excitation. Experiment - solid lines; Theory - dashed lines. 1,2,3 –structures represented in Table 1.

Thus, the interference conditions, when changing depth of location, could be altered from constructive to destructive, making it possible to alter the sign of the optical response. This was the reason why the responses shown in figure 2 were so much different.

To separate the contribution caused by heating into the response, a one-dimensional thermoelastic problem was solved. Elastic properties of diamond and the graphitized layer were assumed to be isotropic. As a result the space-time distribution of temperature and elastic stresses in the structure were obtained. As the thermal-conductivity coefficient of diamond at room temperature is extremely high, we did not neglect the thermal conductivity as usually is used while calculating elastic stresses. Besides, the finite length of the laser pulse was taken into account. The variation of the reflection coefficient was calculated similarly to [9]. As structure 3 was semi-transparent (transmission ~15%), the reflection from both sides of the graphitized layer was taken into account. Results of calculating the response caused by nonstationary heating of the structure, are depicted in figure 2 by dashed curves. The calculations have shown that not only temperature dependence of refraction indices but also the thermal expansion of the sample considerably affected the thermal component of the response.

The best fit was achieved with photothermal coefficients being $\partial n/\partial T=5.5\cdot10^{-6}+j5.3\cdot10^{-6}$ K$^{-1}$ for the graphitized layer and $\partial n/\partial T=9.7\cdot10^{-6}$ K$^{-1}$ for diamond, which varies from published [10] by 8%. In addition, thermal-conductivity coefficients of graphitized layers of different thicknesses were defined (see table 1). The decrease of thickness appeared to cause the increase of the thermal-conductivity coefficient. After subtracting thermal components, we obtained the responses caused by propagation of the elastic waves (coherent phonons) (figure 3). It should be noted that all responses have well pronounced rippling. For example, curve 3 has 11 ps-period oscillations at the beginning of the response, curve 2 - spikes repeating every ~30 ps, and curve 3 - groups of 17 ps-period spikes repeating after ~70 ps.

However, we failed to get a quantitative fit between the calculated responses and the experimental ones. In our opinion, this can be explained in two ways. First, because the diamond plane (100) was misoriented against the sample surface (and against the interface between the graphitized layer and diamond), the laser pulse excited a shearing component of deformation [11]. Preliminary calculations taking into account the cubic structure of diamond showed that the maximum value of the shearing deformation component was about 50 times less than the volume component, with mode conversion at the sample surface taken into account. It was found out, however, that the diamond wafer used for implantation is birefringent. That means that the shearing component of deformation could affect the optical response. Second, the graphitized layer is likely to be inhomogeneous depthward. Dependence of the thermal-conductivity coefficient on the graphitized layer thickness testifies in favor of this assumption. After ion implantation distribution of radiation defects in the near-surface layer had its pronounced maximum.

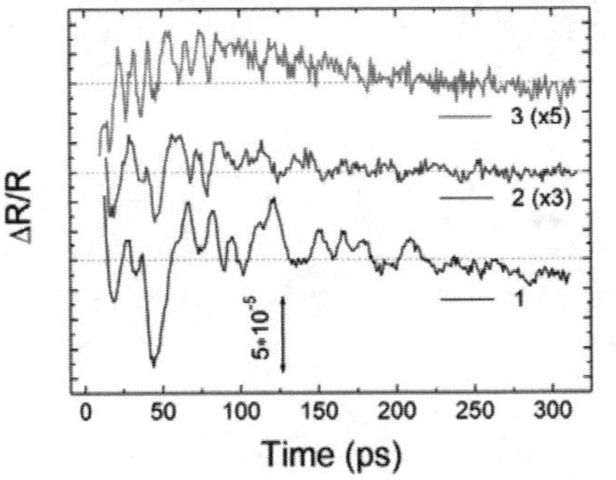


**Figure 3.** Responses of the sample with buried graphitized layers caused by propagating stress pulse. 1,2,3 –structures represented in Table 1.

During the annealing process, areas with defect concentration lower than critical restored the crystalline structure while areas with defect concentration higher than critical were graphitized. It seems that those areas with lower defect concentration (yet exceeding critical) appear to be more fully restructured during annealing than those with higher defect concentration. Therefore, elastic properties of the graphitized layer may also be inhomogeneous depthward.

## 4. Acknowledgement

We are grateful to V.A.Dravin for diamond implantation, V. Martovitsky for X-ray diagnostics and fruitful discussion, and T.I.Galkina for her interest and remarks. This study was supported by the Russian Foundation for Basic Research (projects No 08-02-01012, 10-02-00825-a) and by the Russian Academy of Sciences Program of «Basics of Fundamental Research of Nanotechnologies and Nanomaterials».

## References

[1] Krauss A R *et al.* 2001 *Diamond and Related Materials* **10** 1952
[2] Gippius A A, Khmelnitskiy R A, Dravin V A and Tkachenko S D 1999 *Diamond and Related Materials* **8** 1631
[3] Khomich A V, Kovalev V I, Zavedeev E V, Khmelnitskiy R A and Gippius A A 2005 *Vacuum* **78** 583
[4] Galkina T I, Klokov A Yu, Sharkov A I, Khmelnitskii R A, Gippius A A, Dravin V A, Ral'chenko V G and Savel'ev A V 2007 *Physics of the Solid State* **49** (4) 654
[5] Galkina T I, Sharkov A I, Klokov A Yu, Bonch-Osmolovskii M M, Khmelnitskii R A, Dravin V A and Gippius A A 1996 *Pis'ma Zh. Eksp. Teor. Fiz.* **64** (4) 270
[6] Thomsen C, Grahn H T, Maris H J and Tauc J 1986 *Phys. Rev. B* **34** 4129
[7] Khmelnitskiy R A, Dravin V A and Gippius A A 1996 *J. Chem. Vapor Deposition* **5** 121
[8] Khomich A V, Khmelnitskiy R A, Dravin V A, Gippius A A, Zavedeev E V and Vlasov I I 2007 *Physics of the Solid State* **49** (9) 1661
[9] Matsuda O and Wright O B 2002 *J. Opt. Soc. Am. B* **19** 3028
[10] Fontanella J, Johnston R L, Colwell J H and Andeen C 1977 *Appl. Optics* **16** 2949
[11] Pezeril T, Ruello P, Gougeon S, Chigarev N, Mounier D, Breteau J-M, Picart P and Gusev V 2007 *Phys. Rev. B* **75** 174307

Journal of Physics: Conference Series **278** (2011) 012020 doi:10.1088/1742-6596/278/1/012020

# Picosecond ultrasonics on a single micron carbon fiber

**D. Ségur, Y. Guillet, B. Audoin**

Université de Bordeaux, CNRS, UMR 5469, Talence, F-33405, France

E-mail: b.audoin@lmp.u-bordeaux1.fr

**Abstract.** A pump-probe picosecond technique is used to reveal the opto-acoustic response of micrometric carbon fibers. Two carbon fibers having the hexagonal symmetry are experimentally studied. A dedicated experimental setup allows to generate acoustic waves that propagate in a cross section of the fiber. The elastic properties in the transverse direction of a single carbon fiber are thus evaluated. Complex optical refractive index is also measured. The results provide interesting perpectives for the non-destructive evaluation of elastic and optical properties at a micron scale.

## 1. Introduction

Carbon reinforced composites are widely used materials in various fields such as transport, civil engineering, sports, etc. Carbon fibers are of particular interest owing to their outstanding strength to weight ratio. It is mainly composed of carbon atoms forming microscopic crystals that are more or less aligned with the fiber axis, giving the fiber a hexagonal symmetry.[1] Such carbon fibers have diameters of only a few microns. They are assembled in millimetric yarns before weaving the carbon plies that are finally embedded in a matrix. Usually, manufacturers measure the mechanical properties of those millimetric sized yarns by a traction test. At the same time, the knowledge of these properties at the scale of a single fiber is of strong interest for a better understanding of the micromechanical behaviour of composite materials. A recent work reports experimental results of a micro-traction test on a single micron carbon fiber assessing the elasticity in the direction of its axis.[1] The elasticity in a normal direction to the fiber axis has not yet been investigated essentially due to the experimental difficulties. We propose the first application of the picosecond ultrasonics technique [2] to generate and detect high frequency acoustic waves in such a single micron carbon fiber. Owing to its non contact generation and detection process, this optical technique is particularly well suited for the non-destructive evaluation of cylindrical shaped targets. A description of the experimental setup used in this study is first given; then its application to the non-destructive evaluation of the elastic and optical properties of carbon fibers is detailed.

## 2. Experimental setup and samples

The pump-probe picosecond ultrasonics technique has been successfully applied since the end of the 1980s to analyze plate structures with sub-micrometric thicknesses.[2] The absorption of a subpicosecond laser pulse (pump) suddenly heats the surface material that releases the incoming

Published under licence by IOP Publishing Ltd

Journal of Physics: Conference Series **278** (2011) 012020 doi:10.1088/1742-6596/278/1/012020

energy through a picosecond acoustic pulse propagating back and forth in the sample. A second light pulse (probe) is used to detect the change of the optical reflectivity of the sample caused by the propagating transient strain pulses. This optical technique allows non-contact acoustic generation and detection with very high resolution in time and space, making it particularly well suited for the non-destructive evaluation of cylindrical targets with micron diameters.

Two samples made of carbon fibers are used in this study. A high elastic modulus UMS40 carbon fiber (from Tenax Co.) made from a polyacrylonitrile precursor (ex-PAN) and a low elastic modulus XN05 carbon fiber (from Nippon Graphite Fiber Co.) produced from an isotropic pitch precursor also called an ex-brai precursor . These fibers have respectively a mean diameter of $d = 5\ \mu$m and $d = 10\ \mu$m and show a very regular cylindrical section and a small surface roughness. A small piece of a given fiber of one millimeter length is stretched at its two ends in a dedicated experimental setup to avoid free vibrations.

Experiments are performed with a conventional picosecond ultrasonics setup.[2, 3] Pump and probe beams are collinearly focused on the sample surface by the same objective with a 50X magnification and a high numerical aperture NA = 0.8. The widths of the pump and probe beams on the surface of the fibers are estimated to be 1.5 $\mu$m and 3 $\mu$m, respectively.

## 3. Experimental results and characterization of the carbon fibers

### *3.1. Elastic and acoustic properties*

Transient reflectivity changes measured on the UMS40 and the XN05 carbon fibers are shown in Fig. 1 (left) and Fig. 1 (right) respectively. The absorption of light creates a sudden temperature rise followed by a low thermal decay. Acoustic signal is then superimposed at corresponding times for which the reflected acoustic wavefronts reach the probed surface. The thermal background is removed to emphasize the acoustic signature of the signal as shown in Fig. 1 (right) for the XN05 fiber, where two acoustic echoes are detected corresponding to successive reflections of the longitudinal acoustic wave at the free surface of the sample. These echoes are shown with more details in the insets in Fig. 1 (right). Their dispersive feature appears clearly, with the low frequency components, about 15 GHz, of the transient echoes arriving earlier to the probed surface than the high frequency components. This negative dispersion cannot be due

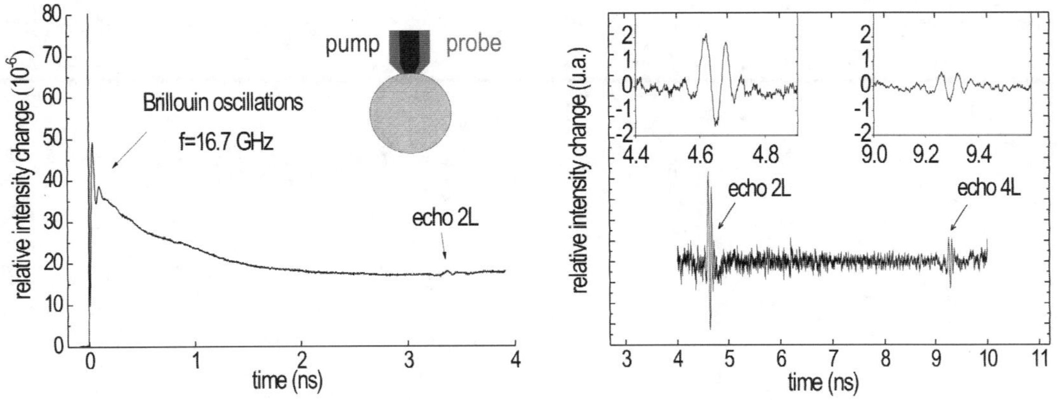


**Figure 1.** Reflectivity change for the UMS40 carbon fiber of 5 $\mu$m mean diameter (left). A sudden change in the reflectivity of the sample is followed by a slow thermal decay. The thermal background is substracted to reveal the acoustic content measured for the XN05 fiber (right). The two upper insets focus on each acoustic echo.

Journal of Physics: Conference Series **278** (2011) 012020 doi:10.1088/1742-6596/278/1/012020

to the structural relaxation (e.g. caused by visco-elasticity), since then the dispersion would be opposite, with high frequencies propagating faster than low frequencies.[4] This phenomenom results actually from the intrinsic geometric scattering of the acoustic waves in the cylindrical cavity.[3]

The acoustic displacements generated by the absorption of a laser pulse along a radial direction in a cylinder have been predicted theoretically [5] and have permitted to successfully describe experimental results observed at the surface of a glass cylinder with a diameter of 5 mm.[6] The wavelength of the bulk acoustic waves approximately equals the optical penetration depth. In the case of carbon fibers, acoustic diffraction is weak since the acoustic wavelength ($\sim 0.2\ \mu$m) is small with respect to the source width, and hence only bulk waves with a longitudinal polarization are detected. Their successive arrivals are observed at the times $\tau_{2L} = 4.65$ ns and $\tau_{4L} = 9.3$ ns for the XN05 fiber whereas only one echo has been observed for the UMS40 at 3.4 ns. From the round trip of the acoustic pulses we find a sound velocity $c_L$ in the cross section equal to $4.3 \pm 0.4$ km s$^{-1}$ and $2.9 \pm 0.3$ km s$^{-1}$ for the XN05 and UMS40 fibers, respectively. The mean diameters $d$ of 10 $\mu$m are provided by the manufacturer with standard deviation of 1 $\mu$m.[7] The elastic coefficients $c_{11}$ can be evaluated as follows: $c_{11} = \rho c_L^2 = 30 \pm 6$ GPa (where the density is $\rho = 1.65$ g cm$^3$) for the XN05 fiber and $c_{11} = 15 \pm 3$ GPa (where $\rho = 1.79$ g cm$^3$) for the UMS40 fiber. These elastic properties are in the same range of order whereas their elastic moduli in the axial direction are drastically different: E = 53 GPa and E = 380 GPa, respectively.[1] This can be understood from the microstructure of such carbon fibers. The XN05 fiber is made from an ex-brai isotropic pitch that leads to a low-modulus fiber in the axial direction with a weak anisotropy between axial and transverse directions. This is essentially due to axial traction forces during the graphitization process. The UMS40 fiber, made from a PAN pitch precursor, is a highly anisotropic carbon fiber where microcrystals, made of graphene sheets, are preferentially aligned in the axial direction leading to a high-modulus fiber.[8] Acoustic attenuation can also be measured for the XN05 fiber from the amplitudes of the two successive echoes.[3]

### *3.2. Optical properties*

Interest is now focused on the Brillouin oscillations detected in the first picoseconds, Fig. 2. The

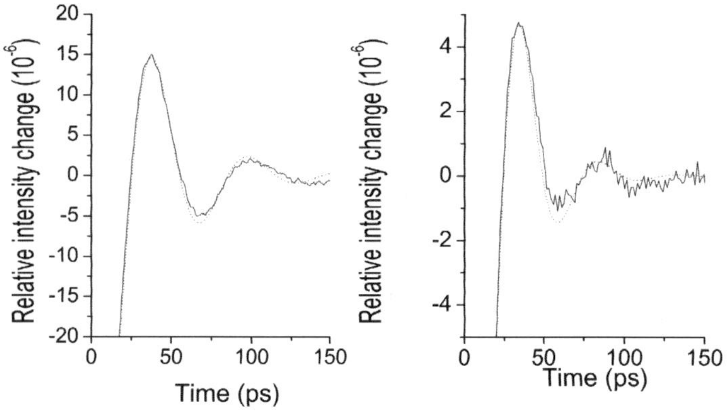


**Figure 2.** Brillouin oscillations observed in the first picoseconds (solid lines), with frequencies of 16.7 GHz and 20 GHz for UMS40 (left) and XN05 (right) fibers, respectively. The thermal background has been substracted. The corresponding curve fitting (dashed lines) is also reported.

transient reflectivity changes are sensitive to the modulation of the optical properties created by the in-depth temperature rise and strain. Part of the probe light penetrates into the material and is scattered by acoustic phonons provided that their wave number $k = 2q\cos\theta$, where $q$ stands for the probe light wave number and $\theta$ is the angle between the two wave vectors. The scattered light interferes with the part of the beam reflected at the sample surface giving rise to the so-called Brillouin oscillations. Their frequency is

$$f_B(\theta) = 2\,n\,c_L \cos(\theta)/\lambda, \tag{1}$$

where $\lambda$ is the wavelength of the probe light and $c_L$ is the sound velocity. Using Eq. (1), the real part of the optical refractive index $n$ at a wavelength of 796 nm can now be deduced from $f_B(0)$ at normal incidence of the probe beam: $n$=1.9 ± 0.2 and $n$=2.3 ± 0.2 for the XN05 and the UMS40 fibers respectively. In addition, the lifetime of the Brillouin oscillations can be the result of either acoustic or optical absorption. Since the acoustic echoes are detected after propagation along a long distance of 40 and 10 $\mu$m for the XN05 and UMS40 fibers respectively, the decrease of the first Brillouin oscillations comes mainly from the optical absorption. The experimental signal can be fitted by a dumped oscillation function. The value of the optical penetration depth at 796 nm, defined as the inverse of the absorption coefficient ($\alpha^{-1}$), can then be calculated using $\xi = C\tau$, giving $\xi$ = 91 ± 10 nm and $\xi$ = 94 ± 10 nm for the XN05 and UMS40 fibers, respectively. It yields the values for the imaginary part of the complex refractive index of the fiber as $k$ = 0.7 and $k$ = 0.6 for the XN05 and UMS40 carbon fibers, respectively. The measured refractive index for the carbon fibers compares well with the complex refractive index $\tilde{n}$ given in [9] for graphite at 800 nm, $\tilde{n}$ = 2.6 + 1.5 $i$. The measured physical parameters are reported in Tab.1.

| fiber | $d$ ($\mu$m) | $\rho$ (g cm$^{-3}$) | $c_L$ (cm s$^{-1}$) | $c_{11}$ (GPa) | $\alpha^{-1}$(nm) | $n$ | $k$ |
|---|---|---|---|---|---|---|---|
| XN05 | 10 | 1.65 | 4.3 ± 0.4 | 30 ± 6 | 91 ± 10 | 1.9 ± 0.2 | 0.7 ± 0.07 |
| UMS40 | 5 | 1.79 | 2.9 ± 0.3 | 15 ± 3 | 94 ± 10 | 2.3 ± 0.2 | 0.6 ± 0.06 |

**Table 1.** Physical parameters for the XN05 and UMS40 carbon fibers: the mean diameter $d$, the longitudinal sound velocity $c_L$, the elastic constant $c_{11}$, the optical penetration length $\alpha^{-1}$ at 796 nm, and the real and imaginary parts $n$ and $k$ of the refractive index.

## 4. Conclusion

The picosecond ultrasonics technique has been applied to the non-destructive evaluation of two carbon fibers of 5 and 10 $\mu$m diameter. Detection of the acoustic echoes propagating back and forth through the fiber cross section has led to measurement of the fiber mechanical properties. In consequence, theses data completements tensile tests that provide the elasticity only in the axial direction of the fiber. In addition, the complex optical refractive index has been measured locally. Such measurements may find interesting applications for the non-destructive evaluation of elastic anisotropy at a micron scale.

## References

[1] Sauder C, Lamon J and Pailler R 2004 *Carbon* **42** 715 – 725
[2] Thomsen C, Grahn H T, Maris H J and Tauc J 1986 *Phys. Rev. B* **34** 4129–4138
[3] Ségur D, Guillet Y and Audoin B 2010 *Appl. Phys. Lett.* **97** 031901
[4] Litovitz T and Davis C 1965 *in Physical Acoustics* vol 2A p.281 (New York: W.P. Mason Academic)
[5] Ségur D, Shuvalov A L, Pan Y and Audoin B 2009 *J. Phys.: Conf. Ser.* **195** 012007
[6] Ségur D, Shuvalov A L, Pan Y and Audoin B 2010 *J. Acoust. Soc. Am.* **127** 181
[7] Pradere C and Sauder C 2008 *Carbon* **46** 1874 – 1884
[8] Guigon M, Oberlin A and Desarmot G 1984 *Fibre Sci. Technol.* **20** 177
[9] Taft E A and Philipp H R 1965 *Phys. Rev.* **138** A197–A202

# Dynamics of specific heat and other relaxation processes in supercooled liquids by impulsive stimulated scattering

**J Fivez[a], R Salenbien[b], M Kuriakose Malayil[b], W Schols[b] and C Glorieux[b,c]**

[a]HUB, Stormstraat 2, B-1000 Brussel, Belgium
[b]Laboratorium voor Akoestiek en Thermische Fysica, Department Natuurkunde en Sterrenkunde, Katholieke Universiteit Leuven, Celestijnenlaan 200 D, B- 3001 Leuven, Belgium
[c]Corresponding author- email: christ.glorieux@fys.kuleuven.be

**Abstract**. Laser based impulsive stimulated scattering or transient grating excitation in a heterodyne diffraction scheme is a powerful method to extract information about different relaxing properties from different signal contributions. Longitudinal acoustic waves are detected simultaneously with thermal expansion and thermal diffusion. Careful fitting of the time-domain density response at different temperatures makes it possible to obtain the various relaxing physical parameters, and to construct Arrhenius plots for the respective relaxation processes. In this work we focus on the influence of the specific heat capacity C on the slower part of the density response function $S_\rho(t)$, and, inversely, on the possibility to extract from experimental $S_\rho(t)$ data the relaxation behaviour $C(\omega)$. The specific heat capacity is relevant for both the initially rising part of the impulsive stimulated scattering signal (together with the time and frequency dependent thermal expansion $\gamma(t)$), and for the thermal diffusion dominated decrease of the signal at later times after the excitation. By simulating $S_\rho(t)$ data in different scenarios, we address the feasibility of unravelling the impulse response functions C(t) and $\gamma(t)$ (and via Fourier transform also $C(\omega)$ and $\gamma(\omega)$) by careful fitting of the signal. This approach offers a unique possibility to extend the 100 kHz bandwidth of current dynamic calorimetric techniques determining $C(\omega)$ (photopyroelectric spectroscopy) to the sub-GHz range.

## 1. Introduction

Laser based impulsive stimulated scattering (ISS) or transient grating excitation in a heterodyne diffraction scheme is a powerful method to characterize glass forming liquids [1,2]. Via heating by partial optical absorption, followed by thermal expansion (impulsive stimulated thermal scattering or ISTS) and electrostrictive effects (impulsive stimulated Brillouin scattering or ISBS), two crossed pulsed laser beams, forming a pulsed optical interference pattern with a controllable grating spacing, act as an impulsive transient grating source for longitudinal or shear acoustic waves. Besides the acoustic contributions, the density response after laser excitation also contains contributions of thermal expansion (immediately after the deposition of optical energy) and thermal diffusion (which is slowly washing out the laser induced thermal grating). The dynamically changing density grating $\rho(T)$ can be detected via the accompanying refractive index grating $\delta n(x,t)\sim\rho(T)$ by monitoring the resulting diffraction of probe laser beams crossing the excitation region. By carefully selecting the polarization of the excitation and detection laser beams, it is also possible to extract information about the rotational relaxation that is triggered by the viscosity-driven aligning effect of the launched acoustic waves [3] or of the electrostrictive effect (ISBS [1,2]). Usually, transient grating experiments on glass forming liquids are performed in a so-called transmission configuration, by crossing laser beams in the bulk of the typically semi- transparent sample. However, by using a reflection configuration, containing an interface with a solid [4] or with air [5], it is also possible to extract information about the relaxation behaviour of the shear velocity.

In the transmission ISTS configuration, careful fitting of the time-domain density response makes it possible to determine the acoustic velocity and damping (related to the oscillatory part of the signal) and the thermal expansion dynamics and thermal diffusivity (related to the slowly varying part of the signal). Performing the experiment at different temperatures for different transient grating wavelengths (and thus for different frequencies) it is then possible to determine the Debye-like behaviour of the longitudinal acoustic velocity and of the thermal expansion dynamics, and to get the temperature dependence of the thermal diffusivity. From that, the temperature dependence of the relaxation time or frequency of the longitudinal acoustic velocity and of the structural relaxation time can be assessed and analysed in an Arrhenius plot.

In this work we focus on the influence of the relaxation of the specific heat capacity $C$ on the slower parts of the density response function $S_\rho(t)$, and, inversely, on the possibility to extract from experimental $S_\rho(t)$ data the relaxation behaviour of the heat capacity, $C(\omega)$. By simulating $\rho(t)$ data in different scenarios for a supercooled material with physical behaviour similar to the one of the well-known glass former glycerol, we address the feasibility of unravelling the relaxing part of $C(t)$ and $\gamma(t)$ (and via Fourier transform also $C(\omega)$ and $\gamma(\omega)$) by careful fitting of the ISS signal.

## 2. Impulsive stimulated scattering in a transient thermal grating configuration for supercooled liquids with relaxing thermo-elastic behaviour: theoretical model and simulation results

A typical ISTS experiment is performed by exciting the sample with an impulsive, quasi-1D light pattern (acting as a heat source via the process of optical absorption) that is very short in time, and quasi-periodic in space. This is achieved by crossing two laser first diffracted and then recombined beams in the $yz$ plane, which results in an interference pattern with wave number $q=2\pi/\lambda$ and spatial period $\lambda$ along the x-axis. In order to analyse the density response $\rho(t)$, driven by the temperature response $T(t)$ of the sample, we first solve in frequency domain the thermal diffusion equation and the longitudinal acoustic wave equation with thermal source term, that are given by:

$$\frac{\partial^2 T}{\partial x^2}-\frac{\rho C}{\kappa}\frac{\partial T}{\partial t}=-\frac{Q}{\kappa} \qquad \frac{\partial^2 u}{\partial x^2}-\frac{1}{c_L^2}\frac{\partial^2 u}{\partial t^2}=\gamma\frac{\partial T}{\partial x} \tag{1}$$

with $\rho$ [kg.m$^{-3}$] the static part of the density, $C$ [J.kg$^{-1}$.K$^{-1}$] the specific heat capacity, $\kappa$ [J.s$^{-1}$.K$^{-1}$.m$^{-1}$] the thermal conductivity, $\gamma$ [K$^{-1}$] the thermal expansion, $c_L$ [m.s$^{-1}$] the longitudinal acoustic velocity and $Q$ [J.s$^{-1}$.m$^{-3}$] the power density amplitude. For the spatially periodic part of an impulsive grating $Q(x,t)=Q_0\delta(t)\cos(qx)$ the solutions for the temperature $T(x,\omega)$ and the displacement $u(x,\omega)$ then read:

$$T(x,\omega)=\frac{Q_0\cos(qx)}{2\pi\left(\kappa q^2+i\omega C\right)} \qquad u(x,\omega)=\frac{-q\gamma c_L^2 Q_0\sin(qx)}{2\pi\rho C\left(\alpha q^2+i\omega\right)\left(\omega^2-q^2c_L^2\right)} \tag{2}$$

In the classical case where all physical properties are frequency independent, the displacement response $u(x,t)$ can then be calculated by performing a double Fourier transform, obtaining:

$$u(x,t)=\frac{\gamma Q_0\sin(qx)}{\rho Cq\left(1+\frac{q^2\alpha^2}{c_L^2}\right)}\left(\exp\left(-q^2\alpha t\right)-\cos(qc_L t)+\frac{\alpha q}{c_L}\sin(qc_L t)\right)\theta(t) \tag{3}$$

The specific heat clearly plays a role in the pre-factor, and, via the thermal diffusivity $\alpha=\kappa/(\rho C)$, in the exponentially decaying tail of the signal.

For a supercooled liquid, one can roughly model the frequency dependent dynamics by assuming Debye behaviour for $c_L(\omega)$, $C(\omega)$ and $\gamma(\omega)$:

$$c_L(\omega) = c_{L,\infty} + \frac{c_{L,0} - c_{L,\infty}}{1 + \mathrm{i}\dfrac{\omega}{\omega_{c_L}}} \qquad C(\omega) = C_\infty + \frac{C_0 - C_\infty}{1 + \mathrm{i}\dfrac{\omega}{\omega_C}} \qquad \gamma(\omega) = \gamma_\infty + \frac{\gamma_0 - \gamma_\infty}{1 + \mathrm{i}\dfrac{\omega}{\omega_\gamma}} \tag{4}$$

This leads to the following exact solution for the transient grating impulse response for temperature and strain $\Delta\rho_q/\rho$:

$$T(x,t) = \frac{Q_0 \cos(qx)}{\rho C_\infty \left(\omega_1^* - \omega_2^*\right)} \left( \left(\omega_1^* - \omega_C\right) \exp(-\omega_1^* t) - \left(\omega_2^* - \omega_C\right) \exp(-\omega_2^* t) \right) \theta(t) \tag{5}$$

$$\frac{\Delta\rho_q(x,t)}{\rho} = -\frac{\partial u(x,t)}{\partial x}$$

$$= \frac{q^2 Q_0 \cos(qx) \gamma_\infty c_{L,\infty}^2}{\rho C_\infty} \left( A_1 \exp(-\omega_1^* t) + A_2 \exp(-\omega_2^* t) + 2\Re\left( A_3 \exp(i\omega_3 t) + A_4 \exp(i\omega_4 t) \right) + A_5 \exp(-\omega_\gamma^* t) \right) \theta(t)$$

In. Eq. 5 (details given in Ref. [6]), the damping parameters $\omega_{1,2}^*$ are related to the thermal diffusion damping constant $q^2\alpha_\infty = q^2\kappa/(\rho C_\infty)$ and to the relaxation parameters of the heat capacity. The angular frequencies $\omega_j$ are related to the sound velocity and damping. $\theta(t)$ is the Heaviside function. In order to calculate simulations of $T(x,t)$ and $u(x,t)$, relaxation parameters were taken from literature for glycerol, and where necessary extrapolated to the frequency and temperature range of interest by using the Debye model fit for the frequency dependence and a Vogel-Fulcher-Tammann (VFT) fit for the temperature dependence. For the sake of conciseness, we only describe the temperature response in words: due to the time- and frequency dependence of the specific heat capacity, the temperature response, which contains an additional exponential function induced by the relaxation of the heat capacity, has an overshooting behaviour, with the overshoot duration increasing with decreasing temperature. This can be explained as follows. At short times, the supercooled liquid is too slow to take up heat in its slow degrees of freedom, so that the heat capacity is low and the temperature rise is high. Only at later times, energy can flow towards the cooperative rearrangements of the amorphous network, so that the heat capacity becomes higher and the temperature lower. The impact of this non-trivial temperature response on the ISTS density response (figure 1) depends on several factors, the main one being the extent to which $\tau_C$, the relaxation time of the heat capacity, is comparable with the thermal diffusion time $\tau_{thermal} \cong 1/(q^2\alpha_\infty) \cong \rho C_\infty/(q^2\kappa)$. Typically $\tau_C$ is of the order of the acoustic relaxation time $\tau_{ac}$ , and of the structural relaxation time $\tau_\gamma$. This renders it difficult to separately extract the heat capacity dynamics and the thermal expansion dynamics from the initial rise of the signal. As a matter of fact, till now, the effect of both relaxations (exponential functions in equation 5 with decay constants $\omega_1^*$, $\omega_2^*$, and $\omega_\gamma$), on the density response was modelled in a combined way by assuming a stretched exponential behaviour for the signal [1,2,3,7]. Separating the two effects could be greatly enhanced by exploiting the influence of the specific capacity on the thermal diffusion tail. By choosing a short wavelength $\lambda = 2\pi/q$, the characteristic time of the thermal diffusion tail, which is sensitive to the heat capacity relaxation via the thermal diffusivity, can be brought to timescales similar to $\tau_C$. By combining recorded signals obtained with different wavelengths, the q-dependent effect of the heat capacity relaxation can be untangled from the $q$-independent thermal expansion contribution to the signal. Details of the calculation and validation of the approach by experimental data [6], as well as a direct fast thermometry approach based on fluorescence spectroscopy [8] will be published elsewhere. Note that also non-Debye behaviour can be simulated by numerically Fourier transforming equations 2, provided a very large of number of points are considered in frequency domain, in order to cope with the very large dynamic range of the temporal response (typically from 10 ms for the thermal part of the signal to 1 ns for the acoustic part of the signal).

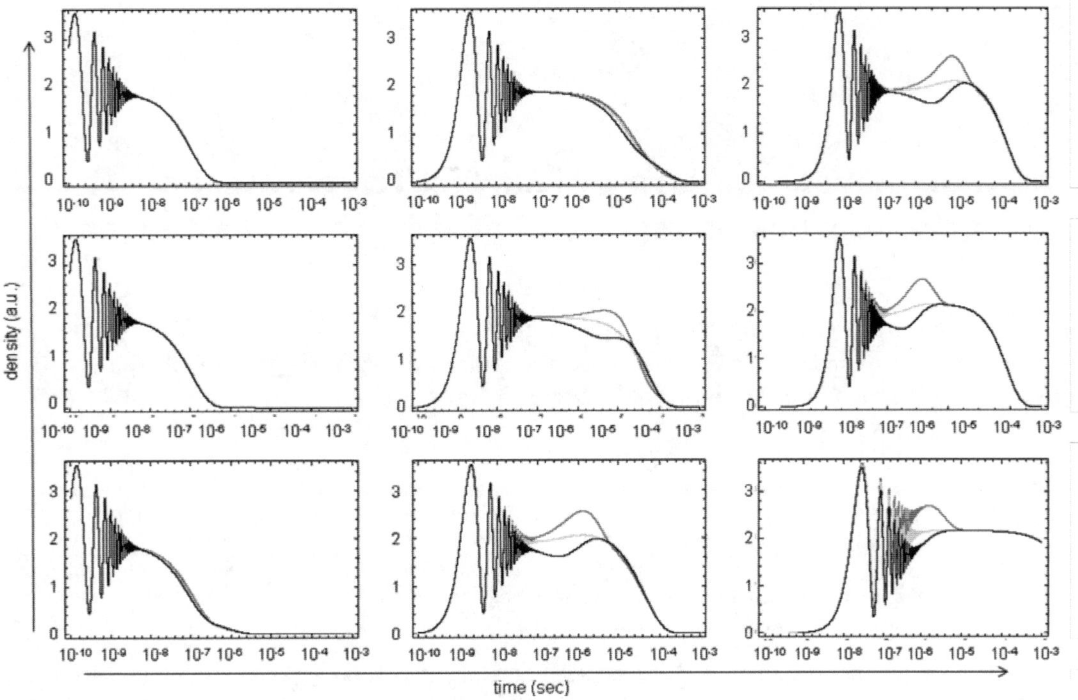


**Figure 1.** Simulated transient grating (negative) density response (a.u.) versus log(t), taking into account relaxing behaviour (based on data in literature [detailed description in Ref. [7]) of the longitudinal acoustic modulus, the specific heat capacity and the thermal expansion coefficient. From left to right: transient grating wavelengths 1, 15, 200 μm. From top to bottom: $\omega_\gamma$=$10^4$,$10^5$,$10^6$.$s^{-1}$. The simulations show the case-dependent effect of varying the relaxation frequency of the heat capacity ($\omega_C$=0 (upper red curve), $\omega_C=\omega_\gamma$ (middle green curve), $\omega_C=3\omega_\gamma$ (lower black curve)).

## References

[1] Yan YX, Nelson KA, *J. Chem. Phys.* **87**(11), 6240-6256 (1987)
[2] Yan YX, Nelson KA, *J. Chem. Phys.* **87**(11), 6257-6265 (1987)
[3] Glorieux C, Nelson KA, Hinze G and Fayer MD, *J. Chem. Phys.* **116**(8), 3384-3395 (2002)
[4] Glorieux C, Van de Rostyne K, Goossens J, Shkerdin G, Lauriks W, and Nelson KA, *J. Appl. Phys.* **99** , 013511(2006)
[5] Sermeus J, "*Capillarity and viscosity at liquid surfaces using impulsive stimulated scattering*", Master thesis Physics, K.U. Leuven 2010
[6] Schols W, "*Study of the relaxation behavior of the thermal expansion coefficient and the specific heat capacity of glass forming liquids by impulsive stimulated scattering*", Master thesis Physics, K.U. Leuven 2010
[6] Paolucci WA, Nelson KA, *J. Chem. Phys.* **112**(15), 6725-6732 (2000)
[8] Creten S, "*Stroboscopic fluorescence thermometry in glass forming liquids*", Master thesis Physics, K.U. Leuven 2010

## Acknowledgements

This work has been supported by the FWO (Project 'AVISCO' nr. G.0230.07) and the Research Fund of K. U. Leuven, Belgium. The authors would like to thank Keith Nelson (MIT, Cambridge, USA) for very helpful advice and fruitful discussions concerning the ISTS technique and relaxation properties of supercooled liquids.

# Two-dimensional monitoring of surface temperature distribution of a heated material by laser-ultrasound scanning

**I Ihara, H Yamada and M Takahashi**
Department of Mechanical Engineering, Nagaoka University of Technology, Nagaoka, Niigata 940-2188, JAPAN,

E-mail: ihara@mech.nagaokaut.ac.jp

**Abstract.** A non-contact method with a laser-ultrasonic technique for measuring two-dimensional temperature distribution on a material surface is presented. The method consists of a laser-ultrasonic measurement of a one-dimensional temperature distribution on a material surface and its two-dimensional area mapping. The surface temperature is basically determined from a temperature dependence of the velocity of the surface acoustic wave (SAW) propagating on a material surface. One-dimensional surface temperature distributions are determined by an inverse analysis consisting of a SAW measurement and a finite difference calculation. To obtain a two-dimensional distribution of surface temperature on a material surface, SAW measurements within the area of a square on the surface are performed by a pulsed laser scanning with a galvanometer system. The inverse analysis is then applied to each of the SAW data to determine the surface temperature distribution in a certain direction, and the obtained one-dimensional distributions are combined to construct a two-dimensional distribution of surface temperature. It has been demonstrated from the experiment with a heated aluminum plate that the temperature distributions of the area of a square on the aluminium surface determined by the ultrasonic method almost agree with those measured using an infrared camera.

## 1. Introduction

Temperature is one of the fundamental and important parameters to be measured in the various fields of materials science and engineering. This is basically because the properties and behaviour of materials often show any temperature dependence and therefore, measuring temperature and its comprehensive understanding are indispensable for developing and processing of materials. Since a temperature state of a material being heated or cooled is transiently changed, it is often required to monitor not only the temperature at a certain point but also its distribution in a certain area of the material. Although thermocouple techniques are widely used for temperature measurements, they are not always acceptable for obtaining the spatial distribution of temperature in a material because of its limitation of installation. In addition, the thermocouple technique may not be appropriate for monitoring a transient variation of temperature because of its relatively slow time response in measurement. To overcome such problem, an infrared radiation technique is used for measuring temperature distribution and its variation on a material surface. Since this method enables noncontact measurements, it is convenient for on-line measurements of surface temperature. In this method, however, accurate temperature measurements are often hindered by the different emissivity and reflection of infrared radiation from other heat sources. Such influences due to the emissivity and reflectivity result in the deterioration of measurement accuracy.

Ultrasound, because of its high sensitivity to temperature, is expected to be an alternative means for measuring the temperature of materials. Because of advantages of ultrasonic measurements such as non-invasive and faster time response, some works on the application of ultrasound to temperature estimations have been made extensively [1]–[6]. In our previous works [7] [8], an ultrasonic pulse echo method with an inverse analysis was developed and used to measure internal temperature distributions of a thick plate being heated. It was shown that one-dimensional temperature distributions inside materials being heated can be successfully determined by the ultrasonic inverse method coupled with a one-dimensional finite difference calculation [8]. The advantage of the method is that no information on the thermal boundary condition at the heating surface is needed for inversion. Recently, the ultrasonic method is modified to be adapted to surface temperature determination by surface acoustic wave (SAW) [9].

In this work, a new method with a laser-ultrasonic technique for measuring a two-dimensional temperature distribution on a material surface is presented. The method consists of a laser-ultrasonic measurement of a one-dimensional temperature distribution on a material surface and its two-dimensional area mapping by a pulsed laser scanning. To demonstrate the practicability of the method, surface temperature distribution for aluminum plate being heated is investigated. A laser interferometer based on photorefractive two-wave mixing [10] is used to measure the SAW of the plate during heating.

## 2. Method

### *2.1. Ultrasonic determination of temperature distribution*

It is known that the velocity of ultrasonic wave propagating through a material changes with the temperature of the material. The principle of temperature determination by ultrasound is based on the temperature dependence of the ultrasonic wave velocity. Assuming a one-dimensional temperature distribution on a material, the transit time of ultrasonic wave propagating in the direction of the temperature distribution can be given by

$$t_L = \int_0^L \frac{1}{v(T)} dx, \tag{1}$$

where $L$ is the propagation distance, $v(T)$ is the ultrasonic velocity which is a function of temperature $T$. It is noted that the temperature $T$ may change with location $x$. When the material is heated, the temperature distribution in the material can be given as a function of location $x$ and time $t$. Such a temperature distribution $T(x, t)$ is subjected to the thermal boundary condition of the heated material. On the basis of equation 1, if an appropriate inverse analysis with a proper boundary condition is employed, it could be possible to determine the temperature distribution from the transit time $t_L$ measured for the heated material. In fact, the validity of such ultrasonic determination of internal temperature distribution was successfully demonstrated through an experiment with a heated silicone rubber plate in our previous work [7]. In addition, an improved ultrasonic inverse method coupled with a one-dimensional finite difference calculation was developed to determine one-dimensional temperature distributions of materials being heated [8].

### *2.2. Quantitative evaluation of surface temperature distribution*

Recently, the ultrasonic inverse method mentioned above has been modified to be adapted to surface temperature determination by SAW [9]. In the modified method, one-dimensional unsteady heat conduction with a constant thermal diffusivity is considered for the surface of a flat plate whose single side is uniformly heated. Assuming that there is no internal heat source in the plate, the one-dimensional equation of heat conduction in a certain direction on the surface can approximately be defined. The temperature distribution can be estimated by solving the heat conduction equation under a certain boundary condition. In actual heating processes, however, the thermal boundary condition at the heating side is often unstable and unknown, and therefore, the temperature distribution is hardly determined by solving the heat conduction equation analytically. To overcome the problem, we developed an effective method consisting of a SAW measurement and an inverse analysis coupled with a one-dimensional finite difference calculation. The advantage of using the method is that no information on the thermal boundary condition at the heating side is needed for inversion. It was demonstrated through an experiment with a heated plate that the method with SAW is useful for measuring surface temperature distribution in a certain direction [9]. In this work, the method has been applied to two-dimensional mapping of surface temperature distribution.

## 3. Experiment and result

Figure 1 shows a schematic of the experimental setup used. This system provides non-contact measurements of SAWs on a heated plate using a laser-ultrasonic system. SAWs are generated at different positions from $E_1$ to $E_{13}$ consecutively by pulsed laser scanning irradiation (Nd:YAG, $\lambda$=1064 nm, energy 200 mJ/pulse, pulse width 3 ns) using a two-dimensional galvanometer scanner, and each SAW is detected at position D using a laser interferometer based on photorefractive two-wave mixing (Nd:YAG, $\lambda$=532 nm, energy 200 mW). An aluminum plate of 30 mm thickness is used for a specimen. The center frequency of the SAW is about 2 MHz. Because of such pulsed laser scanning, SAWs are measured within the square of 60 mm x 60 mm as shown in figure 1. Since the scanning time from $E_1$ to $E_{13}$ is about 0.9 s, the scanning irradiation can be performed every 1 s, while the left end of the plate is being heated by contacting with a heater of 300 °C. The transit time of each SAW is precisely determined by taking the autocorrelation of the detected signal of SAW during heating, and then used for the inverse analysis to determine the one-dimensional surface temperature distribution in the direction of SAW propagation. The obtained thirteen temperature distributions are combined together to construct the surface temperature distribution of the square at transient moment. An infrared camera is used to measure the reference data of the surface temperature distribution for comparison purpose.

Figure 2 shows the estimated surface temperature distributions and their variations with the elapsed time after heating starts, where the ultrasonically estimated results (upper) are compared with those measured using the infrared camera (lower). It can be seen that both temperature distributions

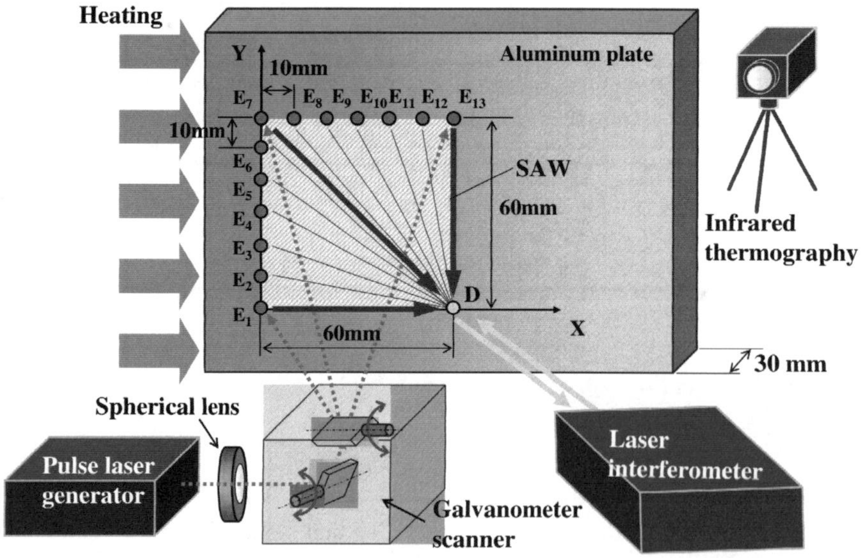


**Figure 1.** Schematic of the laser ultrasonic system with a pulsed laser scanning used for measuring SAWs in the area of a square on the surface of a heated plate.

determined by the ultrasonic method and the infrared camera almost agree with each other. It is noted that the temperature dependence of the Rayleigh wave velocity (m/s) of the aluminum, $v = -0.7557T + 2981.7$, where the temperature $T$ is in degree Celsius, is used for the inverse analysis.

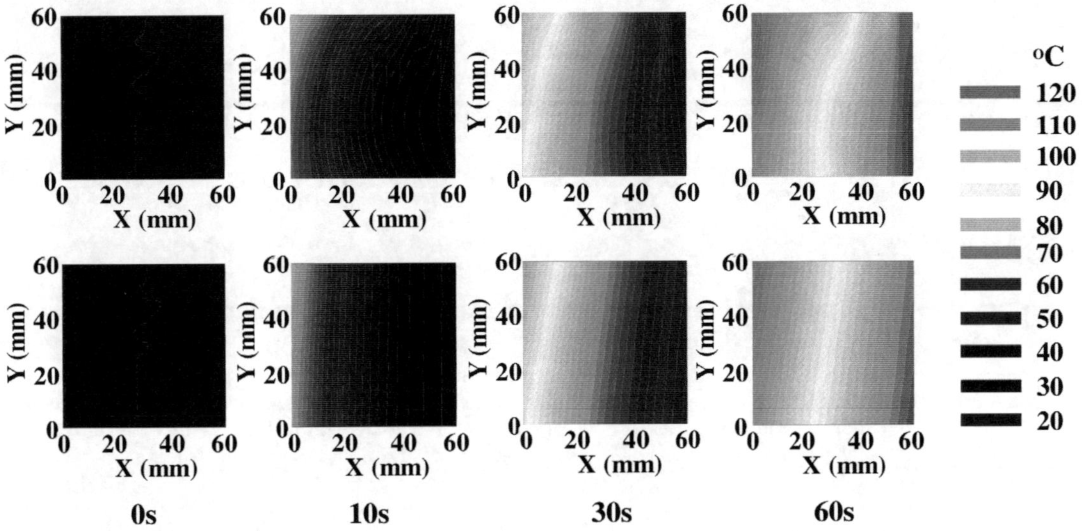


**Figure 2.** Variations in surface temperature distribution of an aluminum plate with heating time, estimated by ultrasound (upper) and infrared radiation (lower).

## 4. Conclusions

A non-contact method with a laser-ultrasonic technique for measuring two-dimensional temperature distribution on a material surface is presented. The practicability of the proposed method is demonstrated through the experiment with aluminum plate heated up to 120 °C. Although further improvements in the analysis and the resulting accuracy are necessary, it is highly expected that the method will be effective in the on-line monitoring of the transient variation in the surface temperature of the material processed at high temperatures.

## Acknowledgment

Financial supports by the Grant-In-Aid for Scientific Research (B22360304) from the JSPS, and Toyota Motor Co. are greatly appreciated.

## References

[1] Degertekin F L, Pei J, Khuri-Yakub B T and Saraswat K C 1994 *Appl. Phys. Lett.* **64** 1338
[2] Simon C, VanBaren P, and Ebbini E 1998 *IEEE Trans. Ultrasonics, Ferroelectrics, and Frequency Control* **45-4** 1088
[3] Chen T-F, Nguyen K-T, Wen S-S and Jen C-K 1999, *Meas. Sci. Technol.* **10** 139
[4] Balasubramainiam K, Shah V V, Costley R D, Boudreaux G and Singh J P 1999 *Review of scientific Instruments* **70-12** 4618
[5] Huang K N, Huang C F, Li Y C and Young M S 2002 *Review of scientific Instruments* **73-11** 4022
[6] Mizutani K, Kawabe S, Saito I and Masuyama H 2006 *Jpn. J. Appl. Phys.* **45** 4516
[7] Takahashi M and Ihara I 2008 *Jpn. J. Appl. Phys.* **47** 3894
[8] Ihara I and Takahashi M 2009 *Int. J. Intelligent Systems Technologies and Applications* **7-1** 80
[9] Takahashi M and Ihara I 2009 *Jpn. J. Appl. Phys.* **48-7** GB04-1
[10] Pouet B, Breugnot S and Clémenceau P 2006 *AIP Conf. Proc.* **820** 233

2nd International Symposium on Laser-Ultrasonics - Science, Technology and Applications IOP Publishing
Journal of Physics: Conference Series **278** (2011) 012023 doi:10.1088/1742-6596/278/1/012023

# Thermomechanical properties of single crystals evaluated by impulsive stimulated thermal scattering technique

**M Landa[1], B Verstraeten[2], J Sermeus[2], R Salenbien[2], P Sedlák[1], H Seiner[1], C Glorieux[2]**

[1]Institute of Thermomechanics, Academy of Sciences of the Czech Republic, Dolejškova 5, Prague 8, CZ 182 00, Czech Republic

[2] Laboratory for Acoustics and Thermal Physics, Department of Physics and Astronomy, Katholieke Universiteit Leuven, Celestijnenlaan 200D, B3001 Heverlee, Belgium

E-mail: ml@it.cas.cz

**Abstract.** This paper describes the application of Impulse Stimulated Thermal Scattering (ISTS) for measurement of surface acoustic wave (SAW) velocity and thermal diffusion along a free surface of a strongly anisotropic material. The motivation for this work stems from the study of thermoelastic properties of individual phases of ferroelastics; experimental results were obtained on a single crystal in the austenitic phase a Cu-Al-Ni alloy (bcc single crystal having elastic anisotropy factor of about 12). The measured SAW velocities in specific directions are in a good agreement with the values calculated for the elastic constants obtained by other ultrasonic methods. Similarly, the evaluated thermal diffusivity coefficient $(22 \pm 2).10^{-6}$ $m^2/s$ of the austenite is consistent with the data in the literature. The proposed approach has also a potential for characterization of thin films grown on anisotropic substrates.

## 1. Introduction

Ferroelastic materials are materials able to form fine, geometrically ordered microstructures consisting of differently oriented regions (ferroic domains) separated by twin boundaries. When the collective orientation of such a microstructure is changed by application an external field (temperature, mechanical loadings, magnetic field) the material macroscopically exhibits a reversible pseudoplastic strain magnitude up to of 10%. The elasticity of individual phases and their behavior in the vicinity of phase transitions are fundamental characteristics of the ferroelastics. However, due to the strength of the elastic anisotropy, the determination of the all independent elastic constants of these materials fall beyond the applicability limits of conventional pulse-echo methods [1] and of the resonant ultrasound spectroscopic (RUS) method [2]. Complementarity of these two methods for bulk elasticity measurements is exploited in the combined method introduced in [3], which enables a reliable determination of the sought elastic coefficients even if the both methods fail when applied separately.

The Surface Acoustic Wave (SAW) propagation can be understood as a surface resonator [6], with properties sensitive to individual components of the elastic tensor or their combinations. This resonant approach enables the sensitivity analysis to be carried out in a similar way as in the case of RUS and the pulse echo methods [3]. Since the displacement and stress fields of SAW are confined to a shallow region between the free surface with controllable penetration depth, their use has a particular potential for the evaluation of thin films. It is obviously desirable that of the elastic properties of the substrate are known.

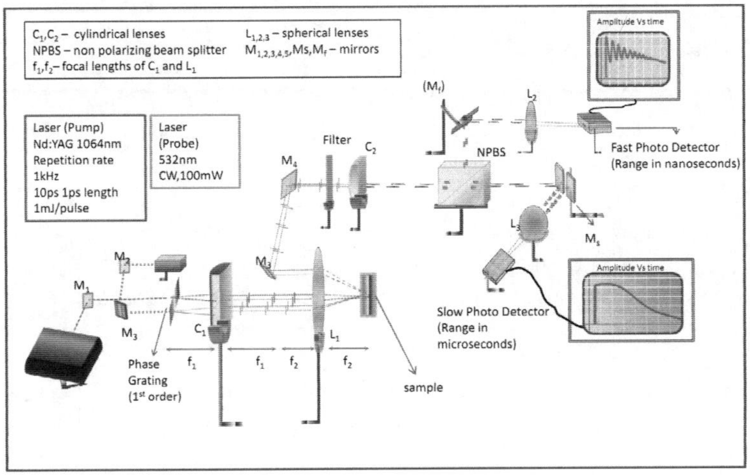


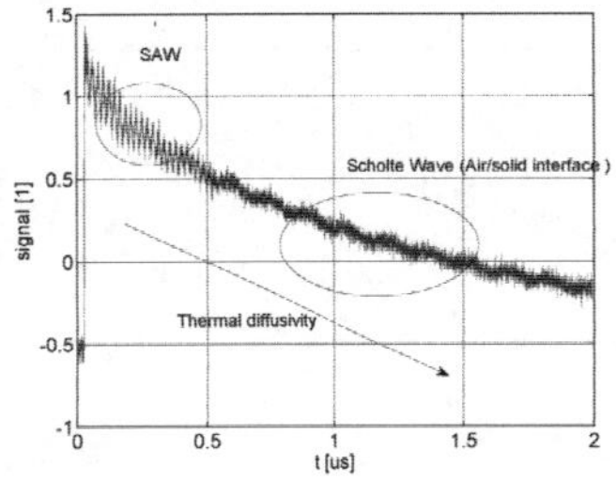


**Figure 1.** Heterodyne experimental set-up and typical detected signal.

## 2. Description of the method

Impulse Stimulated Thermal Scattering (ISTS) is a spectral method [4], using spatially narrow band excitations (figure 1). Grating excitation is particularly useful when the short wavelengths are to be excited, which is of interest for the investigation of shallow material regions and thin coatings. In principle, grating like SAW packets can be detected using laser beam deflection. However, the requirement that the probe beam must be focused into a width that is smaller than the SAW wavelength can be quite demanding as far as the instrumentation is concerned. The heterodyne diffraction detection scheme avoids this complication, since it involves no strict requirements on the width of the probe beam. The used design scheme makes it quite easy to detect short wavelength SAW with a high sensitivity allowing to detect even very small SAW displacements. In this differential variant, the optical phase difference between the diffracted and reference probe beams is adjusted and photodiode signals detecting the respective mixed light beams are electronically subtracted. This optical system was equipped by a rotational table for angular positioning of a specimen with respect to the direction of SAW propagation. The specimen was adjusted such that the surface under study was perpendicular to the optical axis of the ISTS system.

An example of a typical ISTS signal detected by heterodyne diffraction is shown in figure 1, showing contributions of SAW (propagating in the solid surface) and Scholte-Stoneley waves (interfacial wave propagation in air along the interface) may be seen. The decay of the signal amplitude with time (as a DC part of the signal) representing characteristics of thermal diffusivity may be separated by an additional low-pass detector (figure 1).

## 3. Sample preparation

A single crystal of Cu-14.3% Al-4.1% Ni (wt.%) alloy was prepared by the Bridgman method. The crystal structure at room temperature is b.c.c. (cubic austenite $\beta$-phase), because the transition temperature ($M_s \sim 193$K) to the martensite phase is much lower. A cube-like specimen (dimensions equal to approximately $10\times10\times10$ mm$^3$) was cut with faces perpendicular to the [110], [$\bar{1}$10] and [001] crystallographic axes. By cooling below $M_s$, the austenite phase

| $\rho_o$ | $C_{11}$ | $C_{12}$ | $C_{44}$ | $A$ | $C_L$ | $C'$ |
|---|---|---|---|---|---|---|
| [g/cm$^3$] | [GPa] | | | [1] | [GPa] | |
| 7.055 | 142.80 | 126.84 ±0.15 | 95.90±0.30 | 12.02 | 230.73 | 7.98 |

$A = 2C_{44}/(C_{11} - C_{12})$, $C' = (C_{11} - C_{12})/2$, $C_L = (C_{11} + C_{12})/2 + C_{44}$

**Table 1.** Elastic constants of the CuAlNi alloy

can be transformed into the martensite of orthorhombic 2H structure. The elastic constants of this material were measured by conventional pulse echo method [5] and the results (table 1) were then used for the following SAW analysis.

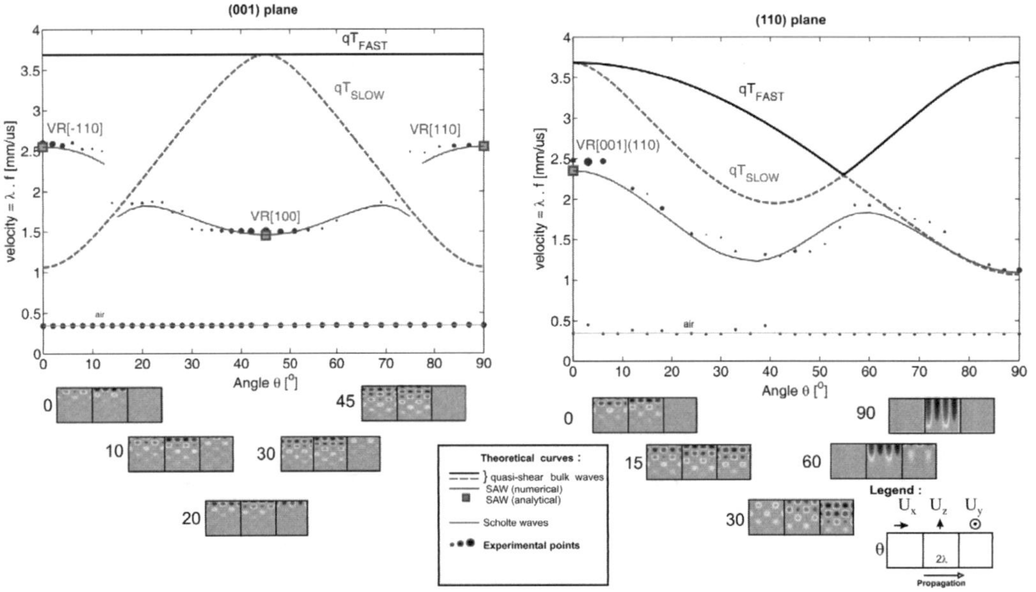


**Figure 2.** Results of ISTS measurements on CuAlNi single crystal.

## 4. Experimental results and analysis

Experiments were carried out on the laser ultrasonic system designed in ATF KUL, Belgium. The ISTS signals were obtained during sample rotation on both the (001) and (110) surfaces of the CuAlNi single crystal at ambient conditions (RT = 25$^o$C). The nominal optical wavelength of the grating was 50 $\mu$m. The fast component of each signal, obtained from a fast (AC) photodetector was averaged 4096 times, while the longer time evolution of the signal, obtained by a slower (DC) photodetector was averaged 2024 times. The angular step was 2.5 or 5$^o$.

The frequencies of the periodical SAW and Scholte wave packets were determined from the peaks in the frequency spectra of the recorded signals obtained by FFT. The effective (imaged) wavelength ($\lambda_{eff}$ = 55 $\mu$m, typically) of the SAW grating waves, which can slightly deviate from the nominal value of the phase mask, was calibrated by comparing the found Scholte velocity with the literature value of the sound velocity of air at the temperature of the ambient. In figure 2, the resulting SAW velocities are represented by points, the size of which reflects the relative magnitude of the spectral peak. Using material properties from Table1, the phase velocity of quasi transverse waves and planar SAW were calculated and plotted in the diagram. The "complete" theoretical angular dependence of SAW was calculated step by step from a numerical model of a surface resonator of length 2$\lambda$ with periodical boundary conditions [6]. For each step, the matrix of elastic constants was recalculated for a given general crystallographic orientation. The results of this model were verified by the solutions of the analytical formulas [7] for the SAW velocities in special directions. The "analytical" points are denoted by squares and labeled by "VR" with the propagation direction. In spite of the strong anisotropy, good agreement was found between the experiment and the theoretical prediction. The dependence of the SAW penetration depth on the orientation can be evaluated from the displacement fields $(u_x, u_z, u_y)$ mapped for $\theta = 0, 10, 20, 30, 45^o$ in the plane (001) and for $\theta = 0, 15, 30, 60, 90^o$ in the (110) plane. The penetration depth significantly increases far from the principal crystallographic directions. Near the [-110] direction in the (110) plane, a transition from the SAW mode to the $qT_{slow}$ mode can be observed.

Since the examined material is cubic, the thermal diffusivity (as a second rank tensor [8]) should be isotropic, i.e. representable by a single scalar quantity. This quantity can be estimated from the DC part of the ISTS signal (figure 1). When fitting this signal by $\text{erfc}(-\beta t)$, the time decay coefficient $\beta$ can be expressed as $\beta = D_T(2\pi/\lambda_{eff})^2$, which enables the thermal diffusivity coefficient $D_T$ to be calculated [9]. The averaged value of the thermal diffusivity obtained from the all angular measurements was $D_T = (22 \pm 2).10^{-6}$ m$^2$/s. This value is slightly higher than estimates (about $16.10^{-6}$ m$^2$/s) obtained from typical thermal properties of CuAlNi alloy [10], indicating a good crystalline quality. The given measurement uncertainty is based on the standard deviation between measurements taken under different angles. As expected, no systematic variation of the thermal diffusivity with angle was found for this cubic crystal.

## 5. Conclusion

Thermomechanical properties of a strongly anisotropic cubic single crystal were determined by ISTS methods. The angular dependence of both the SAW velocity and the thermal diffusivity was determined in the (001) and (110) planes. The experimentally obtained SAW velocities were compared with theoretical predictions based on a numerical model. In spite of the strong anisotropy, a satisfactory agreement between experimental and theoretical results was found, which proves the reliability of the numerical algorithm used for the calculation of the SAW velocities, and shows that this algorithm could be possibly used for inverse determination of the elastic constants. For the thermal diffusivity measurements, the results are, as well, in an acceptable agreement with the literature data. Both the experimental determination and theoretical analysis of thermal diffusivity in single crystals need, however, further investigation.

### Acknowledgement

The authors would like to express their thanks to dr.Václav Novák from the Institute of Physics of the Academy of Sciences of the Czech Republic for the sample preparation. This work was financially supported by the Czech Science Foundation (project No.101/09/0702) and by the research program of IT ASCR (CEZ:AV0Z20760514), and by the Research Council of the K. U. Leuven (project No. OT/07/035).

## References

[1] Sedlák P, Seiner H, Landa M, Novák V,Šittner P and Manosa Ll 2005 *Acta Materialia* **53** 3643-3661
[2] Landa M, Sedlák P, Seiner H, Heller L, Bicanová L,Šittner P and Novák V 2009 *Appl. Phys. A* **96** 557-56
[3] Landa M, Seiner H, Sedlák P, Bicanová L, Zídek J and Heller L 2009 Resonant Ultrasound Spectroscopy Close to Its Applicability Limits *Horizons in World Physics Vol. 268* ed Everett M and Pedroza L (Nova Sci Publ) chpt.3
[4] Rogers JA, Maznev AA, Banet MJ and Nelson KA, 2000 *Ann. Rev. Mat. Sc.* **30** 117-157
[5] Landa M, Novák V , Sedlák P and Šittner P 2004 *Ultrasonics*, **42** 519-526
[6] Westafer RS, Mohammadi S, Adibi A and Hunt WD 2009 Proceedings of the Comsol Conference (CD-ROM)
[7] Musgrave M J P 1970 *Crystal Acoustics* (Holden-Day Inc, SF)
[8] Nye JF 1985 *Physical properties of Crystals* (Oxford Sci Publ)
[9] Käding OW, Skurk H., Maznev AA and Matthias E 1995, *Appl.Phys A* **61** 253-261
[10] `http://www.nimesis.com/anglais/ourproducts/nisaloy.htm`

Journal of Physics: Conference Series **278** (2011) 012024 doi:10.1088/1742-6596/278/1/012024

# Temperature dependant thermal and mechanical properties of a metal-phase change layer interface using the time resolved pump probe technique

**V Schick[1], J-L Battaglia[1], A Kusiak[1], C Rossignol[2], C Wiemer[3]**

[1]Laboratoire de **T**ransferts **É**coulements **Fl**uides **É**nergétique (TREFLE), Esplanade des Arts et Métiers – 33405 TALENCE Cedex, France
[2]**L**aboratoire de **M**écaniques **P**hysique (LMP), Esplanade des Arts et Métiers – 33405 TALENCE Cedex, France
[3]Laboratorio "Materials and Devices for Microelectronics", Italian Institute for the Physics of Matter, via C. Olivetti 2, 20041 Agrate Brianza (Milan), Italy.

vincent.schick@bordeaux.ensam.fr

**Abstract**. Time Resolved Pump Probe (TRPP) technique has been implemented to study the thermal and mechanical properties of $Ge_2Sb_2Te_5$ (GST) film deposited on a silicon substrate. According to the knowledge of the thermal properties of the GST layer, the temperature dependant Thermal Boundary Resistance (TBR) at the metal-GST interface is evaluated. Measuring the acoustic oscillation and more particularly its damping leads to characterize the adhesion at the metal - GST interface. This quantity can be efficiently related to the temperature dependent TBR in the 25°C – 400°C range. The TBR increases with temperature and follows the changes of the crystalline structure of materials. A linear relation between the acoustic reflection coefficient and the logarithm of the thermal boundary resistance is found.

## 1. Introduction : Phase Change Memory

Phase change materials are extensively studied due to their promising applications in the framework of Phase Change Memory (PCM) or Ovonic Unified Memory [1,2]. PCM functioning involves chalcogenide materials that are allotropic semi conducting elements and alloys belonging to the IV, V and VI group of the periodic classification. They can be reversibly brought from the amorphous to the crystalline state, so that the corresponding different electrical properties between these two states (electrical resistance decreases over 4 decades) can be used for data storage. $Ge_2Sb_2Te_5$, commonly denoted GST, is one of the most popular chalcogenide alloys [3], since it is stable at room temperature in the amorphous and hexagonal crystalline phases (*hcp*), and metastable in the face centered cubic phase (*fcc*) [4,5]. The amorphous/*fcc* phase transition occure at 130°C whereas the *fcc*/*hcp* one at 350°C. The melting temperature is approximately 600°C. The transformation between crystalline and amorphous phases is reversible: heating the amorphous GST to a temperature slightly above the glass-transition temperature leads to the crystalline phase; subsequent heating to a temperature close to the melting temperature with fast quenching permits retrieving the amorphous phase.

Optimization of the phase change memories operation rests on the knowledge of the thermal properties of the phase change material. The value of specific heat per unit volume of the GST-225, which does not depend on the crystallographic configuration at high temperature, is: $\rho\, C_p = 1.28\text{x}10^6$ $J.K^{-1}.m^{-3}$. We measured the thermal conductivity of the GST in previous study by using modulated photothermal radiometry in the [$10^3$-$10^5$] Hz frequency range [8]. This alloy has, according to the crystalline phase, the following densities [6]: 5.87 $g.cm^{-3}$ for the amorphous phase, 6.27 $g.cm^{-3}$ in cubic phase and 6.39 $g.cm^{-3}$ in hexagonal phase. The longitudinal speeds of sound were measured by acoustic picoseconds techniques [7]. The longitudinal velocity is 2.25 $nm.ps^{-1}$, 3.19 $nm.ps^{-1}$ and 3.3 $nm.ps^{-1}$ for amorphous, *fcc* and *hcp* crystalline phases respectively. It appears that the thermal resistance at the scale of the GST thickness (some tenth of manometers) is equivalent to the Thermal Boundary Resistance (TBR) at the interface between contacting layers. This paper aims to measure the TBR at the GST (210 nm)-Al (20 nm) interface from room temperature up to 400°C. It is therefore necessary to generate a short thermal excitation and to carry out the measurement within a sub-nanosecond time range.

## 2. Experimental procedure

Time Resolved Pump Probe (TRPP) was performed with a picoseconds acoustic experimental setup initially designed to study the acoustic wave propagation phenomena at the picoseconds and nanometer scale [9]. With this technique the response of a material to an ultra short disturbance is determined by measuring the thermally induced variations of the optical reflectivity at the surface. Under the linearity assumption, the relative variation of reflectivity ($\Delta R/R_0$) is proportional to the temperature variation at the surface of the sample. A Ti:sapphire laser produces a 200 fs pulse at 800 nm wavelength with a $f_m$=82 MHz repetition rate. The laser beam is then splitted into 2 separated beams using a polarizing cube. The so-called pump beam heats the surface of the sample. The pump power received by the sample is about 2 mW, the spot having a 20 μm diameter. Before arriving on the sample, the pump beam is differentiated from the probe beam by traversing a non-linear crystal that divides its wavelength by two. The probe beam and the pump beam are both focused at the same location on the surface of the sample. The reflected part of the probe will depend on the surface temperature. A silicon photodiode is used to measure the temperature dependent relative variation of the reflectivity ($\Delta R/R_0$). The probe pulse is shifted temporally compared to the pump by increasing the distance covered by the probe using an optical delay line. This delay line allows us achieving a $T$=12.5 ns time resolution for the recording signal. This signal can be viewed as the impulse response of the sample in a front face configuration. The probe power received by the sample is about 0.12 mW, the spot having a 20 μm diameter. The pump is modulated with an acousto-optics modulator at $f_m$=330 kHz. A Stanford Research SR 830 Lock-In amplifier used this modulation signal as the reference and makes it possible to accurately extract the $\Delta R/R_0$ signal in terms of real and imaginary parts. The signal/noise ratio is improved by carrying out several translations of the delay line. Nevertheless, this repetition will lead also to a cumulative phenomenon on the signal that must be taken into account within the model [10].

The measured sample was formed by 210 nm thick amorphous GST layer, deposited by sputtering on a Si (100) substrate covered with a 100 nm thick thermal $SiO_2$ layer.; The GST layer is capped with a 20 nm thick Al layer. The sample is heated using a turntable LINKAM TS 1000 furnace under inert atmosphere (Ar) between 20 and 400°C.

## 3. Heat transfer model

Previous work [9] based on the two temperature model has demonstrated that the electron-phonon thermallization in the Al is reached close to $\tau$=10 psec. Time greater than 10 ps are therefore necessary in order to apply the Fourier's law for the GST/Al system The GST layer behaves as a semi infinite medium between two successive pulses. The size of the laser spot being larger than the heat penetration depth at the modulation of the pump, the heat transfer is considered one-dimensional. No heat loss is assumed on the front face of the sample and the temperature is imposed by the furnace on the rear face. Figure 1 presents the model in a synthetic form by using the thermal impedances schema.

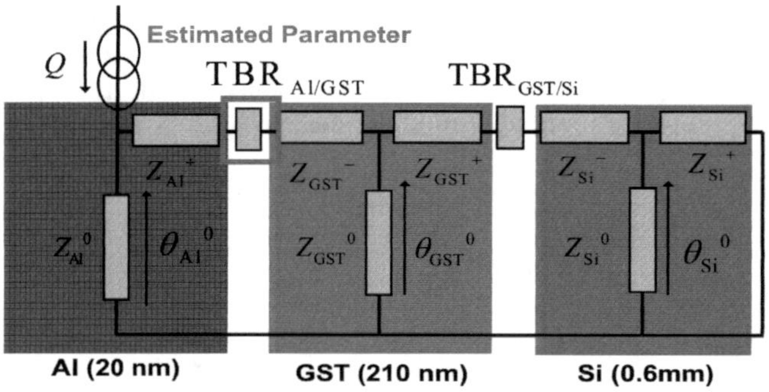


**Figure 1.** Schematic representation of the heat transfer model using the thermal impedances

The repetition rate of the pump is taken into account in the model since the temperature does not have time to go back to its initial value when the next pulse arrives, *i.e.* the cumulative thermal effect. In agreement with the reference [11], the normalized measured signal related to the real part of the lock-in is then:

$$S(\tau)=\sum_{k=-\infty}^{\infty}\theta_{Al}\left(i2\pi\left(\frac{k}{T}+f_m\right)\right)\exp\left(\frac{i2\pi k\tau}{T}\right)$$

where $\tau$ varies from 0 to $T$.

## 4. Results and discussion

The estimated TBR at the Al/GST interface are reported on figure 2 (The estimated values at Au (30 nm)-GST interface is also reported). The Levenberg - Marquardt algorithm based on measurements made up to 6 ns on the 25- 400°C temperature range is used to estimate TBR.

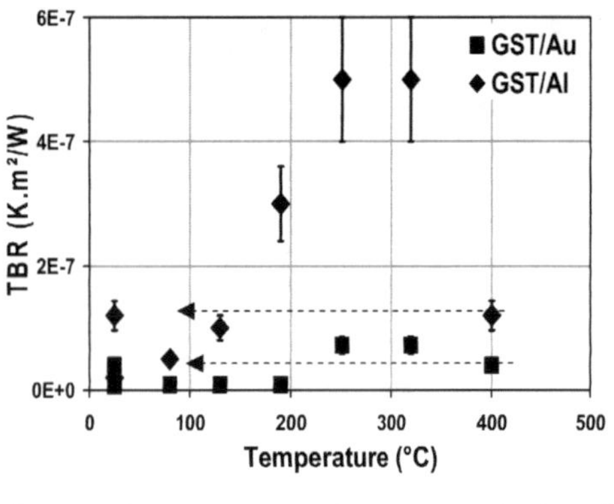


**Figure 2.** TBR at the GST-Al interface. Comparison with that at GST-Au interface.

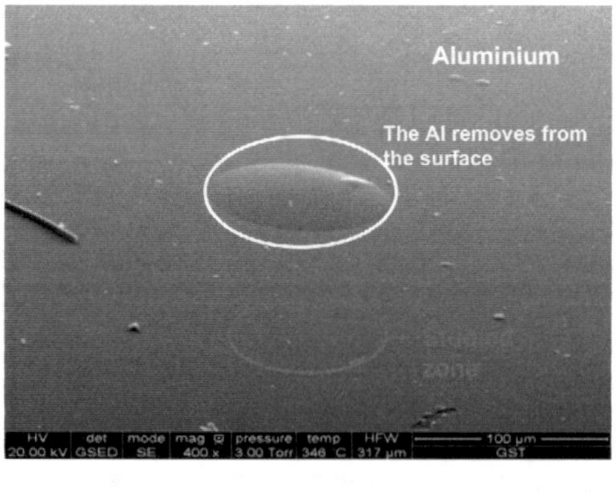


**Figure 3.** Scanning Electron Microscopy (SEM) at the Al at 340°C.

The values of the TBR increase strongly with temperature and thus number of crystals at GST/metal interface. As represented in figure 3, SEM measurement beyond 250°C showed a dissociation of the Al layer. This observation agrees with the high value of TBR starting from 250°C. A degradation of the aluminium film is also observed at 400°C that can explain the fast decrease in the TBR value. Indeed, we observed the Al layer collapsed of the GST layers at this temperature. Calculation with Diffuse Mismatch Model [11] of TBR of Al/GST interfaces is about $10^{-10}$-$10^{-9}$.K.m².W$^{-1}$ for *fcc* phase. By focusing on the short times (0-50ps), an acoustic wave is observed (see

Fig. 4). The period of this oscillation (+-10 ps) corresponds to the first acoustic resonance of the 20 nm aluminium top layer. The attenuation of these vibrations decreases with temperature.

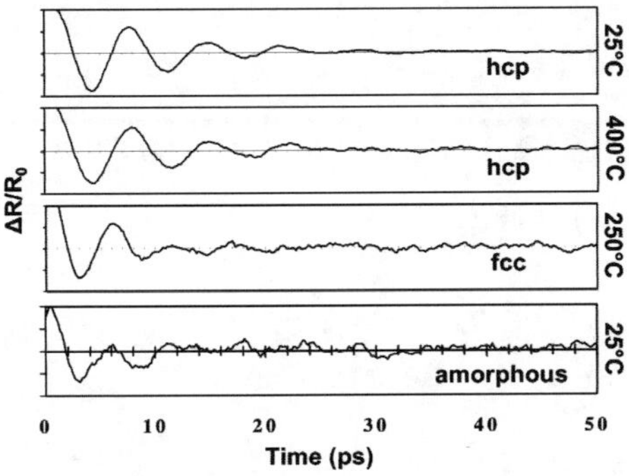


**Figure 4.** Measured acoustic waves in Al according to the sample temperature and crystalline structure of GST

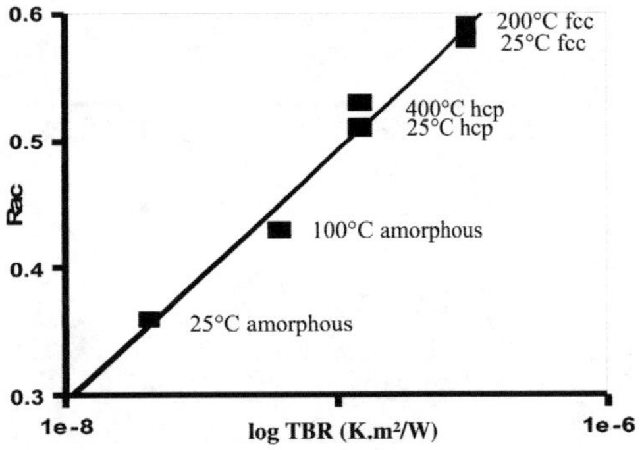


**Figure 5.** Measured GST-Al acoustic reflection coefficient versus TBR.

The acoustic reflection is $|R_{ac}| = \exp\left(-\alpha_{\exp} / f_{\exp}\right)$ where $\alpha_{\exp}$ is the wave attenuation and $f_{\exp}$ is the measured frequency of the acoustic wave (as reported in figure 4). As presented in figure 5, a linear relation between the thermal boundary resistance and the acoustic reflection coefficient is found and is related to the phonon flux transmitted across the interface. A comparable result is obtained for Au on GST but it is more difficult to quantify. However, such a result demonstrates that the TBR can be directly related with the adhesion force. To analyze the acoustic damping results, the Al and GST could be considered as two elastic layers connected by massless springs et the interface or could be connected by a layer of inter-diffusion of element between Al and GST layers [12].

However the standard deviation of estimated TBR and acoustic reflection coefficient is about 20 %. It means that these values need to be improved by making new experiments with better accuracy. In addition, the links between acoustic wave and TBR measurements need to be understood, in particular by studying the evolution of interface structure with chemical and structural analysis.

## References

[1] Wuttig M (2005), *Nature Mater.* **4**, 265
[2] Lankhorst M, Ketelaars B and Wolters R, (2005), *Nature Mater.* **4**, 347
[3] Kolobov A, Fons P, Frenkel A, Ankudinov A, Tominaga J and Uruga T, (2004), *Nature Mater.* **3**, 703
[4] Yamada N and Matsunaga T, (2000), *J. Appl. Phys.* **88**, 7020
[5] Y Park, J Lee, M Youm, Y Kim and H Lee, (2005), *J. Appl. Phys.* **97**, 093506
[6] Njoroge W K, Woltgns H W, Wuttig M, (2002), *J. Vac. Sci. Technol.* A **20**, 230
[7] Lyeo H, Cahill D, Lee B, Abelson J, Know M, Kim K, Bishop S, and Cheong B, (2006), *Appl. Phys. Lett.* **89**, 151904
[8] Battaglia J-L, Kusiak A, Schick V, Cappella A, Wiemer C, Longo M, Varesi E, (2010), *J. Appl. Phys.* **107**, doi. 10.1063/1.3284084
[9] Battaglia J-L, Kusiak A, Rossignol C, Chigarev N, (2007), *Phys. Rev. B* **76**, 184110
[10] Schmidt A-J, Chen X and Chen G, (2008), *Rev. Sci. Instrum.* **79**, 114902
[11] Stevens R, Zhigilei L, Norris P, (2007), *Int. J. Heat Mass Transfert* **50**, 3977
[12] Tas G, Loomis J, Maris H, Bailes A and Seiberling L, (1998), *Appl. Phys. Lett.*, **72**, 18

# Optical detection of ultrasound on rough surfaces using speckle correlated spatial filtering

**M Clark**

Electrical Systems and Optics Research Division, University of Nottingham, University Park, Nottingham NG7 2RD, UK

E-mail: matt.clark@nottingham.ac.uk

**Abstract.**
There are numerous techniques for the optical detection of ultrasound but many of these are of very limited application on optically rough surfaces. Techniques that work on rough surfaces are often complicated, inefficient and expensive. In this paper a new technique for the optical detection of ultrasound on rough surfaces is presented. It is a variation on the knife-edge (KE) technique but adapted for rough surfaces by replacing the knife-edge with a speckle correlated spatial filter (SCSF) which is implemented using a spatial light modulator (SLM). This technique is conceptually and practically simple and may be implemented using relatively low cost components. It is also relatively general and may also be applied to the detection of small displacements in many circumstances.

## 1. Introduction

There are numerous techniques for the optical detection of ultrasound including: KE detection, homodyne and heterodyne interferometry and Fabry-Perot detection. For reviews of these techniques see [1, 2, 3]. All of these techniques operate on optically smooth surfaces but can suffer from greatly reduced performance on optically rough surfaces.

Optically rough surfaces present a particular problem for optical detection because the reflection of laser light from a rough sample surface is speckled rather than specular. Techniques that work on rough surfaces include: adaptive interferometers, Fabry-Perot detectors and photo EMF based receivers. These techniques suffer from a number of drawbacks, for instance, the Fabry-Perot technique requires a frequency stabilised, single mode laser and an etalon with adaptive feedback in order to operate effectively. These are expensive and fragile items. Adaptive interferometers using two wave mixing in non linear crystals or photoEMF devices need to use a lot of the optical power to establish the adaptive pattern in the the crystal or photo-EMF receiver. Consequently they have expensive light budgets. Several commercial rough surface detectors are available on the market at considerable expense; much of which is due to the cost of the detection laser.

In this paper a simple variation of the KE detection technique is demonstrated on optically rough surfaces. It shares the same sensitivity characteristics as a normal KE detector operating on a smooth sample. The system is optically and conceptually simple, robust and inexpensive to implement. The SCSF detector demonstrated in this paper used a 5mW HeNe laser and an SLM removed from an inexpensive ($300) video projector.

Journal of Physics: Conference Series **278** (2011) 012025 doi:10.1088/1742-6596/278/1/012025

## 2. Knife edge detectors

In a normal KE detector the presence of an elastic wave under the illumination beam deflects the reflected beam (see figure 1). The reflected beam is partially blocked by a 'knife-edge'. As the beam moves perpendicular to the knife-edge the amount of light passing it and reaching the photodiode changes. In practice the 'knife-edge' is simply a special case of a general spatial filter and may be implemented in a number of different ways, for instance, by using a split photodiode and differential electronics[4].

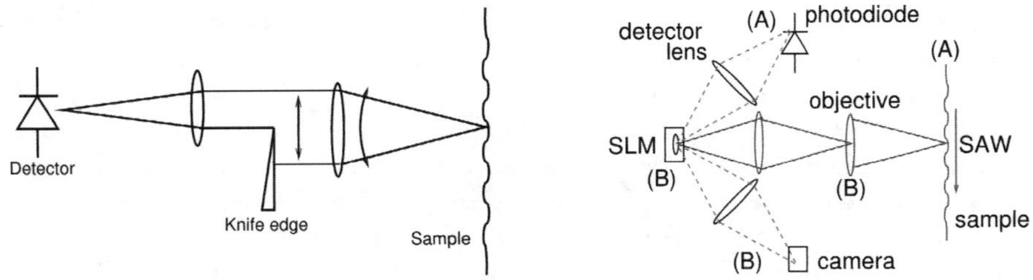


**Figure 1.** (left) Schematic of a KE detector and (right) the SCSF detector. In both cases the illumination optics have been removed for clarity. As the surface is deflected by the elastic wave the reflected beam is subject to angular deflection. This is transformed into an intensity modulation by the knife-edge (left) or spatial filter (right) and picked up by a photodetector. To gain optimal performance the knife-edge is positioned at the point of peak intensity in the reflected beam. In the SCSF detector an 'edge' is placed at the peak of each speckle. Planes marked (A) are conjugates of each other as are planes marked (B), hence the speckle pattern at the objective lens is imaged onto the SLM and the camera.

In order for the KE detector to work effectively the reflection from the sample surface needs to contain a strong specular component. If the sample surface is optically rough this specular reflection is not present, and only a diffuse speckle pattern enters the detector. In this situation the sensitivity is severely dependent on the location of the individual speckles across the knife-edge. Since the light is spread out over a wide range of solid angle, the transfer of optical power across the knife-edge with angular deflection is reduced leading to reduced sensitivity. The light entering the optical system is also likely to be clipped by the entrance aperture so there is a possibility that the response will change sign. As the speckle will change with the position, the sensitivity is not only reduced but also varies randomly with position.

The new detection method presented here uses an ensemble of 'knife-edges', one 'edge' for each speckle. This ensemble of knife-edges is implemented as pattern displayed on an SLM.

This detection method relies on the fact that, for very small displacements of the sample surface, the observed speckle moves rigidly. A carefully designed spatial filter, correlated to the observed speckle pattern, remains highly correlated to the speckle pattern throughout this small movement. The spatial filter needs to be designed so that it divides each speckle in half (across the direction of motion). The light from the sample is then passed through the spatial filter, collected by a lens and focused onto a photodiode. As the sample surface is perturbed by the acoustic wave the rigid movement of the speckles across the spatial filter is turned into an intensity modulation at the photodiode.

The spatial filter may be easily implemented using an SLM which only needs to display the filter as a binary mask. If the SLM is not absorbative and separates the light into a positive and negative images then the light from both of these may be collected on separate photodiodes and the signals subtracted to gain an improvement in the signal levels (and reject common mode noise)[5]. For large movements (compared with the illumination spot size) the speckle pattern

Journal of Physics: Conference Series **278** (2011) 012025 doi:10.1088/1742-6596/278/1/012025

changes its spatial distribution and becomes decorrelated with the spatial filter. If this occurs then the filter ceases to function correctly and the detection sensitivity is lost until the spatial filter is reprogrammed.

### *2.1. Sensitivity of the SCSF detector*

The movement of the speckle pattern for small displacements of the sample surface is geometric and determined by the optics in the same way as a KE detector. The size of the speckles is determined by the numerical aperture (NA) of the optical system delivering light to the sample and the focal length of the objective lens (exactly equivalent to the specular beam size in the KE detector). Neglecting the light that is lost due to scattering at high angles, it is easily shown that the sensitivity of the SCSF detector is exactly the same as the KE detector operating on a smooth sample. In the SCSF detector the light is divided between $N$ speckles crossing $N$ knife-edges (instead of one specular reflection crossing one knife-edge). Since the total amount of light and the distance it moves is the same in both cases and the average speckle size is the same as the specular beam size in the KE detector, the resulting sensitivity of the SCSF detector is the same as the equivalent KE detector operating on a smooth sample.

By electronically dithering the position of the spatial filter and measuring the change in light output, a calibration signal can be generated which, with knowledge of the geometry of the optics, can be used to obtain the absolute displacement.

For small movements of the sample the response is linear. This linear region occurs when the angular deflection caused by the ultrasonic wave is small compared with the NA of the illumination optics (as a rule of thumb $\theta_{\text{deflection}} < \text{NA}/10$). As an example, for 50MHz SAWs the linear region extends to ~50nm amplitude– a substantial amplitude at this frequency.

## 3. Methodology

Figure 1 shows a schematic of the optical configuration used. In this case an inexpensive video projector was used to provide the SLM. This device contained a Texas Instruments digital mirror device with HVGA (480×320) resolution. The projector was stripped of its optics to obtain access to the modulator. These SLMs consist of an array of small mirrors that can be tilted +/-12°. The tilt axis was 45° to the pixel tiling direction, and to keep the optical paths on the horizontal plane the SLM was rotated by 45°. Similarly the camera was also inclined at 45° so that its aperture better matched the SLM aperture. A mapping between the pixels in the SLM and camera was performed using a trapezoidal transform so that the spatial filter could be be accurately mapped to the observed speckle.

### *3.1. Computing the SCSF*

In the normal mode of operation the system needs to determine the SCSF and then display it on the SLM. In order to do this the SLM was first programmed to send a blank image to the camera so that the speckle distribution over the SLM could be determined. Then a SCSF was computed so that each speckle was cut in half by a 'knife-edge' across the expected direction of movement. Once the filter was computed, it was displayed on the SLM. The intensity received by this photodiode is the integral of the product of the spatial filter and the speckle pattern. As the speckle pattern moves (with respect to the spatial filter) the amount of light falling on the photodiode varies. For small movements this variation is proportional to the movement of the sample (see §2.1).

Figure 2 shows the process of the system adapting to a new position on the sample by recomputing the SCSF: (a) an image of the speckle pattern on the SLM was taken with the camera, (b) the spatially correlated spatial filter was computed such that it cuts each speckle in half and (c) spatial filter was displayed on the SLM to filter the speckle pattern. This results

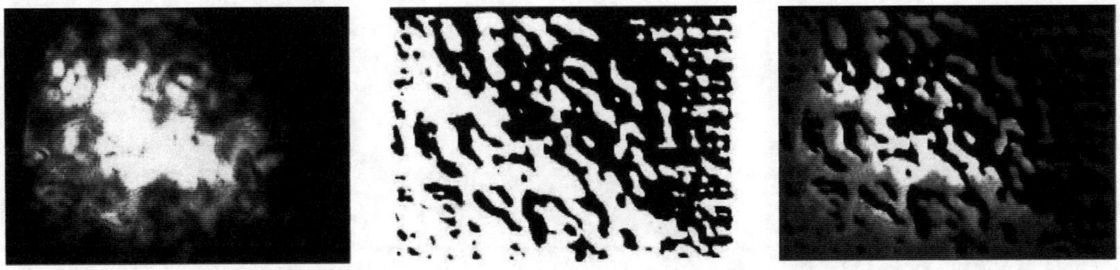

**Figure 2.** (left) (a) Speckle pattern observed on the SLM, (middle) (b) the SCSF computed from the image of the speckle pattern and (right) (c) the overlay of the filter over the speckle pattern. The filter cuts the speckle across the direction of movement (which is 45° left to right in these image due to the inclination of the SLM see §3)

in half the light from each speckle being collected by the detector lens and focused on the the photodiode.

## 4. Results, discussion and conclusion

The SCSF detector can be made to operate as a conventional KE detector by displaying a single edge at the geometric centre of the SLM. Figure 3 shows typical signals taken with a 5mW HeNe laser on an optically rough surface. It compares the signal obtained between the conventional knife-edge mode of detection and the adaptive speckle mode of detection. The b-scans were taken by scanning the detector across the sample, recomputing the spatial filter every point. It can be seen that the SCSF detector exhibits much higher sensitivity than the KE detector. For this system the detector can adapt to new positions on the sample at a rate of ∼12Hz determined by the frame rates of the camera and the SLM (25Hz). Frame rates of well over 1kHz can be achieved with more sophisticated SLMs.

This technique provides a simple, low cost method for the optical detection of ultrasound on optically rough surfaces.

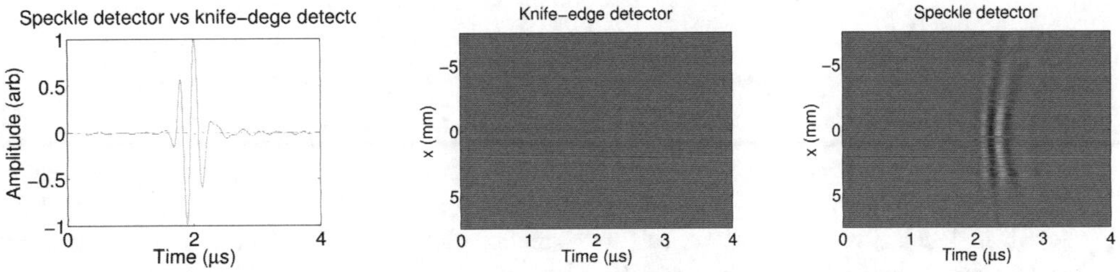


**Figure 3.** (left) Signal trace of KE detector (red, dashed) and SCSF detector (blue) at same position on sample. The sample was a diffusely reflecting, machined block of aluminium and the SAW was excited by a 5MHz transducer mounted on a perspex wedge. (middle) b-scan of the SAW wave taken with the KE detector and (right) with the SCSF detector (both using the same colourmap scale).

[1] Monchalin J P 1986 *IEEE Transactions on Ultrasonics, Ferroelectrics and Frequency Control* **33** 485–499
[2] Dewhurst R J and Shan Q 1999 *Measurement Science and Technology* **10** R139– R168
[3] Scruby C B and Drain L E 1990 *Laser Ultrasonics, Techniques and Applications* (Bristol, UK: Adam Hilger)
[4] Clark M, Sharples S D and Somekh M 1999 *IEEE Transactions on Ultrasonics, Ferroelectrics and Frequency Control* **47** 65–74
[5] Clark M, Sharples S D and Somekh M 2000 *Measurement Science and Technology* **11** 1792–1801

Journal of Physics: Conference Series **278** (2011) 012026 doi:10.1088/1742-6596/278/1/012026

# Frequency spectrum spatially resolved acoustic spectroscopy for microstructure imaging

**Wenqi Li, Steve D Sharples, Matt Clark, Michael G Somekh**

Applied Optics Group, Electrical Systems and Optics Research Division, Faculty of Engineering, University Park, University of Nottingham, NG7 2RD,UK.

E-mail: steve.sharples@nottingham.ac.uk

**Abstract.** The microstructure of a material influences the characteristics of a component such as its strength and stiffness. A previously described laser ultrasonic technique known as spatially resolved acoustic spectroscopy (SRAS) can image surface microstructure, using the local surface acoustic wave (SAW) velocity as a contrast mechanism. The technique is robust and tolerant of acoustic aberrations. Compared to other existing methods such as electron backscattered diffraction, SRAS is completely non-contact, non-destructive (as samples do not need to be polished and sectioned), fast, and is capable of inspecting very large components. The SAW velocity, propagating in multiple directions, can in theory be used to determine the crystallographic orientation of grains. SRAS can be implemented by using a fixed grating period with a broadband laser excitation source; the velocity is determined by analysing the measured frequency spectrum. Experimental results acquired using this "frequency spectrum SRAS" (f-SRAS) method are presented. The instrumentation has been improved such that velocity data can be acquired at 1000 points per second. The results are illustrated as velocity maps of material microstructure in two orthogonal directions. We compare velocities measured in multiple propagation direction with those predicted by the numerical model, for several cubic crystals of known orientations.

## 1. Introduction

SRAS—spatially resolved acoustic spectroscopy—is a nondestructive testing method which maps surface acoustic wave (SAW) velocity on a material surface. A velocity measurement taken using the SRAS technique is local to the area where the waves are being generated. As the measurement is relatively fast the sample may be scanned and an image built up of the velocity over the sample. The velocity at any point is determined by the material elastic parameters at that point and the resulting images clearly reveal the grain structure. This method is used to measure the velocity of the SAW over individual grains on industrial materials such as aluminium and titanium alloys. The variations of velocities, which indicate the orientations of grains, are presented as a colour map.

Microstructure imaging can also be accomplished by other methods such as electron backscatter diffraction (EBSD), which gains crystallographic information by impinging electrons on to the sample and analysing diffraction patterns [1]. Alternatively, the scanning acoustic microscope (SAM)—a contact technique—analyses the phase shift of a narrow band SAW signal. Both these methods have disadvantages such as restriction on sample size, or (in the case of the SAM) ambiguity caused by a periodic phase function. SRAS eliminates those concerns.

Experimental data are compared with theoretical results to determine the direction of single crystals on specific cuts. An advanced fast f-SRAS technique, where the scanning speed is maximised by utilising segmental memory acquisition on the oscilloscope, is also presented.

## 2. Theory

Figure 1 shows a schematic of the system setup. In the figure the pulsed laser is indicated as the excitation source. The pulsed laser beam passes through a mask consisting of a set of stripes. The unblocked light produces SAWs when it is imaged onto the sample surface. In our instrument there are twenty masks which have different curvatures and fringe spacings, to cover a range of different materials and conditions. A smaller distance between the fringes generates SAWs with a shorter wavelength and higher frequency. The results presented here use a straight (rather than focused) grating with an effective period of 45 $\mu$m on the sample.

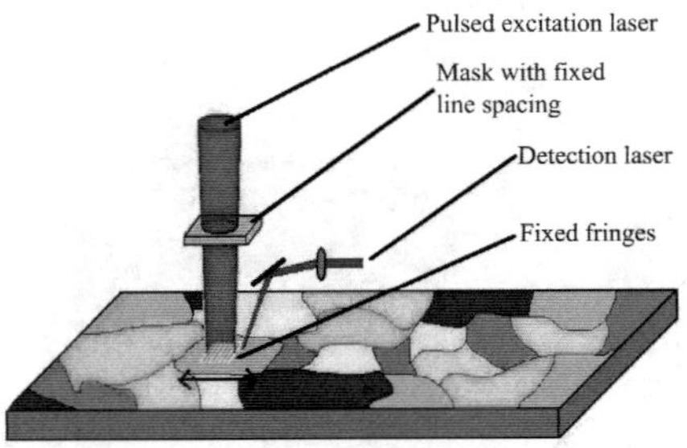


**Figure 1.** Schematic of f-SRAS laser ultrasound system.

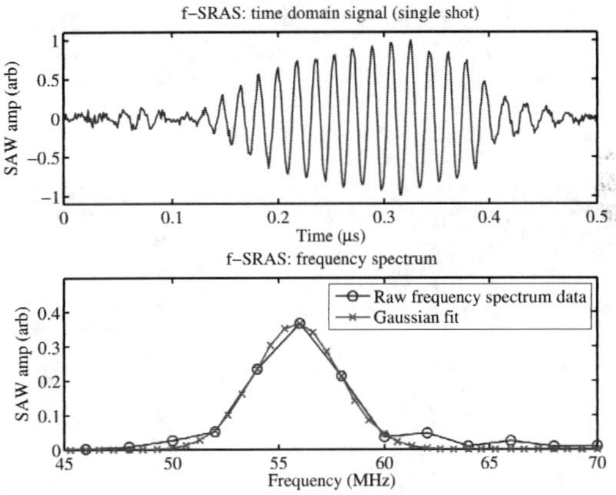


**Figure 2.** Raw data in the time domain (top) and frequency domain (bottom) with Gaussian fit.

The pulse from the Q-switched Nd:YAG laser ($\lambda = 1064$ nm) has a width of approximately 10ns, and has a wide bandwidth from DC to >100 MHz. The pulse repetition is 1 kHz, the energy per pulse onto the sample is 135 $\mu$J. The grating excitation pattern acts as a filter on the generation of the SAWs and is multiplied with the frequency content of the laser. The effective centre frequency of the grating is determined by the grating period and the local velocity ($c = f\lambda$). Since the effective bandwidth of the filtering action of the grating generation is much smaller than the bandwidth of the laser pulse, we can make the assumption that the frequency content of the laser is locally flat across the bandwidth of the grating and hence that the peak of the combined frequency response is at the same location as the peak of the grating filter; thus we can determine the local velocity. This method of determining the velocity—using the frequency spectrum of the SAWs excited using a fixed grating—is termed frequency spectrum SRAS (f-SRAS).

Figure 2 shows a typical SAW signal captured by the system. The spacing of the grating determines the wavelength of SAWs. The top graph is the waveform in the time domain; the amplitude FFT is on the bottom along with a Gaussian curve which has been fitted to the raw FFT data. The frequency where the amplitude is the largest indicates the closest value corresponding to the SAWs travelling on the grain. Note that the velocity measurement relates to the sample area under the generation patch, rather than at the detection point or in between the generation and detection.

## 3. Experimental Results

### *3.1. High Resolution of f-SRAS*

Figure 3 shows f-SRAS experimental results where SAWs are propagating in orthogonal directions on Ti-685. The colour represents the velocity from 2700 m/s to 3500 m/s. Deep red means the surface waves propagate at a higher speed; deep blue shows the lowest velocity. The imaging speed of f-SRAS is restricted by several factors: oscilloscope acquisition speed, data processing time, scanning stage velocity, laser repetition rate, and potentially the number of averages required if the SNR (signal to noise ratio) is poor. Even though the Q-switched laser fires at a repetition rate of 1 kHz, previously [2] the data acquisition rate was limited—by the acquisition architecture—to a maximum of 60 points/sec. Taking advantage of segmented memory acquisition found on modern oscilloscopes, every single laser pulse is now utilised to take a velocity measurement. The SNR is sufficiently high enough to permit single shot measurement the acquisition rate is limited only by the laser repetition rate: 1000 points/sec. The images in Figure 3 were obtained in 40 minutes, approximately 76% of that time is spent on stage flyback and data transfer to the host PC; work is ongoing on to improve scanning efficiency.

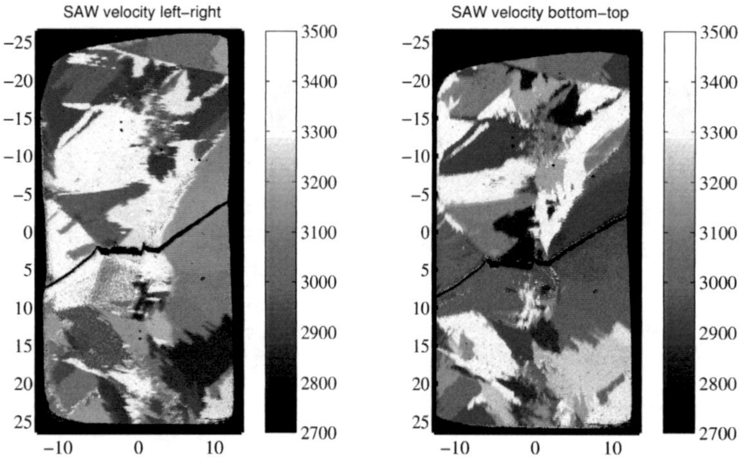


**Figure 3.** SAW velocity map of Ti-685 by fast f-SRAS: SAWs propagating left to right (left graph), top to bottom (right graph). Image size 27×52 mm, pixel size 50×50 $\mu$m.

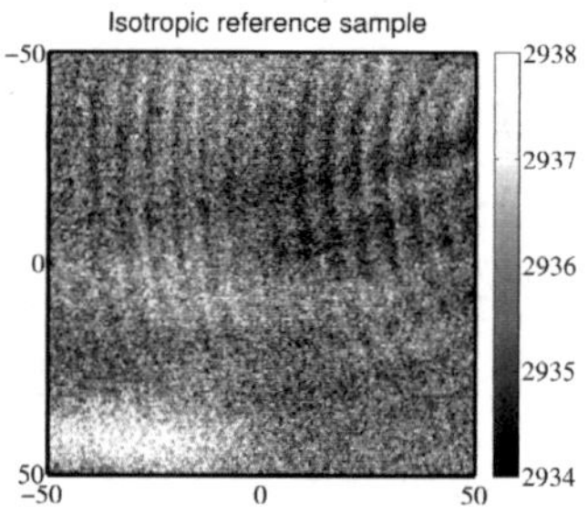


**Figure 4.** Standard deviation of f-SRAS. Image size 1×1 mm, pixel size 5×5 $\mu$m, velocity range 2934 m/s to 2938 m/s, 16 points average.

### *3.2. Noise Analysis of System*

Figure 4 shows the variation in SAW velocity on a nominally isotropic reference sample, consisting of BK7 glass coated with a thin layer of aluminium. The intention was to determine the system noise in terms of the standard deviation of the measured velocity, which is 0.87 m/s (0.03%). We note, however, that spatially coherent features—distinct areas where the velocity is faster or slower than the mean—implies that the noise is probably lower than this value.

### *3.3. Experimental vs. Theoretical: Results on Aluminium Crystals*

Farnell [3] gave a detailed iterative search procedure of how to calculate the SAW velocity on a generic plane, along with the calculation results of a few crystals of different crystal structures in typical cuts. Based on this, the SAW velocity in any arbitrary direction on all the planes of a single crystal can be calculated if the elastic constants are known. Curve fitting could be used to find out what plane of crystal's surface is, when the SAW velocities in a certain limited number of directions are known. Figure 5 illustrates how well the experimental results match the calculated results on three planes—(100), (110) and (111)—on single crystal aluminium. The mismatch on plane (100) is due to the wave mode at a certain angle transforming between a SAW and a pseudosurface wave, which degenerates into a bulk wave [3]. This occurs at 32 degrees from [100] to [110] on aluminium. The experimental results are approximately 100 m/s (3%) faster than the calculated velocities. This could either be due to an error in the size of the excitation grating, or to a small error in the elastic constants used in the theoretical calculations.

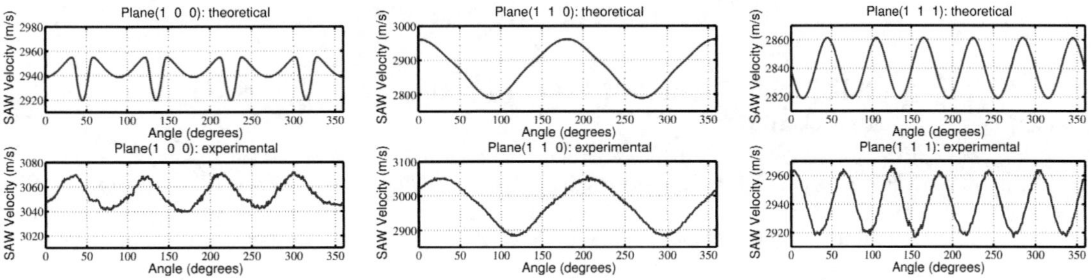


**Figure 5.** SAW velocity curves for all directions on three cuts of single crystal aluminium. The angular offset is due to the randomly selected starting propagation direction.

## 4. Discussion

An acoustic spectroscopy technique called frequency spectrum SRAS has been described in the paper. It is based on laser ultrasonics, and material microstructure can be imaged by determining the local SAW velocity on the sample surface. In the implementation described, the f-SRAS technique can acquire the local velocity of SAWs at 1000 points/sec by taking advantage of segmented memory acquisition, 17 times faster than previously described [2]. The spatial resolution is currently of the order of 400 $\mu$m, which is the size of the grating on the sample. This can be significantly reduced by using different masks with fewer lines at shorter wavelengths. Experimentally, the bandwidth of the excitation laser source used in our instrument limits the maximum frequency SAWs that can be excited (hence the minimum size of the grating). Ultimately—with a laser of sufficiently wide bandwidth— the resolving power of the optics used to produce the grating structure is the limiting factor on the resolution of SRAS. We have demonstrated velocity acquisition on three different planes of aluminium in all directions, and shown how this relates to theoretical calculations of the SAW velocity on these planes. The challenge now is to determine the optimum number of velocity measurements in a limited number of discrete propagation directions that is required to ascertain the crystallographic orientation at each point in multi-grained materials.

This work was supported by the UK RCNDE (Research Centre in Nondestructive Evaluation).

## References

[1] Randle V 2003 *Microtexture Determination and Its Applications* (London: Maney Publishing)
[2] Sharples S D, Li W, Clark M and Somekh M G 2009 *Rev. Prog. in QNDE* **29A**, 279–286
[3] Farnell G W 1970 *Properties of Elastic Surface Waves (Physical Acoustics* vol 6*)* ed W P Mason and R N Thurston (New York and London: Academic Press) chapter 3 pp 109–166

# Electrical excitation and optical detection of ultrasounds in PZT based piezoelectric transducers

**P Babilotte[1], O Diallo[1], L-P Tran Hu Hue[1], M Kosec[2], D Kuscer[2] and G Feuillard[1]**

[1]University Francois Rabelais de Tours, Laboratory Imaging and Brain, Team Ultrasonic Characterisation and Piezoelectricity, ENIVL, Rue de la Chocolaterie, 41034 BLOIS CEDEX, France
[2]Josef Stefan Institute, Jamova cesta 39, 1000 LJUBLJANA, Slovenia

E-mail: philippe.babilotte@univ-tours.fr

**Abstract.** The displacement response of piezoelectric PZT thick films fabricated by means of electrophoretic deposition and laid down an alumina substrate is investigated using coherent optical detection. According to thickness properties determined by electrical impedance measurements, the film presents a resonance around 40 MHz. Other resonance peaks are observed that correspond to eigen modes of the film substrate couple structure. Uniformity of the response of the integrated structure is studied across the surface of the sample when excited by either a continuous or impulse electrical voltage. Results on the amplitude of the detected signal versus the frequency and the input excitation voltage are reported. The optical detection used in these experiments is complementary to conventional techniques of characterization of piezoelectric devices such as electrical impedance measurements and allows getting information on the displacement response of the device.

## 1. Introduction

Sensing of vibrations is a very large technologic and scientific domain. It can be done using different types of techniques, such as acoustic vibrometer, where the vibration measurement is performed through the reflection of a continuous ultrasonic beam on the vibrating surface [1], or using a full optical generation and detection of vibrations [2,3,4,5]. These techniques have been used for characterization of vibrating surfaces in the MHz range in the perspective of non-destructive testing. On the other side, very high frequency measurements can be done by photogeneration and photodetection in the GHz range [6]. Here we report displacement measurements on PZT thick film integrated structures in the 20 MHz - 60 MHz frequency range using laser interferometry.

## 2. Description of samples and experimental set-up

PZT thick films characterized in this study were fabricated by electrophoretic deposition process [7]. Samples are described in figures 1 and 2. The film thickness is 27 $\mu$m for sample (1) and 18 $\mu$m for sample (2). It is laid down on a 600 $\mu$m platinized alumina substrate and covered by a top gold electrode (4 mm diameter). According the structure of the device (heavy backing on one side and air loading on the other side), it operates in a quarter wavelength mode and has a fundamental resonance around 40 MHz. The longitudinal wave velocity previously determined from electrical impedance measurements is 4255 $m.s^{-1}$ [7].

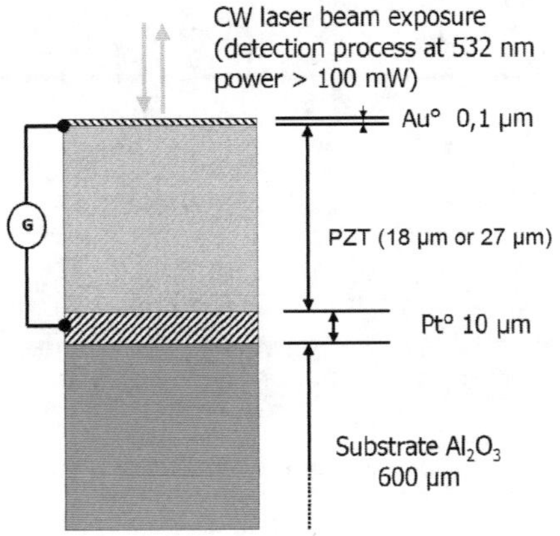


**Figure 1.** Schematic description of samples (1) and (2).

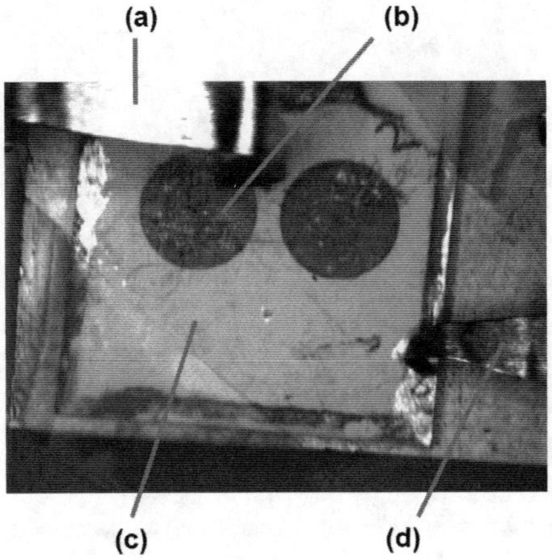


**Figure 2.** Topview of the 27 $\mu$m thick PZT sample. Contact electrode **(a)** touching the circular top electrode **(b)** in *Au* deposited on PZT free surface **(c)** and contact electrode of the bottom electrode **(d)**.

The generators used in the experiments are either an *Agilent 33250 A* or *HP 8116 A* ; optical detection of ultrasound is performed using a 532 nm continuous laser beam. The power of the detection laser beam have been evaluated to 172 mW using a $PM10-3$ *ThorLabs* thermal head sensor. This value is below the damage threshold of the samples.

## 3. Experimental results and discussion

### *3.1. Spectral response of the samples*
Mechanical displacement responses of the sample versus the frequency have been recorded for samples (1) et (2) (figures 3 and 4); clear peaks have been identified at expected values obtained by measurements of the real part $\Re e\{Z\}$ of electrical impedance measured.

Strong resonance peaks are observed for sample (1) at 13.3 MHz and 39.5 MHz. Weaker secondary peaks have been identified, in accordance with electrical input measurements. For sample (2), matching between resonant frequencies measured by electrical input impedance and those determined from optical detection is more clear (4.5 % in relative difference between modes measured by conventional techniques and optical detection), and eigenmode at 25 MHz is more visible for this sample, due to the use of another type of generator.

### *3.2. Dynamic responses versus excitation voltage*
Displacement responses to different level of excitation voltage have been measured for the samples (1) and (2) (figure 5). The linear dependence of the amplitude of vibrations with applied voltage is observed. No *hysteresis* phenomenon is observed after a decreasing and increasing cycle of voltage excitation.

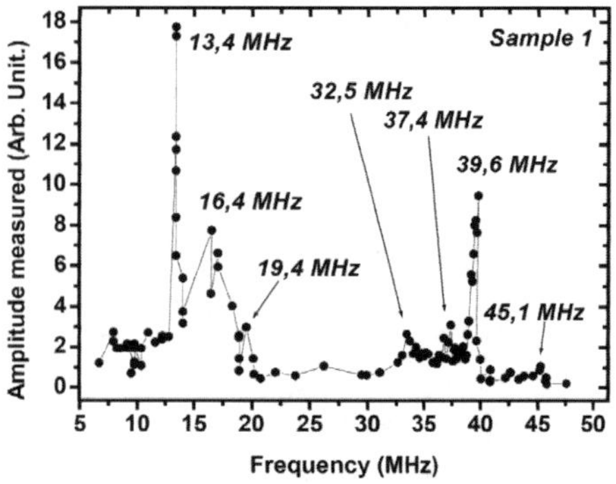


**Figure 3.** Amplitude of response versus frequency of the sample (1) (continuous excitation, 16.00 $V_{PP}$ peak to peak voltage fixed, measurements were carried in 50Ω environment).

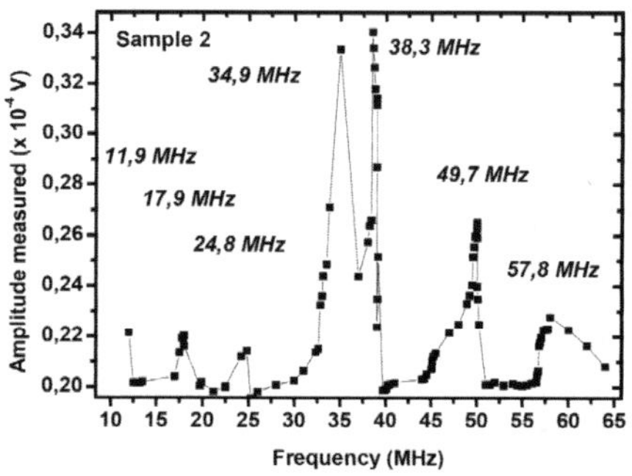


**Figure 4.** Amplitude of response versus frequency of the sample (2) (continuous excitation, 10.00 $V_{PP}$ peak to peak voltage fixed)

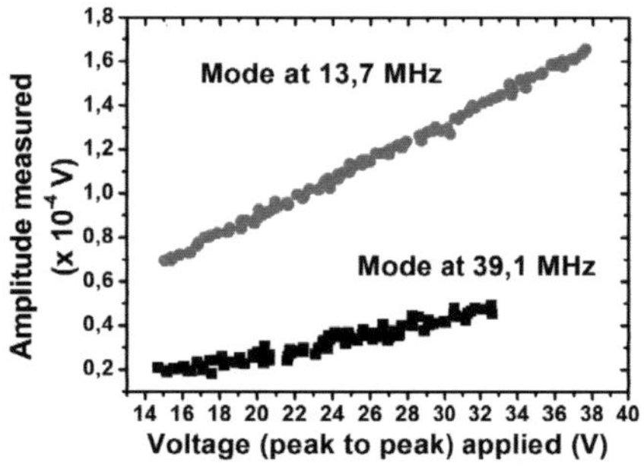


**Figure 5.** Sample (1). Response of sample to a voltage amplitude. Measurements were made using a high impedance environment

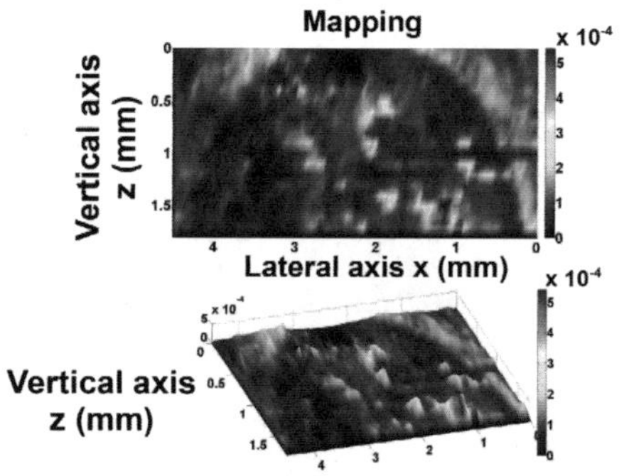


**Figure 6.** Sample (1). Displacement cartography using a 0.05 mm lateral step and a 0.1 mm vertical step, at fixed 32.70 $V_{PP}$ peak to peak amplitude of electrical excitation. Calibration coefficient for all experiments : 94 $mV.nm^{-1}$. Excitation at 39.6 MHz.

### *3.3. Scanning of surface of the sample*

Surface mapping of the sample have been recorded, showing change of optical reflectance coefficient, and consequently showing limit of the disk (area with/without deposition of gold). Electrical generation of vibrations have been performed at 39,6 MHz. Numerical processing have permitted to map the vibrating surface of the sample, exhibiting modes of vibrations at the particular frequency of 39.6 MHz. On all measurements (figure 6), we observe a mechanical displacements on the edge of the active disk area that may due to effect of horizontal (radial) traction.

### *3.4. Temporal responses to an electrical pulse excitation*

Temporal responses to a single electrical pulse excitation delivered by a 50 Ω impedance source have been measured (figure 7).

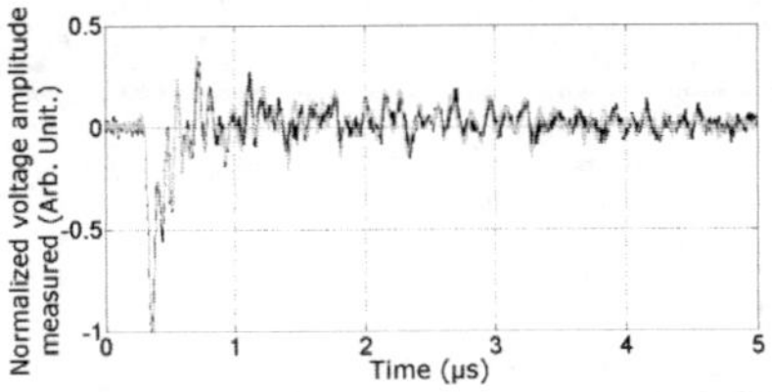


**Figure 7.** Sample (1). Temporal response to an electrical pulse excitation. Amplitude peak to peak of this pulse is 10.00 V, pulse width is fixed at 8.0 ns and edge time fixed at 5.00 ns.

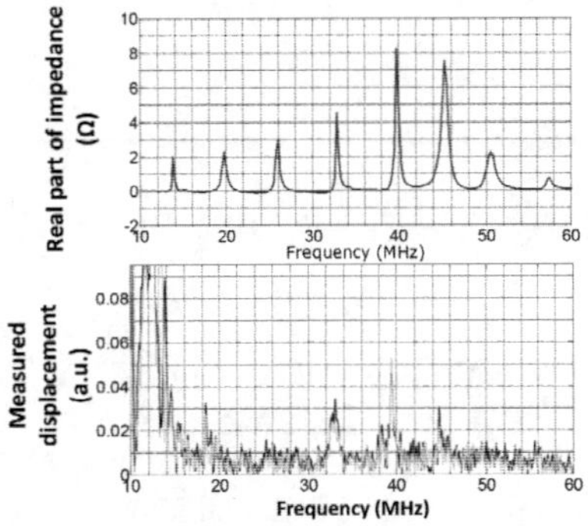


**Figure 8.** Sample (1). Comparison between measurements of electrical impedances (upper) and using optical detection (lower).

Several modes have been detected in the frequency spectrum established by the use of the optical detection. Considering the longitudinal velocity in alumina $V_l(Al_2O_3) = 8860\ m.s^{-1}$ (porous alumina), and a thickness fixed at $\delta = 600\ \mu$m (figure 1), the arrival time of reflected acoustic waves on substrate bottom surface (round trip time into alumina is $\tau \approx \left(\frac{2\delta}{V_l(Al_2O_3)}\right)$). Consequently, we determine fundamental resonant frequency of $f_{ac} \approx 7.4$ MHz, which can be easily found on spectrum, and corresponds to the frequency related to the first return (round trip) of acoustical wave in the alumina substrate. Additional modes due to multiple reflection in the alumina substrate explain the detected mode at entire multiple of 7.4 MHz. The resonances in Au and Pt exist but according the thicknesses of the layers (respectively 10 $\mu$m and 0,1 $\mu$m), their fundamental resonances are in a frequency range higher than 100 MHz and are not observed on figures. The PZT resonance can be observed around 40 MHz on figure 8. When designing a high frequency transducer, round trip echo in the alumina (backing material) should be damped in order the device to operate at the resonance frequency close to that of the piezoelectric ceramic.

In conclusion of the present work, results obtained are consistent with those measured with conventionnal techniques. Eigenmodes have been well-identified and characterisation of the sample has been completed.

## References

[1] Bou Matar O, Remenieras J-P, Bruneel C, Roncin A and Patat F 1998 *Ultrasonics* **36** 391
[2] Levassort F, Lethiecq M, Tran Huu Hue P and Wolny W 1997 *IEEE Ultrasonics Symposium* 947
[3] Royer D and Dieulesaint E 1986 *IEEE Ultrasonics Symposium* 527
[4] Royer D, Noroy M H and Fink M 1994 *J. Phys. IV France, 8th ITMP3 Symposium*
[5] Francini F, Pampaloni E and Innocenti G 1993 *Optics and Laser Tech.* **25** 103
[6] Akhmanov S A and Gusev V E 1992 *Sov. Phys. Usp* **35** 153
[7] Feuillard G, Kuscer D, Tran Huu Hue P, Le Clezio E, Kosec M and Lethiecq M 2009 *IEEE Symposium* 1138

Journal of Physics: Conference Series **278** (2011) 012028 doi:10.1088/1742-6596/278/1/012028

# Superlattice ultrasonic generation

**T E Wilson**[1,4], **M Oehme**[2], **E Kasper**[2] **and H-J L Gossmann**[3]
[1] Department of Physics, Marshall University, One John Marshall Drive, Huntington, West Virginia, 25755, USA
[2] Institute of Semiconductor Engineering, University of Stuttgart, Pfaffenwaldring 47, Stuttgart, 70569, Germany
[3] 29 Brainerd Road, Summit, NJ, 07901, USA

E-mail: wilsont@marshall.edu

**Abstract.** We report the first experimental evidence for the resonant excitation of coherent high-frequency acoustic phonons in semiconducting doping superstructures by far-infrared laser radiation. After a grating-coupled delta-doped silicon doping superlattice is illuminated with ~1 $kW/mm^2$ nanosecond-pulsed 246 GHz laser radiation, a delayed nanosecond pulse is detected by a superconducting bolometer at a time corresponding to the appropriate time-of-flight for ballistic longitudinal acoustic phonons across the (100) silicon substrate. The absorbed phonon power density in the microbolometer is observed to be ~10 $\mu W/mm^2$, in agreement with theory. The phonon pulse duration also matches the laser pulse duration. The absence of any delayed transverse acoustic phonon signal by the superconducting bolometer is particularly striking and implies there is little or no incoherent phonon generation occurring in the process.

## 1. Introduction

Coherent phonons in a variety of materials, including metal films, bulk semiconductors and semiconductor heterostructures, have been studied for some time now using ultrafast pulsed lasers for their generation and detection [1]. Here, we investigate a different approach for coherent phonon generation, that of the direct electromagnetic generation of coherent high-frequency monochromatic acoustic phonons in silicon doping superlattices (DSL) by high peak-power, nanosecond-pulsed far-infrared (FIR) laser radiation. This approach has the potential to produce intense nanosecond pulses of high-frequency narrow-bandwidth coherent acoustic phonons of the desired polarization, with reasonable conversion efficiency, and without concomitant electronic or lattice heating.

## 2. Theory

A number of authors [2,3,4,5] have discussed the possibility of using FIR laser radiation to excite ultrasonic acoustic waves in superlattices in the frequency range 0.1-1 THz, although to date, there has been no experimental confirmation. A *nipi* doping superlattice consists of a periodic variation in the dopant in an otherwise homogeneous semiconductor with resulting layers of spatially separated and periodic space charge. The sinusoidally-varying electric field of incident far-infrared laser radiation will transfer to the crystal momenta of opposite directions in the alternating layers. If the superlattice period is chosen to match the acoustic wavelength, the resulting time-varying momenta in the crystal

[4] To whom any correspondence should be addressed.

will induce a resonant acoustic wave of the same frequency as the electric field. In this case, the fundamental frequency of excitation of the generated phonons is simply given by f=s/d, where *s* and *d* are the phase velocity for phonons and the superlattice period, respectively (an optical phonon mode in the zone-folded scheme). In the case of the DSL fabricated upon the same type of crystal substrate there will be no reflection at the DSL/substrate interface and the wave generated in the DSL will propagate through the substrate.

Longitudinal (LA) or transverse (TA) acoustic phonons couple to electric fields normal or parallel to the DSL layers, respectively. The conversion efficiency $\xi$ is defined as the ratio of the generated acoustic energy to the energy of the incident electromagnetic wave and is given by equation (1) (cgs units) below, where $a$ is the host lattice constant (assumed simple cubic lattice), $n_D^{(2)}$ is the two-dimensional doping concentration, $\kappa_0$ is the dielectric constant, $M$ is the atomic mass, $d$ is the superlattice period, $c$ is the speed of light, $s$ is the phonon phase velocity, $N$ is the number of superlattice periods, and $r \equiv (d/a)$, is the enlarged unit cell:

$$\xi \cong \left[ \frac{\left(4\pi e^2 a^2 (n_D^{(2)})^2 (adN^2\right)}{(\kappa_0 Mdr)(cs\sqrt{\kappa_0})} \right] \quad (1)$$

## 3. Experiment

We have used a delta-doped silicon DSL [6,7] consisting of 30 periods, fabricated upon a (100) float zone silicon substrate, with a period of length $d$ = 34.5 nm, and with individual layers having a two-dimensional doping concentration $n_D^{(2)} = 8.5 \times 10^{12}$ cm$^{-2}$. The conversion efficiency $\xi$ of this DSL is $9.6 \times 10^{-9}$. The period was chosen to match the LA phonon wavelength in silicon at a frequency (245.5 GHz) corresponding to a particularly strong FIR laser line (1.22 mm) produced by a $^{13}CH_3F$ gain medium.

We have used a small-period metallic grating coupler [8] on top of the DSL. Undesirable bulk broadband phonon generation might be expected in the absence of such a grating-coupling scheme, as a result of both multi-phonon- and defect-induced- absorption in the substrate. The grating-coupler also serves to protect the sensitive superconducting bolometer detector from direct exposure to the intense FIR laser radiation, i.e., heat pulse generation in bolometers exposed to intense FIR laser pulses has been observed previously. The grating coupler converts normally incident plane wave radiation, into an evanescent surface wave with a large electric field component perpendicular to the interface for efficient LA phonon generation. We have used a grating period of ~10 $\mu$m, resulting in a decay length of ~1.6 $\mu$m, somewhat larger than the thickness of our DSLs. Niobium is used as the grating material because of its small surface resistance below its superconducting transition temperature.

We have previously developed [9] a cavity-dumped optically-pumped molecular-gas FIR laser resonator, and modified it for use at 1.22 mm as a source of intense ~nanosecond pulses of p-polarized FIR laser radiation for these experiments. We cavity-dump the circulating resonator FIR radiation through the use of an optically-switched intra-cavity intrinsic silicon wafer, oriented at the Brewster angle to the circulating p-polarized far-infrared radiation, through the absorption of above-bandgap radiation. The risetime of the cavity-dumped output pulse is on the order of a nanosecond and the pulsewidth corresponds to the roundtrip time of the circulating FIR – 6 ns. Recent improvements to the FIR laser have included: (1) an in-situ mode-matching Newtonian telescope, employing custom TPX lenses, to couple the output into a 3 m long 246 GHz corrugated waveguide for beam transport to the sample access window of the floor-mounted liquid helium cryostat and, (2) a recycling cryogenic absorption pump for the isotopic methyl fluoride gain medium. After the 246 GHz laser radiation propagates through our corrugated waveguide, we routinely obtain power levels at the cryostat access window of 5 kW in 6 ns pulses, at a pulse repetition rate of 10 pulses per second. The DSL sample is installed at the end of an insertable sample rod and is enclosed in a split-cylinder metal-coated 'rapid-prototyping' polymer jig that also holds a hyper-hemispherical high-purity silicon lens in contact with the grating-coupler/DSL front sample surface. The jig also shields the bolometer from scattered FIR radiation. The focused FIR laser power density at the surface of the grating-coupler/DSL is estimated to be ~1 kW/mm$^2$ in a beamwaist of 3 mm.

The DSL sample assembly (rod and jig) is immersed in low-temperature liquid helium in a cryostat equipped with optical access. The temperature is established by pumping on the helium vapour through a precision computer-controlled butterfly valve with a vacuum pump. The temperature is typically held near 1.750 K (approximately 1/3 up the R-T curve for the superconducting bolometer detector) with a stability of 0.1 mK. Any resonantly excited phonons by pulsed FIR laser radiation are detected on the rear sample face (substrate thickness 0.5 mm) by a superconducting granular aluminium/palladium bilayer microbolometer operating in a constant bias-current mode [10]. The active region of our 100 nm thick granular aluminium bolometer is 10 $\mu$m x 20 $\mu$m and is capped with a 5 nm thick palladium layer. We pattern both the niobium grating-coupler and microbolometer using image-reversal optical lithography, dual-target dc-magnetron sputtering, and liftoff. When biased at 20 $\mu$A and near 1.75 K, our bolometer has a bare time constant (C/G) of 2.9 ns and a calculated responsivity $\mathcal{R}$, at 50 MHz, of 9.8 kV/W. (50 MHz is the bandwidth associated with a 6 ns phonon pulse.) Electrical contact is made to the bolometer with an *MMCX* microminiature coaxial jack that is integrated into the rear half of the split-cylinder sample jig. A 1 GHz bias tee passes the battery-powered dc-bias current through to the bolometer while allowing the ac-coupled bolometer output to be passed to a 1 GHz preamplifier (voltage gain 35 dB). It was necessary to enclose the cryostat and bolometer electronics in a copper Faraday cage in order to shield the experiment from intense electromagnetic noise generated by the pulsed TEA $CO_2$ pump laser and the pulsed frequency-doubled YAG laser (the latter induces the cavity-dumping of the FIR laser resonator).

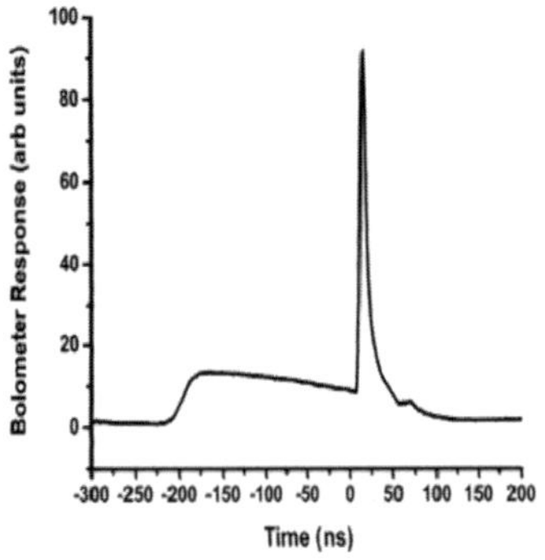


**Figure 1.** Bolometer response (arbitrary units) 25 ns/div

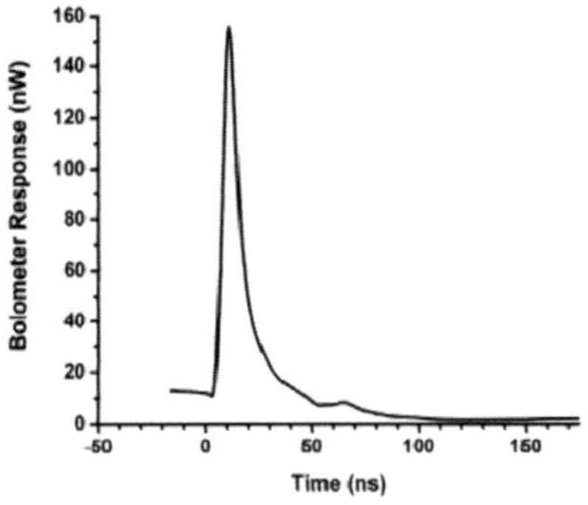


**Figure 2.** Bolometer response (nW) 10 ns/div

Figure 1 above shows the response of the bolometer. We first observe (beginning near -225 ns) pulsed FIR laser radiation leaking through the 500 nm thick superconducting niobium grating coupler (skin depth ~ 39 nm). Next, one observes the onset of the large cavity-dumped FIR pulse near 0 ns (the cavity-dumped pulse has saturated the bolometer; hence, the pulse is seen to be larger in time than 6 ns, the observed FIR pulsewidth in earlier experiments). Finally, one observes a delayed (~65 ns) smaller pulse corresponding to the arrival for ballistic LA phonons that have traversed the 0.58 mm thick (100) Si substrate. Figure 2 displays the bolometer response at higher time resolution; there is no evidence of either transverse ballistic or diffusive acoustic phonons. The absence of a TA phonon signal is to be compared to an earlier heat pulse experiment [10] on identical DSL samples; in that case 0.5 $\mu$J, 532 nm, 1 ns pulsed mini-YAG laser radiation was focused on the Nb grating-coupler/DSL to a beam waist of 50 $\mu$m and a similar (identical geometry and bias conditions) superconducting granular aluminum/Pd bolometer was used as the detector. In the latter case, both LA and larger (due to phonon focusing) TA phonon signals were clearly observed at the appropriate delays relative to the initial pulse of scattered laser light reaching the bolometer. We take this to be evidence that there is very little incoherent heat pulse generation observed in our new results. It is also noteworthy that the absorbed power at the bolometer in the case of figure 2 also matches closely the coherent power predicted by Ruden and Dohler [2] for our 200 $\mu m^2$ active area microbolometer, i.e., ~2 nW.

## 4. Conclusions

We provide evidence for the first experimental demonstration of the coherent generation of acoustic phonons in doping superlattices by pulsed FIR laser radiation. The absorbed phonon power density in the microbolometer is estimated to be ~10 $\mu W/mm^2$, in agreement with theory [2]. The observed phonon pulse duration (~5 ns) in figure 2 matches the FIR laser pulse duration. The absence of any larger and delayed transverse acoustic phonon signal by the superconducting bolometer is particularly striking. These observations taken together provide strong evidence that coherent zone-folded longitudinal acoustic phonons can be generated in a silicon doping superlattice through the resonant absorption of grating-coupled pulsed far-infrared laser radiation, and with negligible associated electronic or lattice heating. In the near future, we plan to increase the thickness of the niobium grating-coupler so as to reduce the FIR laser radiation leakage and thereby allow for Si:B phonon spectroscopy, to use larger bolometers to intercept a larger phonon flux, and to investigate the use of new DSL samples with significantly larger number of periods. We hope to develop a THz cryogenic acoustic microscope [11] with ~1 nm lateral resolution, using the superlattice as a novel high-frequency transducer. We acknowledge the support of the United States National Science Foundation through an ECCS grant 0622060.

## References


[1] Glavin B A, Kochelap V A, Linnik T L, Walker P, Kent A J and Henini M 2007 *J. Phys.: Conf. Ser.* **92** 012010

[2] Ruden P and Dohler G H 1983 *Solid State Communications* **45** (1) 23

[3] Quinn J J, Strom U and Chang L L 1983 *Solid State Communications* **45** (1) 111

[4] Dransfeld K 1985 *Rayleigh-Wave Theory and Applications*, ed E A Ash and E G S Paige (Berlin: Springer) p 10

[5] Wilson T E 1989 *JOSA B* **6** 1058

[6] Gossmann H J 1996 *Delta-Doping of Semiconductors*, ed E F Schubert (Cambridge: Cambridge) p 161

[7] Kasper E and Paul D J 2005 *Silicon Quantum Integrated Circuits* (Berlin: Springer) p 119

[8] Li W J and McCombe B D 1992 *J. Appl. Phys.* **71** (2) 1038

[9] Wilson T E 1989 *U.S. Department of Energy Strategic Defense Initiative's New and Innovative Concepts Program Summary Report*, ed R LeChevalier, (Falls Church VA:TPI Publishing) p 79-80

[10] Wilson T E 2007 *J. Phys.: Conf. Ser.* **92** 012180

[11] Muha M S, Moulthrop A A and Kozlowski G C 1990 *Appl. Phys. Lett.* **56** (11) 1019-1021

Journal of Physics: Conference Series **278** (2011) 012029 doi:10.1088/1742-6596/278/1/012029

# Ultrasound induced by CW laser cavitation bubbles

**N Korneev, P Rodriguez Montero, R Ramos-García, J C Ramirez-San-Juan, J P Padilla-Martinez**

Instituto Nacional de Astrofísica, Óptica y Electrónica, Apt. Postal 51 y 216 CP72000, Puebla, Pue., Mexico

korneev@inaoep.mx

**Abstract**. The generation of ultrasound by a collapsing single cavitation bubble in a strongly absorbing liquid illuminated with a moderate power CW laser is described. The ultrasound shock wave is detected with hydrophone and interferometric device. To obtain a stronger pulse it is necessary to adjust a liquid absorption and a beam diameter. Their influence can be qualitatively understood with a simple model.

## 1. Theoretical considerations

The cavitation generated by a short and powerful laser pulse is a well known phenomenon[1-3 ], but the possibility of producing a cavitation bubble with a medium power CW laser in strongly absorbing liquid is less known [4,5]. Cavitation with CW illumination is interesting both from the fundamental point of view and as a cheaper and safer source of ultrasound shock wave. Here we shortly discuss the conditions for obtaining a shock wave from a collapsing cavitation bubble with high amplitude using the CW laser as a light source.

The bubble formation includes following stages. First, the small volume of liquid (typically 0.1-0.3 mm size) is overheated. This lasts from milliseconds to seconds. When the temperature of the overheated water reaches $T_N$=180-240 *°C*, an explosive phase transition occurs producing a vapour bubble. The bubble growth starts from a random nucleation event. When the bubble grows to the region of a liquid with lower temperature $T_B$, the collapse occurs. At the final stage of collapse, a shock wave is generated. The local temperature returns to its initial value, and under CW illumination the process repeats itself, resulting in nearly periodic train of bursts. We are interested to obtain a shock wave with the maximal energy, thus the bubble has to be larger, because the energy transferred into the wave is proportional to the energy stored in the bubble, and this is roughly proportional to its volume. The factors we can control for CW illumination are the laser power, the beam diameter and the absorption length of the liquid. The qualitative understanding of their influence on the resulting bubble size can be obtained with a simple approximate model.

Let us suppose, that the laser beam with a total power $U$ is absorbed inside a liquid with thermal conductivity $k$ in a sphere with a radius $a$. The steady state temperature distribution in this case depends on the distance from the sphere centre $r$. It is obtained by solving a steady-state heat conduction equation, and the solution is mathematically equivalent to the potential of a uniformly charged sphere [6]:

$$T(r) = \frac{3U}{8\pi a k} - \frac{Ur^2}{8\pi a^2 k}\ , \quad r < a \tag{1}$$

$$T(r) = \frac{U}{4\pi k r}\ , \quad r > a \tag{2}$$

where $T$ is the temperature excess over the liquid initial temperature $T_0$.

If cavitation occurs when the temperature distribution is close to the steady state, the temperature in the centre is:

$$T(0) = \frac{3U}{8\pi a k} = T_N - T_0 \tag{3}$$

and the resulting radius of the bubble is close to the $r_0 = a$. The maximal bubble radius for a given laser power can be estimated as $r_0 = \frac{3U}{8\pi k(T_N - T_0)}$, which gives a correct order of magnitude for experimentally observed bubbles.

If the steady state temperature distribution is not reached when the temperature in the centre becomes sufficient for the start of nucleation, the temperature distribution width is smaller, than in a steady state, which is a known property of the heat conduction equation (it follows from the Green function for this equation [6] ). This leads to a smaller bubble, and smaller shock wave energy. Thus, making a laser power bigger with a fixed beam diameter results in diminished shock wave (because of smaller bubble size), and higher repetition rate (because of faster heating to the temperature of nucleation). To obtain stronger shock wave with a higher laser power, it is necessary to make a beam diameter bigger, and to adjust the liquid absorption to the final bubble size. For higher laser power the optimal absorption coefficient diminishes. If the laser spot at the input is too big, or absorption is too low, the temperature of nucleation cannot be reached, and no cavitation is observed.

## 2. Experiment

To investigate the influence of the spot size, light absorption and laser power on the shock wave amplitude we used the cell with a water solution of copper nitrate ( $Cu(NO_3)_2$), figure 1. The beam of semiconductor laser with maximal output power 1.5 W was focused inside a cell with a 10X objective. The shock wave was detected with a hydrophone placed at a distance 10 mm from the bubble spot.

In figure 2 we present the results. It is seen, that to obtain bigger signal for higher laser power it is necessary to increase the beam diameter. When the diameter becomes too big, cavitation stops. Note, that for a *fixed* spot diameter raising power generally results in lower signal. If the absorption length is much smaller, than the final bubble diameter, one can expect stronger deviations from the theory above. This situation is clearly seen for 1120 mW laser power and 90 $cm^{-1}$ absorption coefficient. In this case the pressure diminishes when the beam diameter grows from 120 to 220 microns, and grows again for bigger diameters till the disappearance of cavitation. If the absorption becomes smaller, the beam diameter for which the signal maximum is obtained becomes bigger, and the maximal obtainable signal grows.

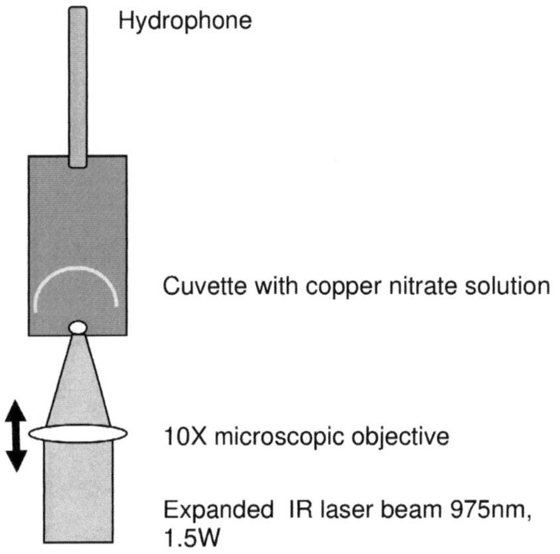


**Figure 1.** The experimental setup. The focused laser beam illuminates the cuvette from the bottom. The spot size is varied by objective displacement.

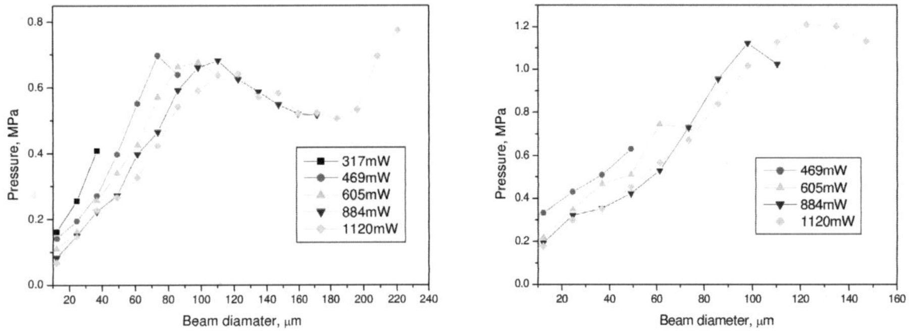


**Figure 2.** Shock wave pressure as function of the input beam diameter for different incident light power and liquid absorption coefficients 90 $cm^{-1}$ (left) and 45 $cm^{-1}$ (right). Permanent illumination, every point represents an average over 200 collapse events.

For our studies of adaptive optical detection [7] we are interested in a simple method of producing high-frequency ultrasound shock waves in solids. The cavitation cell coupled to a mirror results in a useful device for this purpose (figure 3). The resulting mirror surface displacements are comparable to the displacements, obtained with a traditional method of laser pulse impinging on the mirror from the back side [7]. Thus, potentially, this excitation method can be used for ultrasound inspection with a CW laser instead of the pulsed one. However, the resulting spikes are wider for cavitation, than for a pulsed laser. Additionally, because of a big number of interfaces in the setup, many spikes due to the wave reflections in a cell appear. If the propagation of wave in a solid is of interest, these additional spikes complicate the interpretation. Also the obvious disadvantage is the need for a liquid layer and a window for light. We have tried to obtain cavitation bubbles near the free surface of absorbing liquid, but the resulting shock wave proved to be much weaker, than for bubbles which form close to the water-glass interface.

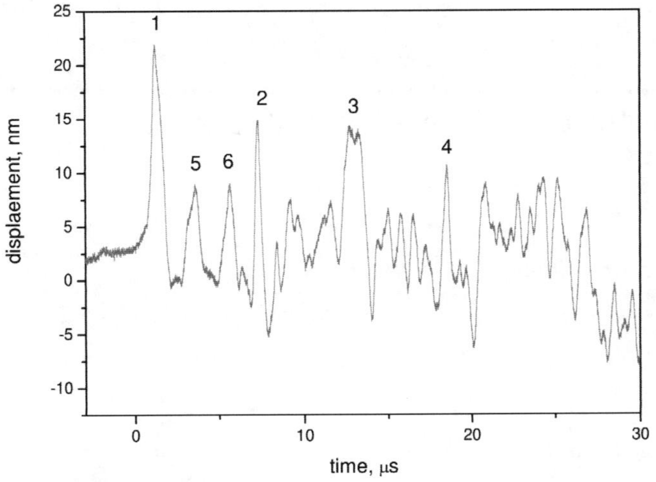


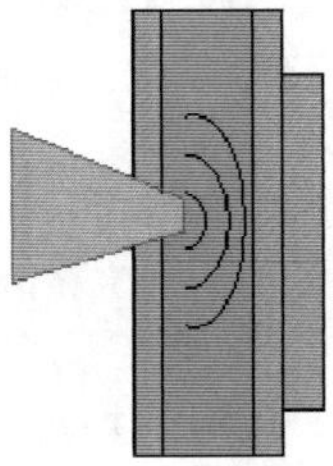

**Figure 3.** Mirror displacement after the bubble collapse. The cell geometry is depicted to the right. The 5 mm absorbing liquid layer is between two 1.7 mm thick glass walls, the 3 mm thick glass mirror ( the last element to the right) is set in contact with a cell wall using a thin oil film. The laser beam with 1.5 W power is chopped with 1Hz frequency, and illumination time is 0.1 s. The single shot measurement.

## 3. Conclusions

In conclusion, the cavitation induced by a moderate power CW laser produces an ultrasound shock wave which can be potentially useful for sample characterization. This is a relatively cheap and safe method of ultrasound generation with light. To obtain the strong wave with a given laser power, it is necessary to adjust the beam diameter and liquid absorption coefficient. A simple theory permits to understand qualitative influence of parameters involved, but for better description more realistic models are required.

## References

[1] Michel J M and Franc J P 2004 *Fundamentals of Cavitation* (Springer)
[2] Lauterborn W 1972 *Appl. Phys. Lett.* **21** 27
[3] Sacchi C A 1991 *JOSA B* **8** 337
[4] Rastopov S F and Sukhodol´sky A T 1990 *Phys. Lett.* A **149** 229
[5] Ramirez-San Juan J C , Rodriguez-Aboyetes E, Martinez-Canton A E, Baldovino-Pantaleon O,Robledo-Martinez A, Korneev N and Ramos-Garcia R 2010 *Opt. Express* **18** 8735
[6] Tikhonov A N, Samarskii A A 1990 *Equations of mathematical physics* (Dover)
[7] Korneev N, Rodriguez-Montero P and Benavides O 2009 *Opt. Lett.* **34** 1964

Journal of Physics: Conference Series **278** (2011) 012030 doi:10.1088/1742-6596/278/1/012030

# Oblique laser incidence to select laser-generated acoustic modes

**S Raetz, T Dehoux and B Audoin**

Université de Bordeaux, Laboratoire de Mécanique Physique, UMR CNRS 5469, France

E-mail: samuel.raetz@u-bordeaux1.fr

**Abstract**. The purpose of this paper is to study the effect of a non-normal optical penetration due to an obliquely incident laser source. It is shown that the loss of symmetry due to such a penetration influences specific bulk acoustic modes. For a given detection position, increase in shear wave amplitude is obtained by orienting the incident laser source.

## 1. Introduction

Since the first quantitative approaches in the 60's [1], a large number of studies has been conducted to better understand the generation of laser-generated acoustic waves. Models for the acoustic generation in the thermoelastic regime, taking into account the optical penetration of the source [2] or the effect of source width [3], were developed. A more general approach based on Green's function formalism also included thermal diffusion in the medium [4]. These works have dealt with the modeling of a circular spot of laser illumination. However, modeling of the acoustic field generated by a line source is of interest, since the signal-to-noise ratio may be increased in this experimental situation [5].

All the works cited above assume a normal incidence of the laser beam with respect to the illuminated surface of the sample. The purpose of this paper is to analyse the effects of oblique incidence on acoustic waves generated by a line-focused laser source. We present a theoretical model accounting for the effects of optical penetration, finite width of the beam and pulse duration of the laser source. The absorbed energy density is first calculated by solving Maxwell's equations. Then the thermoelastic equations are solved to obtain the displacement field through a semi-analytical calculation based on a double Fourier transform in space and time. It is shown that oblique incidence increases shear wave amplitude. Changes of the longitudinal waveforms are also discussed.

## 2. Volume source model for oblique incidence

### *2.1. Electromagnetic energy density distribution*

A plate of finite thickness with parallel-sided slab of an absorbing medium is considered from here on in this paper. The thickness is assumed to be large compared to the optical penetration depth. A Schott NG1 glass film of thickness $h$ = 2.5 mm is considered for calculations. The refractive index is $n$ = 1.51, and the attenuation index is $\kappa$ = 3.39 $10^{-4}$, corresponding to an optical penetration length of $1/\beta$ = 0.25 mm for normal incidence, $\beta$ being the attenuation coefficient.

In order to determine the amount of energy deposited by the laser in the volume of the sample, Maxwell equations are solved [6]. A monochromatic electromagnetic plane wave of frequency $\omega$ is considered and, by convolving the solution with a spatial Gaussian profile, the actual spatial shape of the laser beam is obtained. Let $\theta_i$ be the incidence angle formed by the incident wave vector $\mathbf{k_i}$ and the direction normal to the surface of the sample $\mathbf{x_1}$ (see figure 1). It is convenient to assume that the incident electric field $\mathbf{E_i}$ is perpendicularly polarized:

$$\mathbf{E_i} = A_\perp \, \mathrm{e}^{-j(\omega t - \mathbf{k_i} \cdot \mathbf{r})} \mathbf{x_3} \text{ with } \mathbf{k_i} = k_i \left( \cos\theta_i \mathbf{x_1} + \sin\theta_i \mathbf{x_2} \right). \tag{1}$$

$t$, $\mathbf{r}$ and $A_\perp$ are the time, position vector and the amplitude of $\mathbf{E_i}$, respectively. Since the sample thickness is large compared to the optical penetration depth, optical reflections at the second interface, $x_1 = h$, of the sample are not considered. Because the sample is absorbing, the optical wave vector of the transmitted waves $\mathbf{k_t}^*$ is complex. Optical phase continuity at the interface implies that $\mathbf{k_t}^*.\mathbf{x_2}$ is real. From the continuity for electric and magnetic fields at the interface, one can deduce relations between $A_\perp$ and the amplitude of the transmitted electric wave $T_\perp^*$. This expression is written in terms of complex optical index $n^* = n + j\kappa$ using the optical dispersion relation in an isotropic medium [6]:

$$T_\perp^* = \frac{2\cos\theta_i}{\cos\theta_i + (n + j\kappa)\cos\theta_t} A_\perp = \Upsilon^* A_\perp , \tag{2}$$

where $\Upsilon^*$ is the complex transmission coefficient in amplitude.

The local law of the electromagnetic conservation of energy leads to the following expression of energy density in the sample ($x_1 \geq 0$):

$$Q(x_i, t) = \beta n \left| \Upsilon^* \right|^2 I_0 \, \mathrm{e}^{-\beta x_1} g(x_1, x_2) f(t), \forall x_1 \geq 0 . \tag{3}$$

$I_0 = cA_\perp^2 / 8\pi$ is the incident intensity of the laser beam and $\beta = 2\omega\kappa\cos\theta_t / c$ is the attenuation factor, where $c$ is the light speed in vacuum. $g$ and $f$ represent the spatial and temporal Gaussian distribution of the source, respectively. Assuming a pulse width of 5 ns and a width at half-height of the space distribution of 12.5 $\mu$m, $Q(x_i, 0)$ is shown in figures 2(a) and (b) for $\theta_i$ = 0° and $\theta_i$ = 50°, respectively. A clear effect of the incidence is observed due to the dependence of $\beta$ and $\Upsilon^*$ on the refracted angle $\theta_t$ [and therefore on $\theta_i$, see equations 2 and 3]. Another change in this expression of $Q$, compared with previous models, is the dependence of $g$ on the $x_1$ coordinate due to oblique incidence. To investigate the effect, it can be convenient to express $g$ as a convolution product between a Gaussian function $G$ and the impulse function $\delta$:

$$g(x_1, x_2) = \delta(x_2 - x_1 \tan\theta_t) * G(x_2) . \tag{4}$$

Consequently, the spatial location of the energy along the refracted direction of the laser beam, introduced by $\delta(x_2 - x_1 \tan\theta_t)$, should cause visible change on acoustic waves.

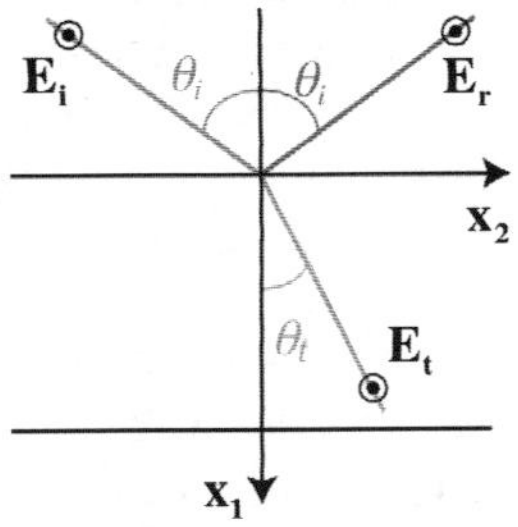


**Figure 1.** Geometry

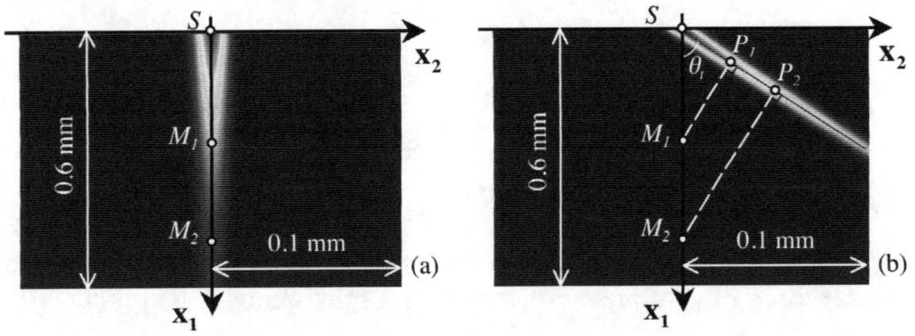


**Figure 2.** Energy density for (a) $\theta_i$ = 0° and (b) $\theta_i$ = 50°

### *2.2. Displacement field*

From the calculated energy density, see 2.1, the temperature distribution $T$ is obtained solving the heat equation. Since the thermal diffusion length is shorter than the optical penetration length in the material of interest at the time scale of the acoustic wave generation, thermal diffusion is neglected. The volume source of expansion is conducted by the optical penetration depth. The heat equation is reduced to:

$$\rho C_p \frac{\partial T}{\partial t} = Q(x_i,t), \tag{5}$$

where $\rho$ is the density of the material and $C_p$ is the specific heat. The temperature field is then obtained by integrating the equation above:

$$T(x_i,t) = \frac{\beta n|\Upsilon^*|^2 I_0}{\rho C_p} \mathrm{e}^{-\beta x_1} g(x_1,x_2) \int_0^t f(u)du \; \forall x_1 \geq 0 . \tag{6}$$

The wave equation is:

$$\nabla \cdot ([C]:\nabla \mathbf{u}) - \rho \frac{\partial^2 \mathbf{u}}{\partial t^2} = [C]:[\alpha]\nabla T , \tag{7}$$

where $\mathbf{u}$ is the displacement vector, $[C]$ the stiffness tensor and $[\alpha]$ the thermal dilatation tensor. A semi-analytical solution is computed, adapting the calculation scheme detailed in Ref. [7].

The effects on the acoustic generation of the dependence of $\beta$ and $\Upsilon^*$ with respect to incidence angle and of the dependence of $g$ on $x_1$ are presented and discussed in the following section.

## 3. Results

Normal displacements within the sample volume along $\mathbf{x_1}$ are plotted vs. time in figure 3. The observation points $M_i$ are located under the source impact point $S$ at the surface (see figure 2). Longitudinal acoustic waves are generated by the in-depth volume source of expansion. Their reflections at the free surface yield bulk longitudinal and shear waves. Arrival times are equal in both the normal and 50° incidence configurations. The amplitude of the normal displacement is weaker in the case of an incident (50°) source due to the decrease of the transmission coefficient $|\Upsilon^*|$ with increasing angle of incidence (equation 2).

Upon oblique incidence, the first longitudinal wave reaching point $M_i$ was generated near the orthogonal projection $P_i$ of this point on the refracted direction [see figure 2(b)]. A step at the beginning of the signal is thus observed, which does not exist in the well known exponential profile of the so-called precursor for normal incidence [8]. This step is obviously less noticeable when the distance between $S$ and $M_i$ increases.

Normal displacements at different points on the rear surface, $x_1 = h$, are displayed in figure 4(a). Film thickness $h$ is large enough to observe longitudinal and transverse contributions to the displacement separately. The longitudinal wave is denoted ($L$) in figure 4(a) whereas the shear wave is denoted ($T$). Shear wave amplitude is more sensitive to the incidence angle than the longitudinal mode [figure 4(a)]. In order to illustrate this effect, shear wave amplitudes in both configurations (normal and oblique incidence) are normalized with respect to the absorbed laser intensity $n|\Upsilon^*|^2 I_0$ to compensate for the fall in transmission coefficient $|\Upsilon^*|$. With $\theta_i$ = 50°, thus $\theta_t \sim$ 30°, it appears in figure 4(c) that, for the same absorbed energy, the shear wave amplitude is larger in the oblique incidence configuration. Furthermore the maximum of amplitude is shifted compared to the one at normal incidence.

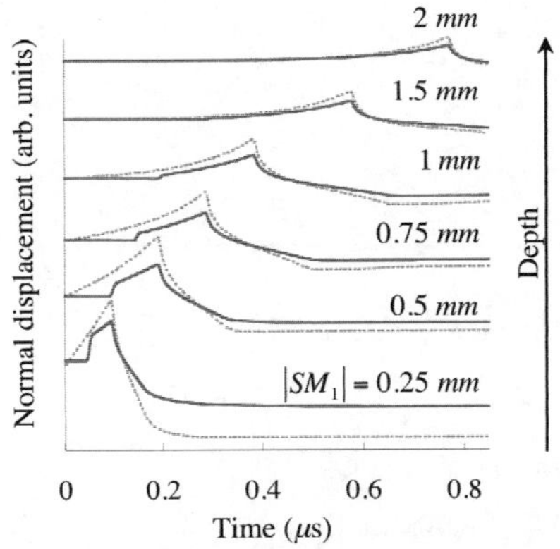


**Figure 3.** Normal displacement at different depths for $\theta_i = 0°$ (dash) and $\theta_i = 50°$ (solid) with $1/\beta = 0.25$ mm

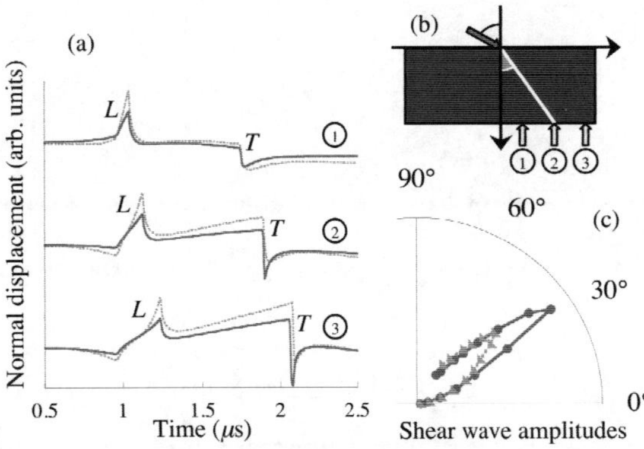


**Figure 4.** (a) Normal displacement at three detection points shown in (b), and (c) shear wave amplitudes in polar coordinates normalized with respect to the absorbed laser intensity for $\theta_i = 0°$ (dash) and $\theta_i = 50°$ (solid)

## 4. Perspectives

A model of the volume source of expansion for oblique incidence has been presented and effects on the thermal normal gradient have been discussed. We have demonstrated that the longitudinal waveforms are changed and that the shear wave amplitude can be increased due to oblique incidence of the laser beam, and thus a specific mode can be enhanced by changing the incidence angle of the source. These predictions will be compared with a pump-probe experiment using nanosecond laser pulses to generate acoustic waves in samples made of colored glasses.

## References

[1] Lee R E and White R M 1968 *Appl. Phys. Letters* **12** 12-14
[2] Dewhurst R J, Hutchins D A, Palmer S B and Scruby C B 1982 *J. Appl. Phys.* **53** 4064-71
[3] Scruby C B, Dewhurst R J, Hutchins D A, Palmer S B 1980 *J. Appl. Phys.* **51** 6210-16
[4] Doyle P A 1986 *J. Phys. D: Appl. Phys.* **19** 1613-23
[5] Arias I and Achenbach J D 2003 *Int. J. Solids Struct.* **40** 6917-35
[6] Born M and Wolf E 2005 *Principles of Optics - 7th ed.* (Cambridge University Press)
[7] Audoin B, Meri H and Rossignol C 2006 *Phys. Rev.* B **74** 214304
[8] Dubois M, Enguehard F and Bertrand L 1994 *Phys. Rev* . E **50** 1548-51

Journal of Physics: Conference Series **278** (2011) 012031 doi:10.1088/1742-6596/278/1/012031

# Ultrasound generated by laser in a coated cylinder

**Y D Pan[1], X H Song[2], Z Zhong[2] and B Audoin[3]**

[1]Institute of Acoustics, Tongji University, 200092, Shanghai, P. R. China
[2]School of Aerospace Engineering and Applied Mechanics, Tongji University, 200092, Shanghai, P. R. China
[3] Laboratoire de Mécanique Physique, CNRS UMR 5469, Université Bordeaux 1, 33405 Talence, France

**Abstract.** An alternative theoretical solution is proposed to predict ultrasound excited by a laser line source in a homogeneous, isotropic cylinder coated with a homogeneous and isotropic material. The spectrum and transient displacement are in agreement to author's previous model for the cylinder without coating. Numerical results are obtained for copper on aluminium and vice versa. It is found for the surface waves that the high frequency component travels faster for hard coatings with velocities faster than that of the substrate, and vice versa for soft coatings.

## 1. Introduction

The inspection of coatings is important for surface engineering, where it is widely used to improve the performance of materials. To access information about a coating by surface waves, ultrasound microscope is often used to obtain the reflection coefficient through the measurement of V(z) curve [1]. It requires the sample to be immersed in water, and thus limits its application on site. Moreover it is difficult to measure a curved sample such as a cylinder. Fortunately, the laser ultrasonic technique, in which ultrasounds are generated and detected by lasers, can measure the acoustic displacement of a curved surface at a distance [2]. Recently, coatings on a plate structure have been studied [3, 4]. However, little work is available on the coated cylinder. The objective here is to develop a model to predict laser-excited surface waves on a coated cylinder.

## 2. Governing equations

Consider acoustic wave propagation in a coated cylinder of diameter $2a$ with infinite length. The coating is homogeneous with a thickness $h$, density $\rho$ and Lamé coefficients $\lambda$ and $\mu$. The rod substrate is homogeneous with diameter $2b$ ($b=a-h$), density $\rho_s$ and Lamé coefficients $\lambda_s$ and $\mu_s$. A system of cylindrical coordinates ($r$, $\theta$, $z$) is chosen. The displacement $\mathbf{U}$ and the stress vector $\sigma$ of the coating are sought in the form:

$$\begin{cases} \mathbf{U}(r,\theta,t) = \mathbf{u}(r)\exp[j\omega(t - pa\theta)], \\ \Sigma(r,\theta,t) = \boldsymbol{\sigma}(r)\exp[j\omega(t - pa\theta)]. \end{cases} \tag{1}$$

$\omega$ being the angular frequency, and $p$ the circumferential slowness. Instead of searching the explicit solution of the authors' recent publication on bi-layered cylinder [5], here the state vector approach [6] is applied to obtain the matrix differential equation as:

$$\partial \boldsymbol{B}(r)/\partial r = i\omega \mathbf{A}(r)\boldsymbol{B}(r) \tag{2}$$

where the state vector $\boldsymbol{B} = (u_i, \sigma_{ri})^T$ ($i=r$, $\theta$) is the displacement –traction and $\mathbf{A}$ is the system matrix depending on the elastic properties of the layered medium, and on $p$ and $\omega$. For an isotropic and layered cylinder, the state vector and the system matrix for P-SV coupled waves are [6]:

$$\boldsymbol{B}(r) = \begin{pmatrix} i\omega u_r & i\omega u_\theta & \sigma_{rr} & \sigma_{r\theta} \end{pmatrix}^{\mathrm{T}} \tag{3}$$

and

$$\mathbf{A}(r) = \begin{pmatrix} -r^{-1}\lambda(\lambda+2\mu)^{-1} & par^{-1}\lambda(\lambda+2\mu)^{-1} & (\lambda+2\mu)^{-1} & 0 \\ pa/r & r^{-1} & 0 & \mu^{-1} \\ \rho & -pa\xi/r^2 & -r^{-1}\lambda(\lambda+2\mu)^{-1} & pa/r \\ pa\xi/r^2 & \rho - p^2a^2\xi/r^2 & par^{-1}\lambda(\lambda+2\mu)^{-1} & -2/r \end{pmatrix} \tag{4}$$

with

$$\xi = 4\mu\frac{\lambda+\mu}{\lambda+2\mu} \tag{5}$$

### 3. Boundary conditions and solution

The matrix $\mathbf{M}(r, r_0)$ is the fundamental solution of Eq. (2) with the form :

$$\boldsymbol{B}(r) = \mathbf{M}(r, r_0)\boldsymbol{B}(r_0) \tag{6}$$

where $r_0$ is a reference point on the $r$-axis. The matrix can be calculated by the Peano expansion with the dimensionless quantity $(r-r_0)/a$ to be small for the truncation to be acceptable:

$$\mathbf{M}(r, r_0) = \mathbf{I} + \int_{r_0}^{r}\mathbf{A}(\zeta_1)d\zeta_1 + \int_{r_0}^{r}\mathbf{A}(\zeta_1)\int_{r_0}^{\zeta_1}\mathbf{A}(\zeta_2)d\zeta_2 d\zeta_1 + \ldots \tag{7}$$

Let the laser excite the coating as a beam focused along the $z$-axis by a cylindrical lens. To model the laser source, the same boundary conditions as that in author's previous model [7], are chosen for the two components $\sigma_{rr}$ and $\sigma_{r\theta}$ of the stress tensor as

$$\begin{cases} \sigma_{rr}\big|_{r=a} = -\sigma_0\delta(t)\delta(\theta) \\ \sigma_{r\theta}\big|_{r=a} = 0 \end{cases} \tag{8}$$

for the ablation generation, where $\sigma_0$ stands for the normal force magnitude in N·μs·m$^{-2}$. The delta functions of time $\delta(t)$ and of space $\delta(\theta)$ are used. We have

$$\boldsymbol{B}(r = a) = \begin{pmatrix} i\omega u_r & i\omega u_\theta & -\sigma_0 & 0 \end{pmatrix}^{\mathrm{T}}. \tag{9}$$

The coated cylinder is assumed to have no defects along its interface. The continuity of the two non-zero components of the displacement and stress [4] are considered at the interface between cylinder and coating. The laser excited transient displacement at $r=a$ is found to be:

$$\begin{pmatrix} i\omega u_r & i\omega u_\theta \end{pmatrix}^{\mathrm{T}} = \left(\mathbf{M}_2 - \mathbf{M}_1\mathbf{M}_3^{-1}\mathbf{M}_4\right)\begin{pmatrix} -\sigma_0 & 0 \end{pmatrix}^{\mathrm{T}} \tag{10}$$

where $\mathbf{M}_1$, $\mathbf{M}_2$, $\mathbf{M}_3$, $\mathbf{M}_4$ correspond to the left-up, right-up, left-down, and right-down off-diagonal block of $\mathbf{M}(a, b)$ as four 2×2 sub-matrices [6].

### 4. Numerical results

First, an aluminium cylinder coated with the same material is considered, and the calculation is compared to author's previous result [7], as shown in figure 1. The density, and the longitudinal and shear wave velocities of aluminium, are chosen to be $\rho$=2700 kg/m$^3$, $V_L$=6400 m/s and $V_S$=3110 m/s. The diameter of the cylinder is 10mm, and the coating is 0.1mm thick through this paper. The dashed white-lines of the dispersion curves by the previous model [7] fit well to the spectrum where energy

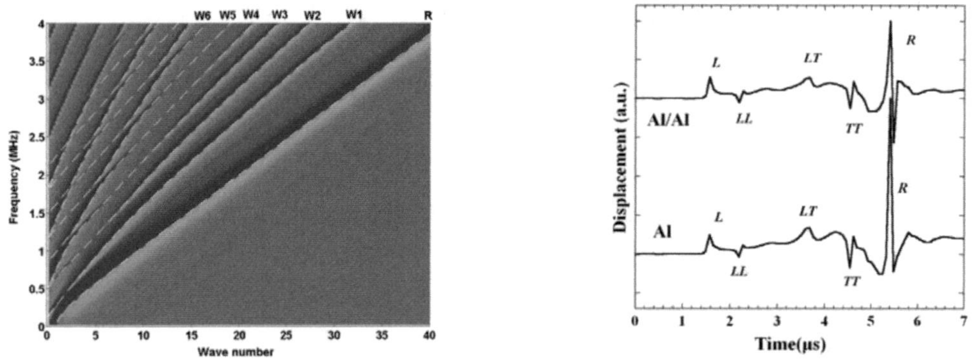


(a) (b)

**Figure 1.** (a) Spectrum and (b) transient displacement excited by laser for a cylinder without coating

exists. Various wave modes are clearly observable, the first mode (*R*) is a surface wave mode, the following modes are Whispering Gallery (*W*1, *W*2,...) waves [8]. The transient displacement is calculated and compared with a previous model [7]. The two waveforms are identical, the direct (*L*) and reflected longitudinal waves, the reflected shear wave (*TT*), the conversion of the longitudinal and transverse waves (*LT*), and the cylindrical Rayleigh wave (*R*) are clearly observable. The agreement illustrates the effectiveness of the proposed model and solution.

Also, an aluminium rod coated by copper and vice versa were chosen for the calculations. The density, and the longitudinal and shear wave velocities of copper, are chosen to be $\rho$=8900 kg/m$^3$, $V_L$=4700 m/s and $V_S$=2260 m/s. The dimension is the same as above. The spectra are obtained for the two coated cylinders in figure 2 (a) and (b). The *R, W*1, *W*2,... waves are clearly observed. As the penetration depth of the surface wave is in the order of wavelength, the slope of the R wave is quite different for both, and it approaches the surface wave velocity of the coated material. However, the waveforms of various *W* waves make no much difference as their penetration depths are in the order of the radius.

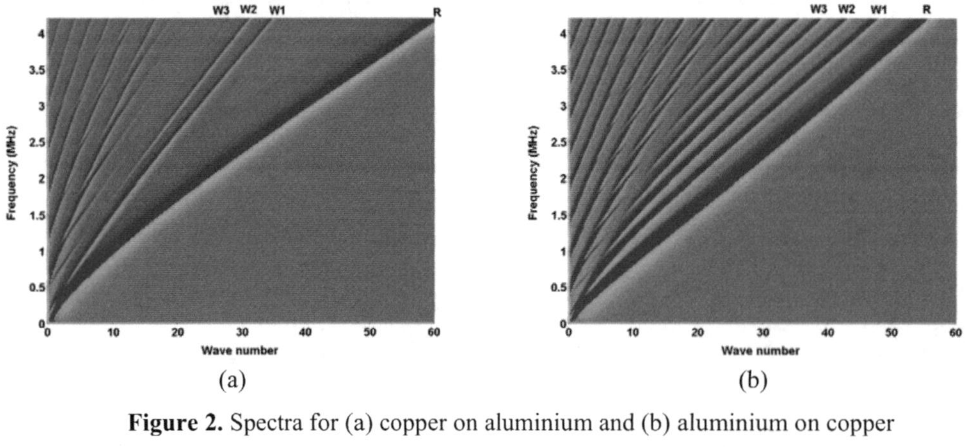


(a) (b)

**Figure 2.** Spectra for (a) copper on aluminium and (b) aluminium on copper

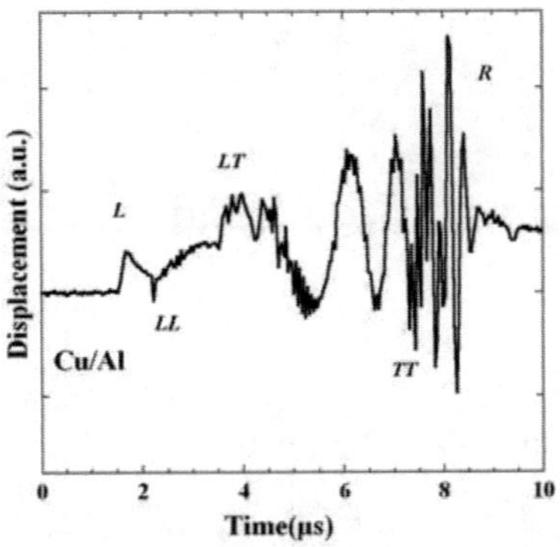


(a)

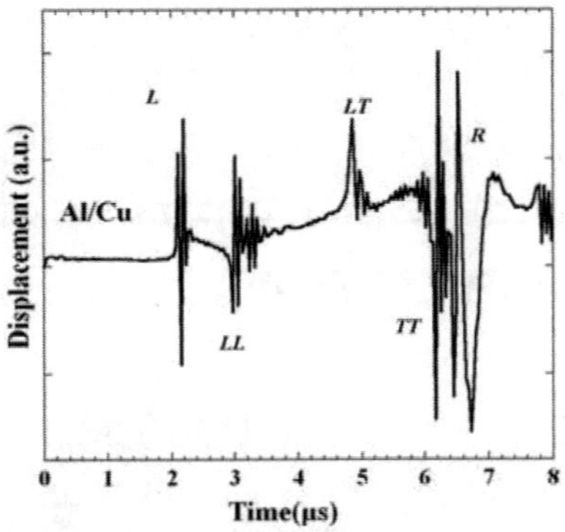


(b)

**Figure 3.** Transient displacements for (a) copper on aluminium and (b) aluminium on copper

Finally, the transient displacements are obtained as shown in figure 3 (a) and (b). The direct (*L*) and reflected longitudinal waves, the reflected shear wave (*TT*), the conversion of the longitudinal and transverse waves (*LT*) are clearly observable. However, the waveforms of Rayleigh (*R*) waves are quite different. For copper on aluminium, the high frequency component travels relatively slower than the low frequency component, and it is vice versa for aluminium on copper. This phenomenon may be explained by the penetration dependence on the frequency. The high frequency component has relatively smaller penetration than the low frequency component. As the thickness of the coating is relatively small, the high frequency component is likely affected by the coating and the counter part by the substrate. The difference of R waves further illustrates the effectiveness of the proposed model and solution.

## 5. Conclusion

An alternative theoretical solution is provided to predict ultrasound generated by a laser line source in a homogeneous, isotropic cylinder coated with a homogeneous and isotropic material. This model is different from previous one to potentially address functionally graded cylinders. The spectrum and transient displacement are in agreement for copper on aluminium and vice versa. It is found for the surface waves that the high frequency component travels faster for hard coatings with velocities faster than that of the substrate, and vice versa for soft coatings.

## Acknowledgments

Y. P. was supported by NSFC under 10872151 and 10834009, and Z. Z was supported by NSFC under 10872150.

## References

[1] Xu W J, Ourak M 1997 *NDT&E International* **30** 75
[2] Scruby C B, Drain L E 1990 Laser Ultrasonics:Techniques and Applications, Adam Hilger, New York,USA
[3] Murray T W, Krishnaswamy S, Achenbach J D 1999 *Appl. Phys. Lett.* **74** 3561
[4] Kawald U, Desmet C, Lauriks W, Glorieux C, Thoen J 1996 *J. Acoust. Soc. Am.* **99** 926
[5] Pan Y, Chigarev N, Audoin B 2010 *J. Appl. Phy.* **107** 023527
[6] Baron C, Poncelet O, Shuvalov A 2007 *Ultrasonics* **46** 1
[7] Pan Y, Rossignol C, Audoin B 2003 *Appl. Phys. Lett.* **82** 4379
[8] Hu W, Qian M, Cantrell J H 2004 *Appl. Phys. Lett.* **85** 4031

Journal of Physics: Conference Series **278** (2011) 012032 doi:10.1088/1742-6596/278/1/012032

# Directivity measurements in aluminum using a laser ultrasonics system

**J M S Sakamoto[1], B R Tittmann[2], A Baba[3] and G M Pacheco[1]**

[1] Instituto Tecnológico de Aeronáutica, Praça Marechal Eduardo Gomes, 50 – Vila das Acácias, São José dos Campos – SP, 12228-900, Brazil
[2] Pennsylvania State University, 212 Earth & Engr. Sciences Building, University Park – PA, 16802-6812, USA
[3] Hitachi, Ltd., 7-2-1 Omika-cho, Hitachi-shi, Ibaraki-ken, 319-1221 Hitachi, Japan

E-mail: sakamoto@ieav.cta.br

**Abstract.** A laser ultrasonics system was setup to measure the directivity (angular dependence pattern) of the amplitude of ultrasonic waves generated in aluminum samples. A pulsed Nd:YAG laser operating at 1064 nm optical wavelength, with typical pulse width (FWHM) of 8 ns, and energy per pulse of 450 mJ, was used to generate the ultrasound waves in the samples. The laser detection system was a Mach-Zehnder interferometer with typical noise-limited resolution of 0.25 nm (rms), frequency range from 50 kHz to 20 MHz, and measurement range from -75 nm/V to +75 nm/V. Two different optical spot sizes of the Nd:YAG laser were used to generate waves in the ablation regime: one was focused and the other was unfocused. Using the obtained data, the directivity graphics were drawn and compared with the theoretical curves, showing a good agreement. The experiments showed the directivity as a function of the optical spot size. For a point ultrasonic source (or focused optical spot), the directivity shows that the longitudinal waves present considerable amplitude in all directions. For a larger ultrasonic source (or an unfocused optical spot) the directivity shows that the longitudinal waves are generated with the higher amplitudes inside angles around $\pm 10^{\circ}$.

## 1. Introduction

The laser ultrasonics technique typically has a pulsed laser that generates ultrasound waves and an interferometer to detect the waves. When inspecting a sample, these two devices can be positioned in the same side of the sample, to perform a single-sided measurement or a single-sided pulse-echo measurement, or they can be placed in opposite sides to make a direct transmission or oblique transmission measurements [1]. In these cases one can consider the relative position between the generation laser spot and the interferometer spot. The directivity or angular dependence is important to show where the generator and the detector should be placed for a better detection when inspecting a sample. Also it is important to characterize the laser ultrasonic source as in the work of Hutchins *et al* [2], Edwards *et al* [3], and Pouet *et al* [4]. In this work, the directivity measurements are presented for a laser ultrasonics source in ablation regime.

## 2. Theory

The theory used to draw the theoretical curves gives the directivity pattern for an ablation laser ultrasonic source [1], [2], [5]. In reference [1], the authors considered that the ablation source was focused to a small area and the finite source size effects were not included. However, in this work, the size of the source is relevant and a function was included to account for the size of the optical spot. The equation that gives the angular dependence of the longitudinal wave in the ablation regime is [1]:

$$U(\theta) \cong \frac{2k^2 \cos\theta \left(k^2 - 2\sin^2\theta\right)}{\left(k^2 - 2\sin^2\theta\right)^2 + 4\sin^2\theta\left(1-\sin^2\theta\right)^{1/2}\left(k^2-\sin^2\theta\right)^{1/2}} \tag{1}$$

where $k$ is the ratio between the longitudinal wave velocity and the shear wave velocity, and $U$ varies with the angle $\theta$ in the farfield. The following equation was used for taking into account the size effect of the laser ultrasonic source [1]:

$$\gamma(\theta) = \frac{J_1\left(2\pi f a \sin\theta / c_{\mathrm{L}}\right)}{2\pi f a \sin\theta / c_{\mathrm{L}}} \tag{2}$$

where $J_1$ is the Bessel function of the first kind, $f$ is the frequency of the ultrasound wave, $a$ is the radius of the source (optical spot) and $c_{\mathrm{L}}$ is the velocity of the longitudinal wave. Multiplying equations (1) and (2), one can obtain the equation that gives the angular dependence, including the optical spot size effects:

$$u_r(\theta) = U(\theta)\gamma(\theta) \tag{3}$$

## 3. Experimental setup

The laser ultrasonics system was composed of a pulsed Nd:YAG laser to generate ultrasonic waves, an optical interferometer to detect the waves, and optical components. The Nd:YAG laser provided optical radiation at 1064 nm wavelength, with minimum pulse width (full width at half maximum - FWHM) of 8 ns, and energy per pulse of 450 mJ. The optical interferometer was a Mach-Zehnder type with typical noise-limited resolution of 0.25 nm (rms), frequency range from 50 kHz to 20 MHz, and measurement range from -75 nm/V to +75 nm/V.

The Nd:YAG laser and the interferometer were positioned in opposite sides of the sample, as shown in figure 1. The Nd:YAG laser beam was firstly deflected by the mirror $M_1$ and after by the mirror $M_2$. These mirrors were used to set the position of the generation laser spot. After the mirror $M_1$ there is an attenuator and a photodetector that detects the stray laser beam and generates a trigger signal that is connected to the oscilloscope trigger input. A converging lens was used in the experiments that needed a reduced (or focused) spot size in the generation side. The interferometer spot was focused over the sample surface, using its own coupled lens set. Its displacement output was connected to the oscilloscope channel.

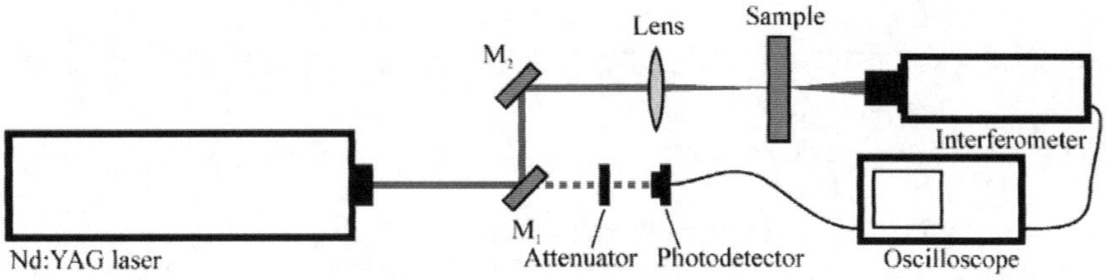


**Figure 1.** Experimental setup.

The Nd:YAG laser beam impinged at the sample surface generating a ultrasound wave that traveled through the thickness of the sample, and it was detected by the interferometer, on the back surface. To detect the ultrasound waves as a function of position (or angle), the interferometer was translated using a micrometer coupled to it. The interferometer spot position (or detection position), $x$, and the thickness of the sample, $h$, are related to the angle, $\theta$, as shown in figure 2.

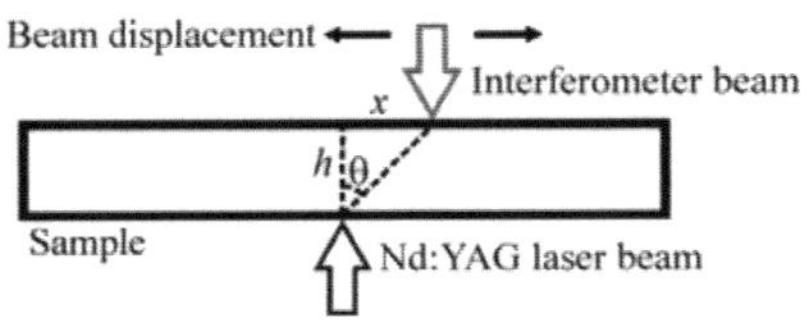


**Figure 2.** Relation between displacement and angle.

The relation used to determine the angle was $\theta = \tan^{-1}(x/h)$ and each waveform detected was related to its corresponding one. The origin of the $x$ axis was the epicenter of the wave. This point was experimentally set through the alignment of the generation spot and the detection spot using an auxiliary semi-transparent sample. A correction factor for the experimental amplitude was included to account for the flat shape of the sample: the experimental amplitude was divided by $\cos\theta$.

Two different optical spot sizes of the Nd:YAG laser were used to generate waves in the ablation regime: one was focused using the lens and the other was unfocused, without any lens.

## 4. Results

The first measurement was the directivity of the ultrasound generated by a focused laser beam, in the ablation regime, in an aluminum 6061 T6 sample with thickness of 4.6 mm. A convergent lens with 200 mm of focal length was used to provide a point source in the generation side. The sample was positioned on the focus. The Nd:YAG laser was operated using an energy of 55 mJ and the pulse width (FWHM) was approximately 17 ns. The laser spot radius was approximately 150 μm so the density of energy was approximately 78 J/cm$^2$. The other surface of the sample, where the spot of the interferometer was positioned to detect the ultrasound, was polished to obtain a better signal to noise ratio at the output of the interferometer. The ultrasonic waveforms were acquired from -13$^\circ$ to +75$^\circ$ and they are shown in figure 3. The data obtained from the waveforms of figure 3 were used to draw the directivity of the longitudinal wave graphic, shown in figure 4. The squares are the experimental data and the solid lines were drawn using equation (3).

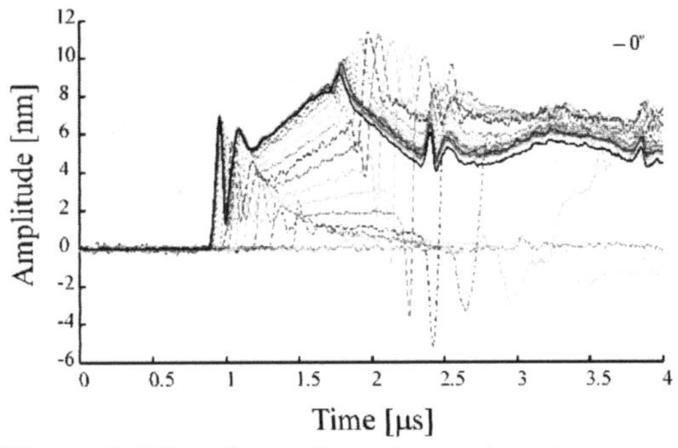


**Figure 3.** Waveforms for a focused spot.

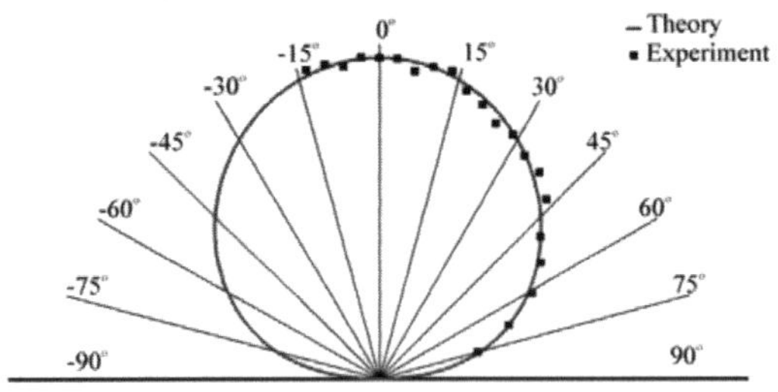


**Figure 4.** Directivity of longitudinal waves.

As can be seen in figure 4, the experimental data present a good agreement with the theory.

The same experimental setup and methods used in the former case were used to measure the directivity of ultrasonic waves generated by an unfocused laser beam, again in ablation regime, in an aluminum 2024 sample, with thickness of 9.6 mm. The energy used was approximately 250 mJ and the pulse width (FWHM) of approximately 12 ns. In this case, the lens was not used and the spot radius was approximately 3 mm at the sample surface so the density of energy at the sample was around 0.9 J/cm$^2$. The waveforms were acquired from -15$^\circ$ to +64$^\circ$, and they are shown in figure 5.

The waveforms of figure 5 have a different shape compared to the waveforms showed in figure 3 due to their different density of energy [1].

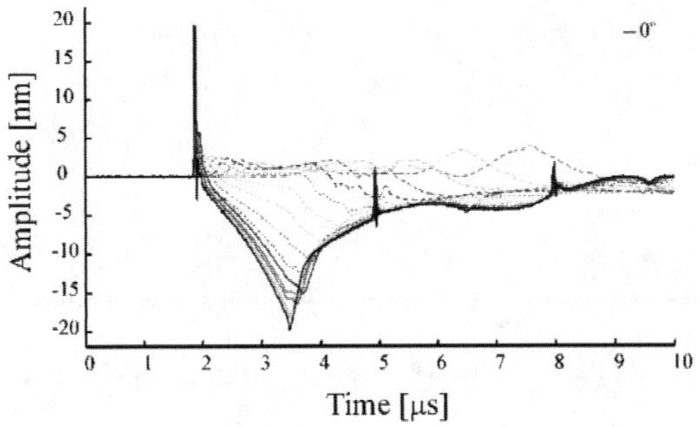


**Figure 5.** Waveforms for a non focused spot.

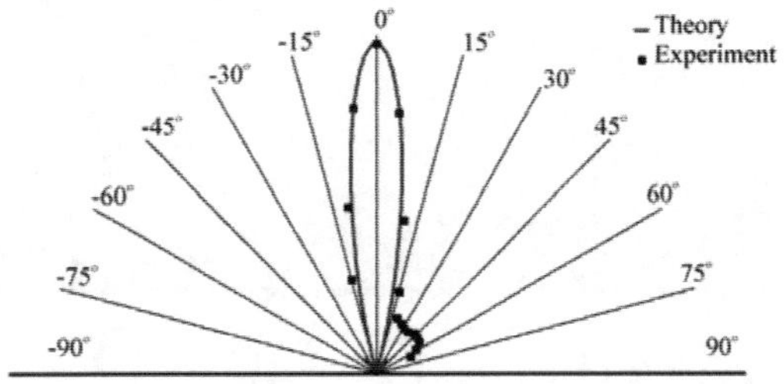


**Figure 6.** Directivity of longitudinal waves.

The data obtained from the waveforms of figure 5 were used to plot the graphic shown in figure 6. Again, the squares are the experimental data and the solid line, the theory. The longitudinal wave directivity for the non focused spot, shown in figure 6, presents the waves with higher amplitude in a direction narrower than the direction for the waves using a focused spot. The experimental data agrees with the theory up to around 20°. The disagreement for angles larger than 20° is attributed to a residual thermoelastic generation. The thermoelastic directivity has a side lobe around 60° [1], similar to the small side lobe shown on the experimental data in figure 6. Although the predominance of the ablation regime, the thermoelastic effect becomes apparent due to the lower optical density of energy used.

## 5. Conclusion

For a point ultrasonic source (or focused optical spot), the directivity shows that the longitudinal waves present considerable amplitude in all directions. For a larger ultrasonic source (or an unfocused optical spot) the directivity shows that the longitudinal waves are generated with the higher amplitudes inside angles around ±10°. This is interesting for the generation of waves in the central direction, to perform a single-sided pulse-echo inspection or a direct transmission inspection. With the unfocused spot, a reduced density of energy is used since the spot area is bigger than the focused spot, nevertheless generating high amplitudes in the central direction. Also, using the non focused optical spot the detector must be positioned near the epicenter to detect the waves. On the other hand, it is not suitable to perform a single-sided inspection or an oblique transmission inspection. In this case the focused optical spot appears to be more suitable. In conclusion, the experiments suggest that the angular dependence pattern can be seen through the Huygens principle, having the waves spreading out at large angles for a small optical spot size compared to the acoustic wavelength. For optical spot sizes greater than the acoustic wavelength, the waves interfere constructively in the region in front of the acoustic source and destructively outside this zone.

## Acknowledgments

The authors would like to thank Dr. Chiaki Miyasaka and the students John Mulry and Robert Cyphers. One of the authors (JMSS) acknowledges the provision of a scholarship from CNPq, process number 142191/2007-8, and an international scholarship from CAPES, process number 4697/08-1.

## References

[1] Scruby C B and Drain L E 1990 *Laser Ultrasonics: Techniques and Applications* (Bristol: Adam Hilger)
[2] Hutchins D A, Dewhurst R J and Palmer S B 1981 *J. Acoust. Soc. Am.* **70 (5)** 1362-9
[3] Edwards C, Taylor G S and Palmer S B 1989 *J. Phys. D: Appl. Phys.* **22** 1266-70
[4] Pouet B, Lefaudeux N, Clemenceau P and Breugnot S 2008 *1st Int. Symp. on Laser Ultrasonics: Science, Technology and Applications* (Montreal), http://www.ndt.net/article/laser-ut2008/papers/Pouet%20LU2008.pdf
[5] Miller G F and Pursey H 1954 *Proc. R. Soc. London*, **A223** 521-41

# Photothermal depth profiling for multilayered Structures by particle swarm optimization

**Z J Chen**[1,2,a)], **J W Fang**[1] **and S Y Zhang**[2,b)]

[1] Department of Physics, Zhejiang Normal University, Jinhua 321004, China
[2] Institute of Acoustics, Nanjing University, Nanjing 210093, China

[a)] E-mail: zjchen@zjnu.edu.cn
[b)] E-mail: zhangsy@nju.edu.cn

**Abstract**. This paper presents a method to reconstruct thermal conductivity depth profile of a layered medium using noisy photothermal data. The method tries to obtain an accurate reconstruction of discontinuous profile using particle swarm optimization (PSO) algorithm and total variation (TV) regularization. The reconstructions of different thermal conductivity profiles have been tested on simulated photothermal data. The simulation results show that the method can find accurately the locations of discontinuities, and the reconstructed profiles are in agreement with the original ones. Moreover, the results also show the method has good robustness and anti-noise capability.

## 1. Introduction

In recent years photothermal techniques have been applied to investigate thermal conductivity depth profiles of inhomogeneous samples [1,2]. In order to characterize such samples one may generate a plane thermal wave at sample surface by using a pump beam modulated at the frequency $f$. The quantity $f$, which is adjustable, drives the penetration depth of the thermal wave used for the investigation. Noninvasive thermal-wave imaging of depth profiles is very appealing in several areas from civil and industrial engineering to nondestructive testing and biomedical applications. In spite of this great importance, most thermal-wave imaging techniques are still far from solving practical problems. In fact, there are theoretical aspects hard to manage, such as, the nonlinear and non convex relationship between the thermal-wave field and the object structure, the robustness of the inversion algorithms against noise. Working on an ill-posed problem, photothermal inversion normally provides either unstable or oversmoothed results. Recently, evolutionary computing methods such as genetic algorithms (GA) [3] and particle swarm optimizations (PSO) [4] have been tried in continuous profile reconstruction problems, and these approaches proved to be able to efficiently find a global optimum in a fitness landscape.

In this paper, reconstructions for discontinuous and step-index thermal conductivity profiles are attempted by using PSO algorithm and total variation regularization. The specific problem regards the reconstruction of thermal conductivity profile of a multilayer slab from measurements of the alternating temperature of the object surface within a certain frequency range. Some details on the formulation of the profile reconstruction problem are given, and several reconstruction simulations are performed to evaluate the capability of the inversion method to find the locations of discontinuities. Moreover, the influences of noise on the reconstruction results are also investigated.

## 2. Particle swarm optimization implementation

The sample considered here is a multilayer slab, which located on a semi-infinite homogeneous substrate. For charactering the thermal properties of the slab, a modulated plane heat source is located at the sample surface. Then the amplitude and phase of surface alternating temperature is measured at $m$ modulated frequencies $f_m$ within the chosen frequency range. The aim is to estimate the thermal conductivity of the slab as a function of the depth coordinate $z$. To solve this problem, the slab is discretized into a finite number $N$ of homogeneous layers. Then the phtothermal signal is calculated as follows: $S=Z_1\times S_{ref}=A\exp(i\phi)$, where $S_{ref}$ is the reference signal of a homogeneous sample, and $Z_1$ is the surface thermal impedance calculated, layer by layer (n= $N$...1), by the following recursive formula [1,4]:

$$Z_n = \frac{\cosh(q_n h_n)Z_{n+1} + \dfrac{1}{k_n q_n}\sinh(q_n h_n)}{k_n q_n \sinh(q_n h_n)Z_{n+1} + \cosh(q_n h_n)}. \tag{1}$$

where $q_n=(i2\pi f\rho c/k_n)^{1/2}$, $k_n$, $h_n$ are thermal wave vector, thermal conductivity, thickness, of the $n$th layer, respectively. $\rho$ and $c$ are density and heat capacity of the slab.

The inversion algorithm has to find the configuration of layers that gives the best fit between the measured and the computed photothermal signals, subject to suitable regularizing constraints. The forward solver implemented in the code evaluates the photothermal signals of any structure composed by a finite number of layers for each of the $m$ frequencies $f_m$. The minimized fitness function to be optimized is the following:

$$F(\boldsymbol{k}) = \left[\sum_{m=1}^{M}\left(S_{teo}(f_m,\boldsymbol{k}) - S_{\exp}(f_m,\boldsymbol{k})\right)^2\right]^{1/2} + \lambda\sum_{n=1}^{N-1}\left|k_{n+1}-k_n\right|. \tag{2}$$

where the first term in equation 2 is a suitable distance between the two $m$-element arrays of the experimental ($S_{exp}$) and the theoretical ($S_{teo}$) photothermal signals (amplitudes or phases). The second term is the total variation (TV) regularization function [5], which is appropriate for problems where we except there to be discontinuous jumps in the model. Hyper-dimensional vector $\boldsymbol{k} = (k_1, k_2, \ldots, k_N)$ represents the reconstructed conductivity profile and $\lambda$ is the regularization parameter. The equation 2 is non-convex, and a global optimization is required to converge to the minimum. A particle swarm optimization (PSO) algorithm [6] is adopted to optimize equation 2 since this approach has proved to be reliable and efficient in the smooth profile reconstruction problems [4].

PSO starts with the random initialization of a population of individuals (particles) in the search space and works on the social behaviour of the particles in the swarm. Each particle is treated as a mass-less and volume-less point in a $D$-dimensional space. The $i$th particle is represented as $\boldsymbol{x}_i = (x_{i1}, x_{i2}, \ldots, x_{iD})$. The best previous position of the $i$th particle that gives the best fitness value is represented as $\boldsymbol{p}_i = (p_{i1}, p_{i2}, \ldots, p_{iD})$. The best particle among all the particles in the population is represented by $\boldsymbol{p}_g =(p_{g1}, p_{g2}, \ldots, p_{gD})$. Velocity, the rate of position change for the $i$th particle, is represented as $\boldsymbol{v}_i = (v_{i1}, v_{i2}, \ldots, v_{iD})$. At every iteration, the velocity and the position of each particle are updated by using the two best values according to the following equations:

$$v_{id}^{t+1} = \omega\cdot v_{id}^{t} + c_1\cdot r_1\cdot(p_{id}^{t} - k_{id}^{t}) + c_2\cdot r_2\cdot(p_{gd}^{t} - k_{id}^{t}), \tag{3}$$

$$x_{id}^{t+1} = x_{id}^{t} + v_{id}^{t+1}. \tag{4}$$

Here $t$ is the iteration number, $d$ = 1, 2, … , $D$, $i$ =1, 2, … , $S$, and $S$ is the size of the population. $\omega$ is inertial weight factor, $c_1$ and $c_2$ are acceleration constants, $r_1$ and $r_2$ are two random numbers uniformly distributed in the interval [0, 1]. In the PSO inversion, a particle represents a possible thermal conductivity profile $\boldsymbol{k}$ ($D = N$), and the choice of $\boldsymbol{k}$ is determined by the intrinsic relationship between the photothermal signal and the thermal conductivity profile. In each epoch of the optimization routine, the cost function is calculated by equation 2, and the repeated application of equation 3 and 4 leads to the evolution of the population towards the best solution.

In this study, the PSO population size $S$ is 30, the inertial weight $\omega$ is linearly decreasing from 0.9 to 0.4 during the iterations, the acceleration constants $c_1$ and $c_2$ are equal to 2, and the maximum velocity $v_{max}$ is set to 0.1 times of the maximum particle position. Moreover, the maximum iteration number (epoch) for training is 4000, which is selected based on the convergence rate of the inverse problem. Moreover, the criterion to select the best regularization parameter $\lambda$ is given by looking at the edge of L curve in the $\chi^{(2)}$ fit [5].

## 3. Discussion of simulation results

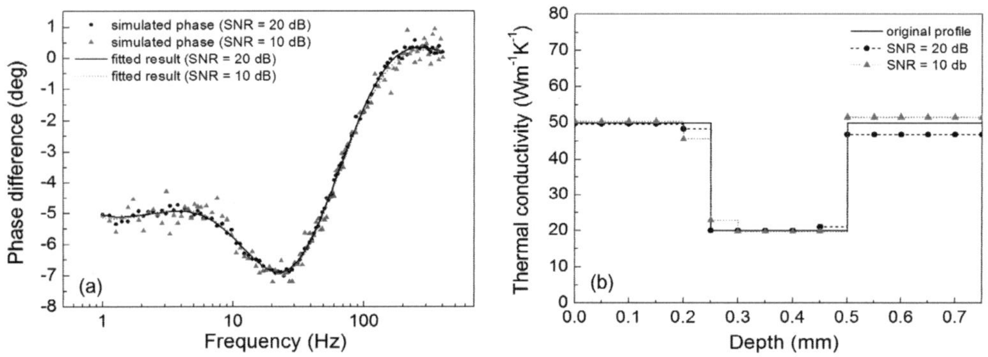


**Figure 1.** Reconstructed results for thermal conductivity profile with the same layer thickness: (a) simulated phase signals and fitted results; (b) reconstructed profiles for different noisy data.

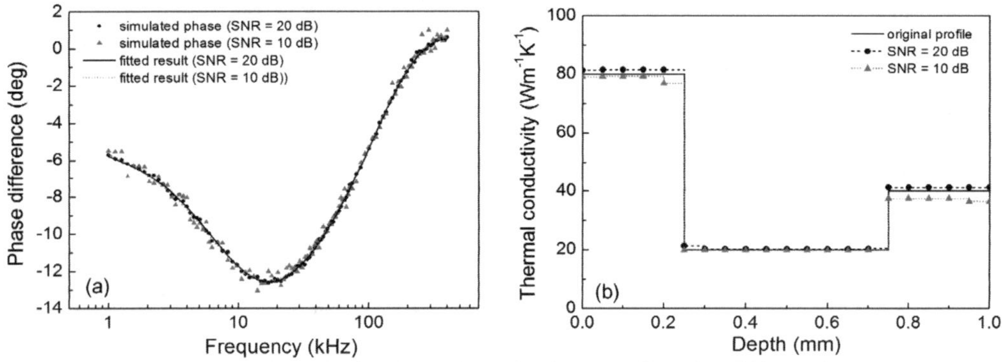


**Figure 2.** Reconstructed results for thermal conductivity profile with different layer thickness: (a) simulated phase signals and fitted results; (b) reconstructed profiles for different noisy data

To assess the inversion procedure we have initially used simulated data to test the reliability of the results from structures with exactly known features. For simplicity, we considered the slab has a depth independent specific heat capacity 480 J $Kg^{-1}$ $K^{-1}$ and density 7700 Kg $m^{-3}$. In the inversion, the bandwidth of the modulation frequency used for profile recognition is of major importance. Here, the method introduced in Ref [7] is used to select a reasonable frequency range. Figure 1 shows a typical result of the inversion method for reconstructing a three-layer profile with the same layer thickness, where $N$ = 15 virtual layers and $M$ = 108 frequencies in the range 1 Hz – 400 Hz is used. Figure 1(a) shows the simulated phase signals of SNR = 20 dB (circle) and 10 dB (triangle) and corresponding fitted results using PSO optimization (solid line: for phase signal with SNR = 20 dB; dotted line: for phase signal with SNR = 10 dB), where good fitted results still can be observed when signal noise is relatively larger. Figure 1(b) shows the reconstructed profiles (dash line of circles: for phase signal with SNR =20 dB; dotted line of triangles: for phase signal with SNR = 10 dB) in comparison with the original profiles (solid line). Figure 1(b) indicates clearly that the reconstruction results are quite satisfactory, where the discontinuous position of profile can be found clearly and the reconstructed profiles are in agreement with the original ones. Moreover, it can be found that the noise increase only decreases slightly the precision the reconstructed profile, so the algorithm is very robust to noise. Figure 2 shows the reconstructed result for a three-layer thermal conductivity profile with different layer thickness ($N$ = 20). It can be found that the results of reconstructed profiles for different noisy data are still quite good.

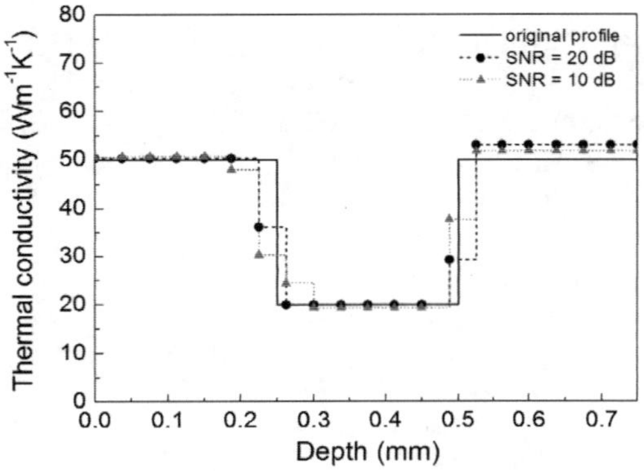


**Figure 3.** Reconstructed results obtained in a case where the discontinuity lines do not match the reconstruction grid.

To show how the choice of virtual layer number is not a critical point, a case when the simulation and the reconstruction grids are different is shown in figure 3 ($N$ = 20), where the true profile is the same to that in figure 1. It can be found that, although the discontinuities in the original profile do not fit the reconstruction lattice cells, the positions of discontinuity still can be located very well within the resolution allowed by the reconstruction grid.

## 4. Conclusion

In this paper, based on particle swarm optimization and total variation regularization, we have presented a reliable inversion method for photothermal depth profiling of multilayered structures. The reconstructions of the thermal conductivity profiles have been tested on different simulated phase data. The simulation results show the method has good resolution and robustness against noise. Moreover, the ability of the inversion method to detect and locate discontinuities is rather satisfactory. It should be mentioned that reconstructing depth profiles of thermal conductivity from photothermal amplitude data is also feasible by using the inversion method.

## References

[1] Opsal J and Rosencwaig A 1982 *J. Appl. Phys.* **53** 4240
[2] Glorieux C, Voti R L, Thoen J, Bertolotti M and Sibilia C 1999 *Inverse Probl.* **15** 1149
[3] Voti R L, Sibilia C and Bertolotti M 2003 *Int. J. Thermophys.* **26** 1833
[4] Chen Z J and Zhang S Y 2010 *Chin. Phys. Lett.* **27** 026502
[5] Aster R, Borchers B and Thurber C 2005 *Parameter Estimation and Inverse Problem* (San Diego: Academic)
[6] Kennedy J and Eberhart R C 1995 *Proc. IEEE Conf. on Neural Networks IV* (Piscataway, NJ) pp 1942
[7] Glorieux C and Thoen J 1996 *J. Appl. Phys.* **80** 6510

Journal of Physics: Conference Series **278** (2011) 012034 doi:10.1088/1742-6596/278/1/012034

# Remote photoacoustic imaging for material inspection

**T Berer[1,2], A Hochreiner[1], B Reitinger[1,2], H Grün[2] and P Burgholzer[1,2]**

[1] Christian Doppler Laboratory of Photoacoustic Imaging and Laser Ultrasonics, Hafenstrasse 47-51, 4020 Linz, Austria

[2] Research Center for Non Destructive Testing GmbH (RECENDT), Hafenstrasse 47-51, 4020 Linz, Austria

E-mail: thomas.berer@recendt.at

**Abstract.** We report on (to our knowledge) the first remote contactless photoacoustic imaging with short excitation pulses on semitransparent solid polymer samples for material inspection. In this work solid semitransparent samples are excited with pulses from a short pulse laser. The local absorption of the electromagnetic radiation leads to generation of broadband ultrasonic waves. Ultrasonic waves arriving on the sample surface are detected with a confocal Fabry-Pérot interferometer. After data acquisition the absorbed energy density is reconstructed by utilizing an F-SAFT algorithm. The work shows the potential of photoacoustic imaging on material inspection of semitransparent solid materials.

## 1. Introduction

In photoacoustic imaging (PAI) - also termed optoacoustic or thermoacoustic imaging - a volume of a semitransparent sample is illuminated with a short pulse of electromagnetic radiation, e.g. a short laser pulse. Depending on the spatially varying optical properties of the sample, the radiation is absorbed and scattered differently. The local absorption in the sample leads to local heating, in sequence to thermal expansion, and finally to the emission of broadband ultrasonic signals. The goal of PAI is to reconstruct the absorbed energy density by measuring the ultrasound signals outside the sample [1]. Thus photoacoustic signals contain information of light absorption at ultrasonic resolution. The used pulses are so short that thermal conductivity can be neglected during pulse time ("thermal confinement"). The imaging modality is therefore different as described in early publications about photoacoustics, e.g. references [2; 3], where photoacoustically generated thermal waves, generated at relatively low frequencies, interact with features of a sample. Although there is frequently confusion about that issue, the term photoacoustic, instead of the alternatively used terms optoacoustic or thermoacoustic, has established in literature.

Various publications about imaging with laser ultrasonic means exist, and also publications of generation of ultrasound waves by laser ultrasonic techniques in (semi)transparent media, e.g. ref [4]. However, up to now photoacoustic imaging with short laser pulses, where the optical absorption at the excitation time is reconstructed in three dimensions, was only applied to biological or medical samples. For these applications the sample is usually immersed in a tank filled with water (or some other fluid). The fluid acts as coupling media between sample and

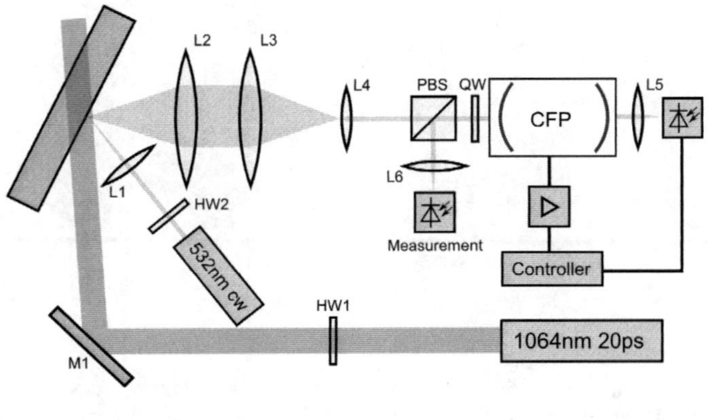


**Figure 1.** Schematic of the setup with the component labels given as follows: M1 mirror; L1 to L6 lenses; HW1 and HW2 half wave plates; PBS polarizing beam splitter; QW quarter wave plate; CFP confocal Fabry-Pérot cavity

detector. Detection of the ultrasonic signals is usually done utilizing piezoelectric transducers, but also detection schemes using interferometric means exist[5–9]. For material inspection, e.g. for in-line process control, a physical contact between sample and detector is a serious limitation. To take advantage of the remote generation of ultrasonic waves also the detection has to be accomplished without physical contact to the sample. In this work we excite ultrasonic waves in solid samples by short laser pulses. For detection we utilize a confocal Fabry-Pérot interferometer (CFPI) which measures the ultrasonic signals directly on the surface of the sample.

## 2. Measurement Setup

A schematic of the measurement setup is depicted in figure 1. A Q-switched Nd:YAG laser operating at a wavelength of 1064 nm generated pulses with a pulse length of 20 ps and a beam diameter of 9 mm. The polarization of the pulses was adjusted to p-polarization with a half wave plate HW1. Subsequently, the beam was deflected by a mirror onto the sample. The angle between the surface normal of the sample and the incident beam was set to coincide with the Brewsters's angle. Thus most of the incident radiation was transmitted into the sample. Absorption of laser pulses inside the sample lead to broadband emission of acoustic waves. These were detected and demodulated at the sample surface utilizing a confocal Fabry-Pérot interferometer. The interferometer employed a 532 nm continuous wave laser with a power of 1 W which was damped with an optical density filter to about 150 mW. After passing through a half wave plate HW2 the (horizontal polarized) beam was focused onto the sample surface with a lens L1. Light reflected from the sample surface was collected and collimated by a 2" lens L2 and subsequently reduced by a Keplarian telescope formed by the lenses L3 and L4. The beam then passed through a polarizing beam splitter PBS, got circularly polarized by quarter wave plate QW and finally entered the confocal Fabry-Pérot cavity. The transmitted beam was focused onto a photo diode, the signal of which was used to stabilize the cavity. Stabilization was done using a digital controller (based on an ARM processor) and an amplifier. Back reflected light from the CFPI experienced another $\lambda/4$ rotation, and was focused onto a fast photodiode after reflection from the polarizing beam splitter. The measured signal of the photodiode corresponds to the surface velocity of the sample. To allow automatic measurements the sample was mounted onto a motorized $x$,$y$-translational stage.

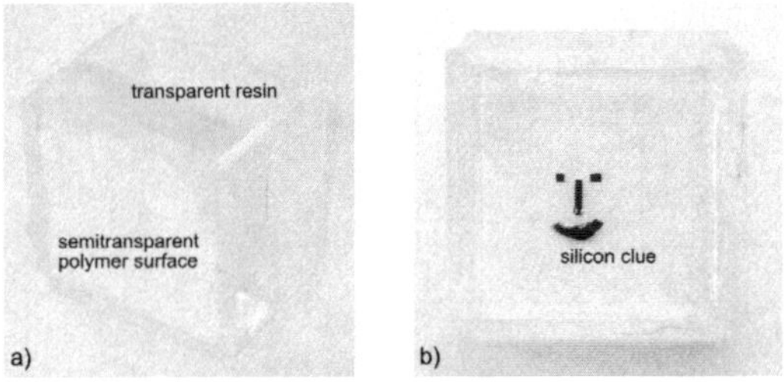


**Figure 2.** Photograph of the tested specimen: a) side view of the sample and b) view from the backside of the sample. The sample consists of a base plate of a semitransparent polymer with a thickness of 3 mm. Onto the plate pieces of a black silicon glue are attached. The back of the object is filled with a transparent resin.

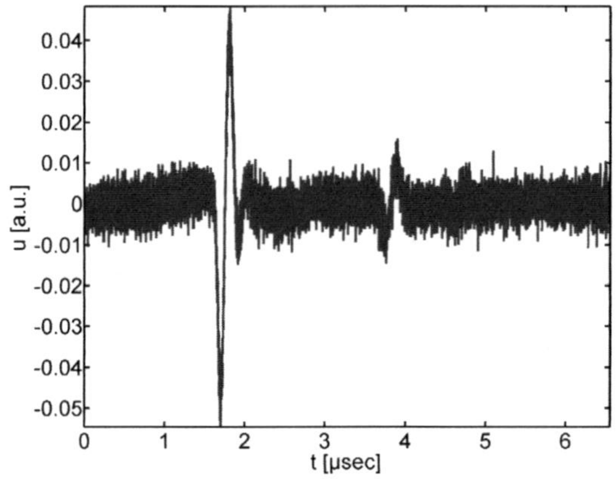


**Figure 3.** Trace measured above a black absorbing silicon clue. The first pulse originates from the glue.

## 3. Measurements

A sample was made from a $50 \times 50 \times 3\,\mathrm{mm}^3$ white semitransparent polymer board (figure 2a). The polymer is commonly used in automotive industries. The transmission of the 3 mm thick polymer sheet in the visible to the near infrared region was measured to be below 5%. On the back of the polymer plane objects made of a black silicon glue, also commonly used in automotive industries, were placed (see figure 2b). To allow ultrasonic waves to propagate properly the back of the plane was filled with a transparent resin.

Ultrasonic waves within the sample were excited with pulses from the generation laser with a pulse energy of 50 mJ at a repetition rate of 10 Hz. The acquired ultrasonic signals were averaged 24 times. A typical trace measured directly above a piece of black glue is depicted in figure 3. The first pulse at a time of $1.5\,\mu$s is the first arriving longitudinal wave. A second pulse arriving at about $3.6\,\mu$s is attributed to transversal waves. The signal of the longitudinal wave which is back reflected at the polymer/resin interface can be identified at $4.5\,\mu$s.

For three dimensional imaging a sample area of $16 \times 16\,\mathrm{mm}^2$ was measured by moving the sample in two directions with the translational stage. The step size was set to $400\,\mu$m for both scan directions. A maximum intensity plot of the measured signals is depicted in figure 4b. To construct the ultrasonic signals at the time of excitation an F-SAFT algorithm [10] was applied on the data. After reconstruction the absorbing structures appear at a depth of 3 mm corresponding to the thickness of the polymer board. Due to the F-SAFT algorithm artifacts from the wave propagation which lead to blurring of the image are removed. A maximum intensity plot of the reconstruction is depicted in figure 4c. For the picture no image enhancement like application of a threshold, filtering of the data or contrast adjustment was applied. As can be seen the F-SAFT reconstruction does not only remove artifacts which arise from the different arrival times of the waves but also increases the signal to noise ratio [11]. For comparison a photograph of the measured area is depicted in figure 4a. To allow a better comparison between photograph, measured data and reconstructed data figure 4a is displayed mirrored at the $y$-axis. All absorbing objects are well resolved. Also small details like the asymmetry in the objects mouth are well reproduced.

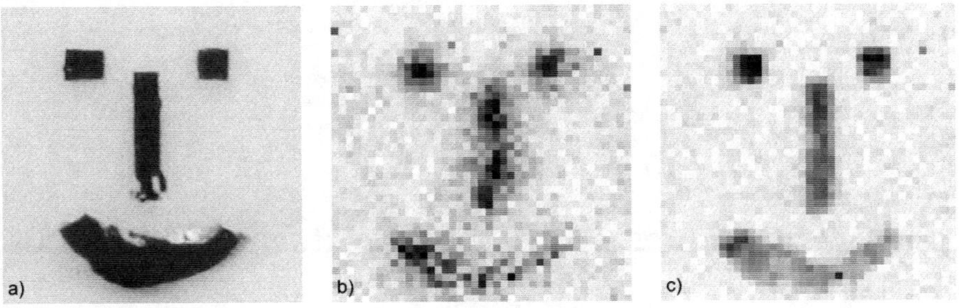


**Figure 4.** a) Photograph of the measured object. b) Maximum intensity plot of the acquired signals. The scan size was 16 mm×16 mm with 41 points on each axis, leading to a step size of 400 $\mu$m. c) Maximum intensity plot of the F-SAFT reconstruction. Note that no threshold, filtering of the data or contrast enhancement was applied.

## 4. Conclusion

We have reported on photoacoustic measurements on solid semitransparent polymer samples. Ultrasonic waves within the specimen were excited with pulses from a picosecond pulse laser. The ultrasonic waves arriving the surface of the sample were detected and demodulated utilizing a confocal Fabry-Pérot interferometer. After acquisition the ultrasonic signals at the time of excitation were reconstructed using an F-SAFT algorithm. All features of the measured object are clearly visible. This work shows that photoacoustic imaging may not only be valuable for medical or biological samples but has also great potential for inspecting semitransparent solid samples.

## Acknowledgments

This work has been supported by the Christian Doppler Research Association, by the Federal Ministry of Economy, Family and Youth, by the industrial partner INPRO Innovationsgesellschaft für fortgeschrittene Produktionssysteme in der Fahrzeugindustrie mbH, by the Austrian Science Fund (FWF), project number S10503-N20, by the European Regional Development Fund (EFRE) in the framework of the EU-program Regio 13, and the federal state Upper Austria

## References

[1] Xu N and Wang L V 2006 *Rev. Sci. Instrum.* **77** 041101

[2] Rosencwaig A and Busse G 1980 *Appl. Phys. Lett.* **36** 725–727

[3] Wong Y H, Thomas R L and Pouch J J 1979 *Appl. Phys. Lett.* **35** 368–369

[4] Matsuda O, Tachizaki T, Fukui T, Baumberg J J and Wright O B 2005 *Phys. Rev. B* **71** 115330

[5] Paltauf G, Nuster R, Haltmeier M and Burgholzer P 2007 *Appl. Optics* **46** 3352–3358

[6] Grün H, Berer T, Nuster R, Paltauf G and Burgholzer P 2010 *J. Biomed. Opt.* **15** 021306

[7] Zhang E, Laufer J and Beard P 2008 *Appl. Opt.* **47** 561–577

[8] Payne B P, Venugopalan V, Mikic B B and Nishioka N S 2003 *J. Biomed. Opt.* **8** 273–280

[9] Carp S A and Venugopalan V 2007 *J. Biomed. Opt.* *12* **12** 064001

[10] Busse L J 1992 *IEEE Trans. Ultrason. Ferroelectr. Freq. Control* **39** 174–179

[11] Blouin A, Lévesque D, Néron C, Drolet D and Monchalin J P 1998 *opt. Express* **2** 531–539

Journal of Physics: Conference Series **278** (2011) 012035 doi:10.1088/1742-6596/278/1/012035

# Design and fabrication of ultrasonic transducers with nanoscale dimensions

**R Smith[1], A Arca[1], X Chen[1], L Marques[2], M Clark[1] , J Aylott[2], M Somekh[1]**

[1]Electrical Systems and Optics Research Division, Faculty of engineering, University of Nottingham, UK.
[2]School of Pharmacy, University of Nottingham, UK

Richard.j.smith@nottingham.ac.uk

The development of nanometre sized ultrasonic transducers is important in both biological and industrial applications. The small size can be important in its own right or necessary in order to generate acoustic waves with nanometric wavelengths. Potential applications of nanotransducers range from embedded sensors through to sub optical wavelength acoustic imaging. In this paper, we show the generation and detection of ultra high frequency acoustic waves using nanometre scaled optical ultrasonic transducers. The optical and mechanical properties of these devices have been modelled using finite element modelling (FEM) and analytical techniques. The models allow the fine tuning of the design parameters to enhance both the acoustic and optical performance of the transducers. The devices were fabricated by evaporating the required metal and transparent layers onto a substrate, and then surface patterning of the device was created by laser machining or photolithography, thus allowing close comparison between model and experiment. We discuss the transducer design process and the effect of the coating parameters and how these affect the operating frequency and efficiency of the devices. We discuss the possibility of using molecular self assembly to produce even smaller devices.

## 1. Introduction

Nanometre sized ultrasonic transducers are important in both biological and industrial applications. The small size is necessary in order to generate nanometric wavelengths or access small structures such as cells. There has been much interest in the vibration of micro[1] and nano[2] objects excited by fs laser pulses and the presented transducers build on this work. The transducers we introduce in this paper act as simultaneous optical and ultrasonic resonators designed so that they couple from the optical to the ultrasonic and vice versa. We have realised these in planar structures using Fabry-Pérot optical resonators and are developing spherical transducers using optically and ultrasonically resonant nanospheres (see figure 1a). There are a number of different approaches to producing these small scale structures and in this paper we shall focus on one method for building the planar devices. At this time the devices are nanoscale in one dimension only and we will discuss methods to reduce the size of the devices to be nanoscale in all dimensions. The devices have been modeled so that the parameters governing their behaviour could be tuned to produce a device that is effective from a mechanical and optical point of view. We have produced devices to be measured in a pump probe picosecond ultrasound system[3], where a pump pulse is absorbed by the sample and generates an acoustic wave packet. The probe beam monitors the sample at different time delays to measure the interaction of the propagating acoustic wave with the sample.

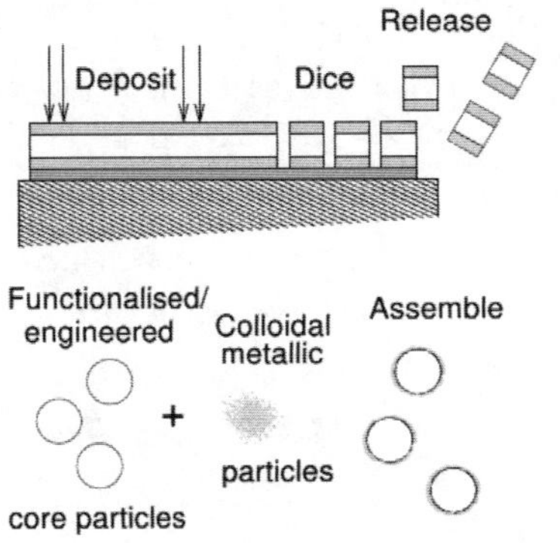


**Figure 1**. Two approaches to producing nano transducers

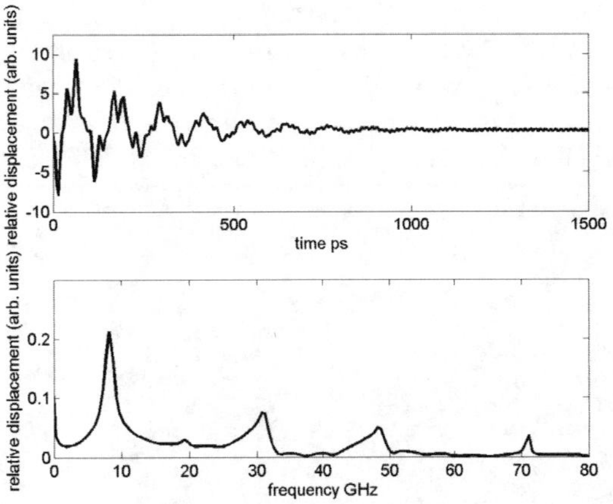


**Figure 2.** FE simulated ultrasound signal generated on nanostructure device

These devices rely on optical resonances and therefore rely on the presence of partially reflecting mirrors sandwiching an optically transparent filling. The incident light reflects and scatters off the partially reflecting boundary layers and interfere. The sensitivity of the device depends on the energy ratio of the reflected beams and the device size. If the size changes (for example due to the presence of an acoustic field) then the phase of the reflected components changes causing a change in the measured intensity at the detector. We have approached the creation of these transducers from a 'top down' direction, where 'top down' means be building the structures from the top, a layer at a time. The produced resonating layer structure that can then be patterned via photolithography or diced up using laser machining or focused ion beam techniques. Once diced the transducers can then be liberated from the substrate into solution.

## 2. Mechanical operation

The transducers generate ultrasound thermoelastically from the light absorbed primarily in the metal layers. Because the absorption in these layers is not uniform we use an EM FE model to predict this. The model calculates where in the structure the laser pulse is absorbed, knowing where and over what time scale the energy is absorbed the model can work out the change in temperature throughout the structure and from that the mechanical motion that arises. For the gold:ITO:gold structure we find that most of the light is absorbed in the top gold layer, which is as expected as gold is very absorbing at the pump laser wavelength of 400nm.

The result from the mechanical model for a structure of gold:ITO:gold:glass substrate (40nm:160nm:40nm:inf) is shown in figure 2 below, the figure shows the difference in displacement between the two gold layers. We observe a decaying oscillating signal composed of 3 or 4 main frequency components. The frequencies that are present are related to the round trip times of the acoustic waves through the structure.

## 3. Optical operation

The devices operate in a manner similar to that of a Fabry-Pérot interferometer. The layer structure is designed such that the top metal layer is a partial reflector allowing the probe beam to penetrate through the structure and reflect off the bottom layer of the structure. The two reflected beams then interfere with a strength that is dependent on the mirror separation. The acoustic field is assumed to change the separation of these partial mirrors, i.e. the optical path difference between the beams, hence creating a difference in the strength of interference. The individual thicknesses of the films are designed to give the optimum change in the reflectivity for a given change in mirror separation/radius, while maintaining desirable acoustic properties. When the transducers are bigger than the probe beam spotsize, they can be modelled analytically under the infinite width assumption using Fresnel coefficients. The result of this analytical optical model is shown in figure 3. The plotted quantity is the sensitivity of the device ($dI/dt$) i.e. the change in the reflectivity with respect to the mirror separation. From figure 3 we can see that for a gold:ITO:gold sandwich the best sensitivity is achieved with a 40nm gold layer and 160nm ITO layer.

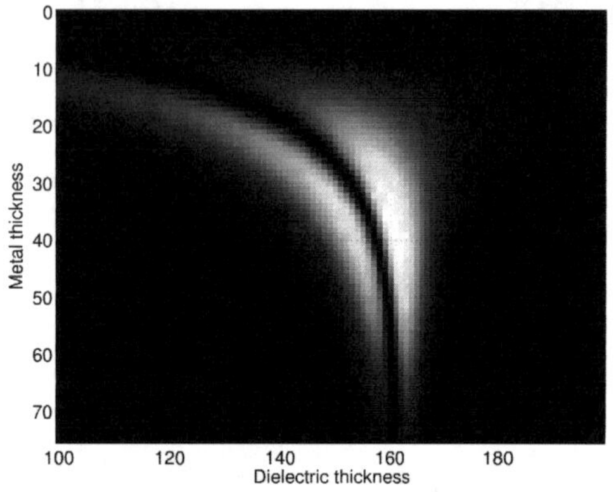


**Figure 3.** Sensitivity from analytical model

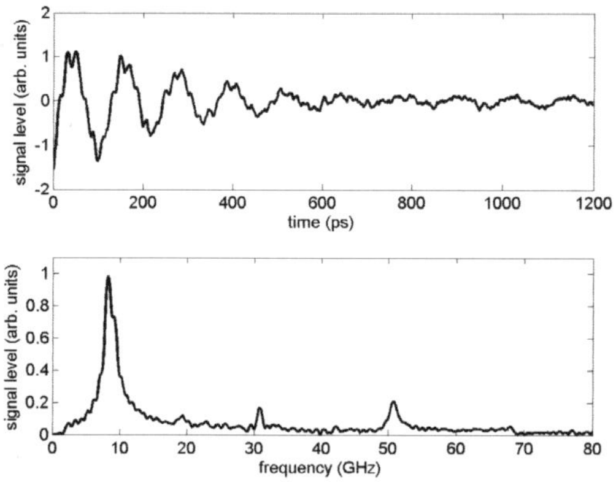


**Figure 4.** Experimental result from 10 micron transducer

## 4. Manufacturing the devices

We have fabricated the devices using standard photolithography techniques to produce patterned substrates of the required films, which can be measured in situ or released from the substrate into solution. This approach allows very fine control over the devices made. It allows easy monitoring of the devices and allows *in-situ* measurements to be made before release to fine tune the design and fabrication process. However, the size and total number of devices made is limited, which in our case leads to a minimum device size of ~ 5 microns and a few million devices per run. Using a better photolithography process or employing a different dicing technique could produce transducers with lateral nanoscale dimensions.

## 5. The first prototype transducers

We have fabricated and tested a set of planar transducers. The experiment is based around an ASOPS [3] laser from Menlo Systems. This uses two femtosecond lasers that are jointly controlled with a set of high speed electronics to control the repetition rates of the two lasers. The electronics allow one laser to be locked at 100 MHz repetition rate and the other laser to be locked at 100 MHz + 10 KHz. This means that the lasers provide the equivalent of a 10 ns delay line sweep every 100 microseconds. The pump and probe beams are focused to the same location on the top of the transducer and probe response is recorded as the delay is swept. We measured a 10 micron transducer on top of a glass substrate that had a polystyrene buffer layer to provide acoustic isolation. The recorded signal is shown in figure 4 after the coincidence peak and background have been removed.

The signal is very similar to the modeled result in figure 2. The differences are likely due to slight variations in actual layer thicknesses and the mechanical properties. The oscillations in the experiment are longer lived than for the modeled result. This is likely due to the model not including the buffer layer and so acoustic energy is lost into the substrate more quickly than in the experimental case.

Figure 5 shows an optical image of the 10 micron patches and a lower magnification image of a mix of 5, 10 and 20 micron patches reattached to a glass slide after being lifted off into solution. This shows that the transducers survive the lift off process and can be reattached successfully.

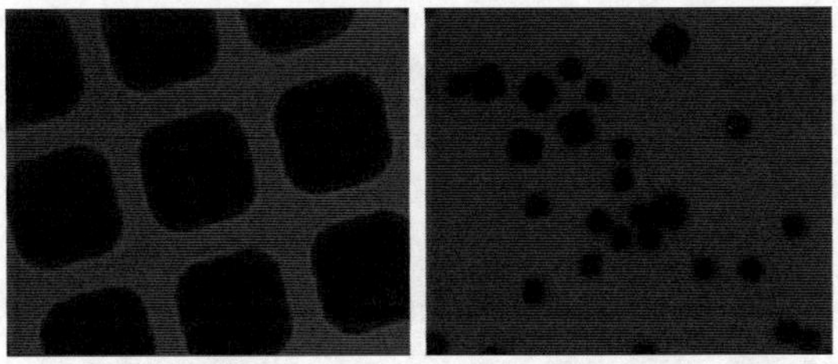

10 micron patches    Assortment of patches re-attached

**Figure 5.** Optical image of transducers in situ and released and reattached on a glass substrate

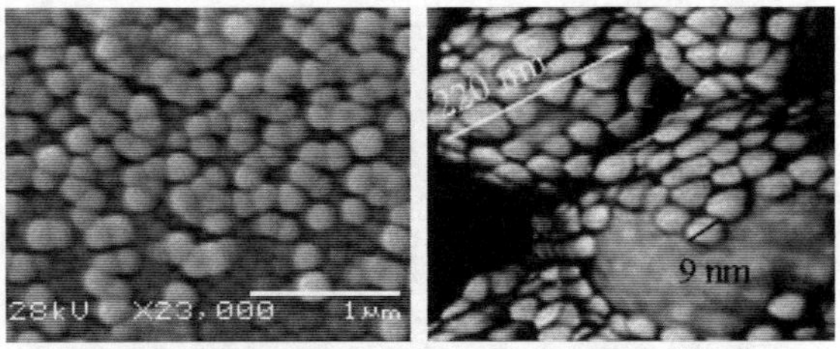


gold coated sillica particles    zoom showing gold coating

**Figure 6.** Gold coated silica nanoparticle transducers

## 6. Future Directions

To reduce the transducer size still further we have started to tackle this problem from another direction – that of self assembled nanoparticles. We have made a series of particles with thin gold coating shells around a transparent core. These transducers are essentially a spherical version of the transducers we have already tested. These have a core of 180nm and a 10 nm gold shell (Figure 6). Testing of these devices is ongoing. In the future we intend to encapsulate and functionalize the particles increasing their utility as transducers.

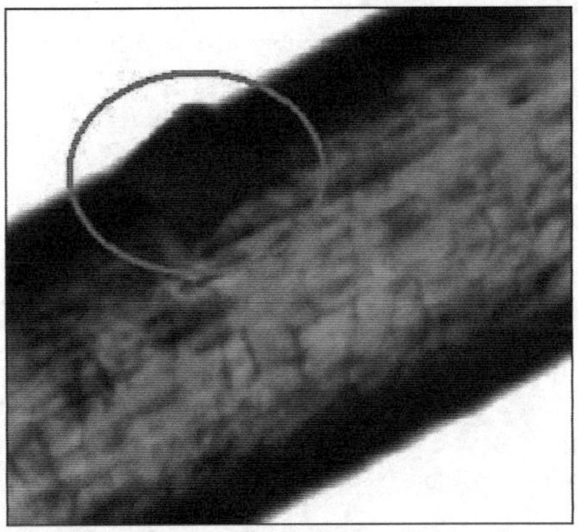

**Figure 7.** Transducer (highlighted) on a human hair

## 7. Conclusion

We have modelled, fabricated and tested an ultrasonic / optical transducer approximately 10-5 microns in size. The transducers can be liberated from the base substrate and attached for non contact measurements (see figure 7 where we attached a transducer to a human hair). These transducers produce and detect ultrasound at frequencies in the 5-30GHz range and they were remotely excited and probed using fs lasers. We have the ability to manufacture even smaller devices using molecular self assembly and have presented out first prototypes of spherical transducers of ~200nm in size.

## References

[1] Dehoux T, Kelf T, Tomoda M, Matsuda O, Wright OB, Ueno K, Nishijima Y, Juodkazis S, Misawa H, Tournat V and Gusev V. 2009 *Opt. Lett.* **34** 3740-42
[2] Guillet Y, Rossignol C, Audoin B, Calbris G, Ravaine S 2008 *Appl. Phys. Lett.* **95** 061909
[3] V. Stoica, Y. Sheu, D. Reis, and R. Clarke, 2008 *Opt. Express* **16** 2322-35

Journal of Physics: Conference Series **278** (2011) 012036 doi:10.1088/1742-6596/278/1/012036

# Low power laser generated ultrasound: Signal processing for time domain data acquisition

**A Cleary[1], I Veres[2], G Thursby[1], C McKee[1], I Armstrong[1], S G Pierce[2] and B Culshaw[1]**

[1]Centre for Microsystems and Photonics, Department of Electronic and Electrical Engineering, University of Strathclyde, Glasgow, G1 1XW, UK
[2]Centre for Ultrasonic Engineering, Department of Electronic and Electrical Engineering, University of Strathclyde, Glasgow, G1 1XW, UK

Email: alison.cleary@eee.strath.ac.uk

**Abstract**. The use of low power modulated laser diode systems has previously been established as a suitable method for non-destructive laser generation of ultrasound. Using a quasi-continuous optical excitation amplified by an erbium-doped fibre amplifier (EDFA) allows flexible generation of ultrasonic waves, offering control of further parameters such as the frequency content or signal shape. In addition, pseudo-random binary sequences (PRBS) can be used to improve the detected impulse response. Here we compare two sequences, the m-sequence and the Golay code, and discuss the advantages and practical limits of their application with laser diode based optical excitation of ultrasound.

## 1. Introduction

The use of low power continuous wave laser diodes for generating ultrasound is an established technique [1,2]. In contrast to the high peak power, short pulse lasers more commonly used in laser generated ultrasound [3,4], laser diodes offer the option of distributing the excitation energy over a longer pulse or a sinusoid to help avoid ablation damage to the sample. The excited ultrasonic waves have very low amplitude and for the case of sinusoidal excitation, lock-in detection is required; the excitation frequency is stepped through the range of interest. For pulsed or toneburst excitation however, lock-in detection is unsuitable and the data is typically averaged and captured on an oscilloscope. In this case, signal processing or coded excitation for noise reduction becomes attractive.

One such method of noise reduction is to use PRBS coded pulses as excitation, as demonstrated by Pierce [5,6] who used 850 nm optical excitation and interferometric detection. Madaras and Anastasi have also demonstrated a similar method, using a contact transducer for detection [7]. Here, we show how coded excitation can be used in conjunction with a laser diode at 1550 nm, where an EDFA can be used for optical amplification.

## 2. PRBS coding

The two types of sequence described here are the m-sequence and the Golay code [8,9]; they use a pseudo-random train of pulses rather than a single repeated pulse. M-sequences consist of a pseudo-random train of $2^N$-1 pulses of magnitude 1 or -1, where N is an integer. The autocorrelation function is an almost perfect delta function with magnitude $2^N$-1. However, as optical pulses cannot be negative, the pulse train must be modified to be a set of 1's and zeroes [10,11]. This modification influences the zero frequency component and reduces the amplitude of the autocorrelation function by a factor of four. Golay codes are more complex, but the autocorrelation function is perfect. They are a set of two complementary codes with magnitudes of 1 or -1. The two codes are then decomposed into two further series that consist of 1's and zeroes, which are suitable for optical excitation. Each code is sent separately to the sample excitation, and then the four measured responses are combined together through cross-correlation [9,13]. Its autocorrelation function has magnitude $2x(2^N)$.

By cross-correlating the ultrasonic signal with the original pulse stream that was sent to the sample, the unique properties of the autocorrelation functions mean that the final recovered signal is the impulse response with increased amplitude. Therefore, the PRBS coding acts as a type of smart averaging where it is not necessary to wait for each individual pulse response to decay before inputting the next, and so the acquisition time required for averaging is considerably reduced.

## 3. Experimental setup

The optical excitation consisted of a distributed feedback (DFB) laser diode which was modulated through a bias tee with an arbitrary function generator, and amplified using an EDFA. The output power from the EDFA was 1 W, and the power incident on the sample was approximately 0.8 W. The output from the EDFA was focussed onto the sample surface using a single element lens, and the smallest $1/e^2$ spot size was measured to be 30 μm. Figure 1 shows the experimental setup.

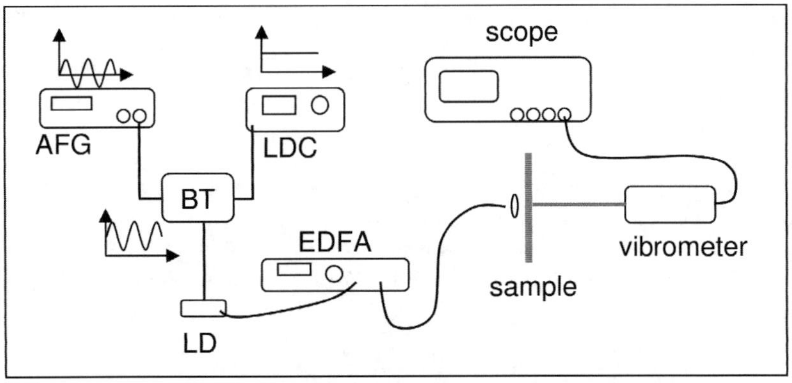


**Figure 1.** Experimental setup, where AFG is arbitrary function generator, LDC is laser diode controller, BT is bias tee, LD is laser diode.

Detection was with a Polytec OFV303 vibrometer, using a displacement decoder with 20 MHz maximum bandwidth. The output from the vibrometer was sent directly to an Agilent Infiniium oscilloscope. The sample under test was a 12.5 μm thick brass plate which was mounted vertically to enable the detection beam to be placed on the opposite side of the plate from the excitation. The source and detection spots were aligned to be within approximately 1 mm of each other, but some drift was observed during the course of the experiments.

M-sequence excitation was provided by sequences of length 2047 ($2^{11}$-1) and 8191 ($2^{13}$-1) bits, and Golay codes of length 2048 bits were used. Single pulse excitation was also used for the purpose of comparing responses from the coded sequences. All measurements were averaged on the oscilloscope 4096 times.

## 4. Results

### *4.1. M-sequences*

A set of measurements under identical excitation and detection conditions were recorded for an individual 5 μs pulse (which is equivalent to one bit of the m-sequence) at 200 kHz and m-sequences of length 2047 and 8191 bits, again at 200 kHz. Control of the frequency content of the pulses is achieved by varying the width of the individual pulses within the sequence. This means that every element (1 or 0) in the m-sequence has a width of (1/200 kHz) or 5 μs. Figure 2 shows a segment of the 2047 bit input m-sequence as recorded on the oscilloscope, and figure 3 shows the response from two different length m-sequences after cross-correlation.

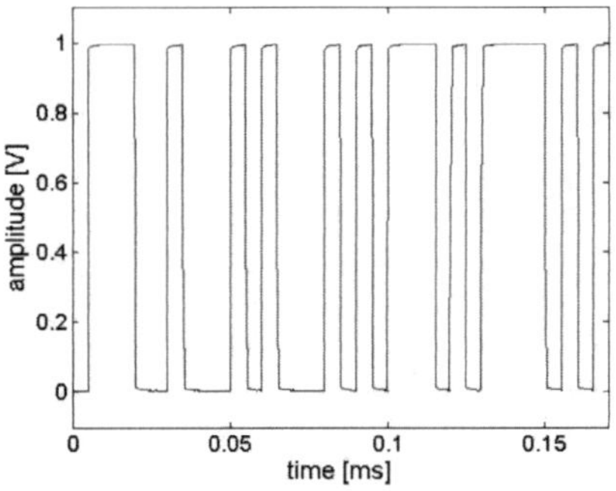


**Figure 2**. Segment of the 2047 bit m-sequence used as input to the DFB laser diode.

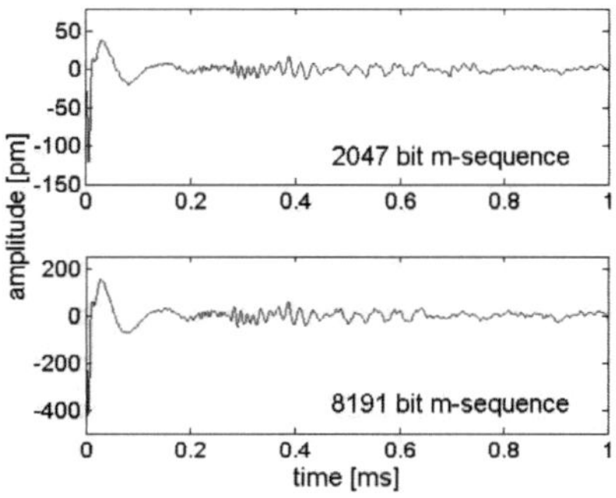


**Figure 3.** Measured waveform after cross-correlation from two different length m-sequence measurements.

The increase in maximum amplitude between the two sequences is approximately fourfold, corresponding to the increase in the number of bits in the m-sequence.

The frequency response of the 8191 bit m-sequence measurement at 200 kHz is shown in Figure 4, showing that the first null occurs at 200 kHz, and that most of the energy is concentrated in the frequencies up to 200 kHz.

### *4.2. SNR enhancement*

If the noise level can be significantly reduced by using coded excitation, there is a chance to detect the small displacement signals associated with low power laser generated ultrasound. The expected SNR enhancement with respect to a single pulse measurement for a perfect m-sequence is $\sqrt{(2^N-1)}$ and for Golay codes $\sqrt{(2^N)}$ [12,13]. This would correspond to enhancements of 45 and 90 for m-sequences of 2047 ($2^{11}$-1) and 8191 ($2^{13}$-1) bits respectively. As we have used imperfect m-sequences with pulses of 1 and zero, the expected SNR is reduced by a factor of four. An increase by a factor of 11 for the 2047 bit and 23 for the 8191 bit m-sequence with respect to the single pulse measurement is then anticipated. The maximum SNR was measured by taking the maximum signal amplitude and dividing by the average noise value in each case. In practice, we found the maximum SNR was improved by factors of 20 and 36 for the 2047 and 8191 bit sequences.

### *4.3. Golay codes*

The SNR obtained with Golay code excitation should be better than that seen with the m-sequences, as a perfect Golay code has a larger autocorrelation function. Figure 5 shows an example of a 2048 bit Golay code measurement at 200 kHz and that of a single 200 kHz pulse response. The SNR improvement between these measurements was expected to be $\sqrt{(2^N)}$ which is 45, but was measured as

102. The greater than expected improvement seen is thought to be due to errors in the noise estimation.

Golay codes are also more suitable than m-sequences for the case where the pulses are used as a trigger for a more complex waveform such as a toneburst, as they consist only of pulses of magnitude ≥0, and therefore the autocorrelation function in practice allows a more complete reconstruction of the original signal. This is beyond the scope of this paper, but will be described in a future communication.

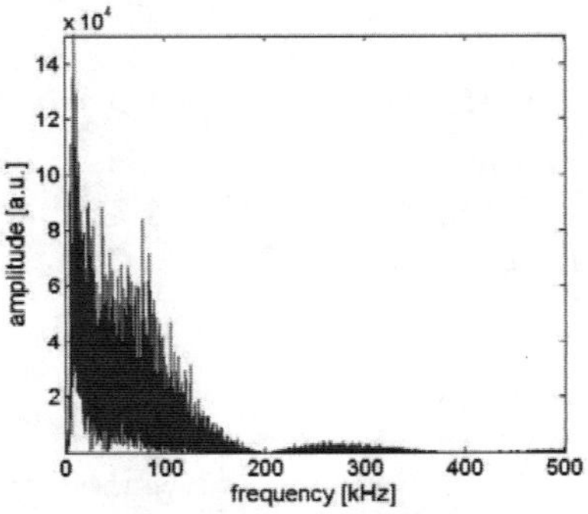


**Figure 4.** Fourier transform of an 8191 bit m-sequence response.

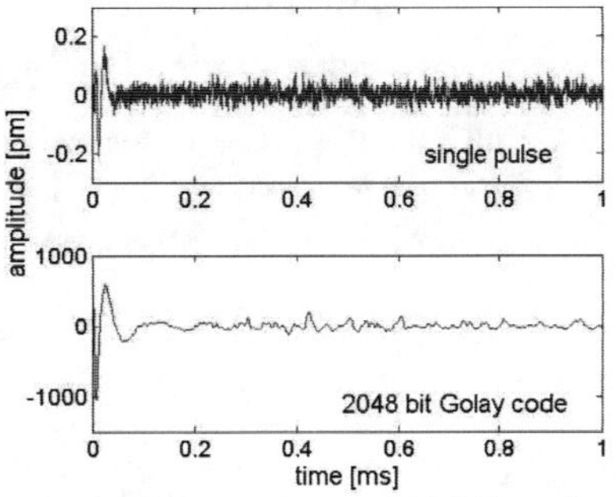


**Figure 5.** Response to 200 kHz pulse, and a 2048 bit Golay code excitation.

## 5. Conclusions

We have shown experimentally how both m-sequences and Golay codes can be used in the laser generation of ultrasound. M-sequences codes have to be modified from their original form for optical excitation, thus reducing the SNR enhancement in comparison to an equivalent length Golay code. The SNR in the case of both m-sequences and Golay codes was improved by increasing the length of the coded sequence and corresponded reasonably well to theoretical expectations.

## Acknowledgements

The authors wish to thank Richard Leach and Nigel Jennett from NPL, and the EPSRC for funding.

## References

[1] Balogun O and Murray T W 2006 *J. Appl. Phys.* **10** 034902
[2] Bramhavar S, Pouet B and Murray T W 2009 *Appl. Phys. Lett.* **94** 114102
[3] Davies S J, Edwards C, Taylor GS and Palmer S B 1993 *J. Phys. D: Appl. Phys.* **26** 329
[4] Sorazu B, Culshaw B and Pierce S G 2005 *Proc. SPIE* **5758** 177
[5] Pierce S G, Culshaw B and Q Shan 1998 *Appl. Phys. Lett.* **72 (9)** 1030
[6] Pierce S G and Culshaw B 1998 *IEE Proc. Sci Meas. Technol.* **145 (5)** 244
[7] Madaras E I and Anastasi R F 2000 *Review of Progress in Quantitative Non Destructive Evaluation 19* **509** 303
[8] Dixon R C 1994 *Spread Spectrum Systems with Commercial Applications* (New York: Wiley) p60
[9] Golay M J E 1961 *IRE Trans. Inf. Theory* **7** 82
[10] Thursby G, Culshaw B, Pierce S G, Cleary A, McKee C and Veres I 2009 *Proc. SPIE* **7293**
[11] Cleary A, Veres I, Thursby G, McKee C, Pierce S G and Culshaw B 2010 *Review of Progress in Quantitative Non Destructive Evaluation 29* **1211** 271
[12] Schroeder M R 1979 *J. Acoust. Soc. Am.* **66 (2)** 497
[13] Foster S 1986 *Proc. IEEE Int. Conf. On Acoustics, Speech and Sig. Proc.* **11** 929

Journal of Physics: Conference Series **278** (2011) 012037 doi:10.1088/1742-6596/278/1/012037

# Ultrafast opto-acoustics applied to the study of material nanostructures

**T J Grimsley[1], F Yang[1], S Che[1], G A Antonelli[2], H J Maris[1], and A V Nurmikko[1,3]**

[1] Physics Department, Brown University, Providence RI 02912 USA
[2] External Research & Development, Novellus Systems, Albany NY 12203 USA
[3] School of Engineering, Brown University, Providence RI 02912 USA

Email: Humphrey_Maris@brown.edu

**Abstract**. The propagation of ultra-short sound pulses in water has been studied using an ultrafast opto-acoustic technique. A pulse time-of-flight technique for measuring the depths of deep channels in Si-based nanostructures was demonstrated. We report in these proof-of-concept ultrasonic experiments how spatial profile information of nanostructures can be acquired, where sound pulses propagate down narrow channels in patterned nanostructures. We have been able to detect acoustic echoes for sound propagating along a channel as narrow as 35 nm with depth to width ratios exceeding 10:1.

## 1. Introduction

Ultrafast opto-acoustic methods [1] provide a way to generate short (~1 ps) acoustic pulses for fundamental studies and metrology of small structures. An ultrashort (~100 fs) pump light pulse is absorbed at the surface of a sample, and the optical reflectivity of the sample is then measured with a probe pulse whose arrival is time-delayed with respect to the pump pulse by a variable optical path. The pump pulse excites the sample and generates an elastic stress; as this stress relaxes it launches an acoustic pulse into the sample. For example, the technique of picosecond ultrasonics has been used to study the vibrational modes of nanostructures directly [2]. Picosecond ultrasonics has also been used to generate sound pulses and scatter them off of a nanostructure on the surface of a silicon wafer [3]. In this paper we demonstrate an acoustic microscope in which sound pulses are generated optically, transmitted through a coupling fluid (water), and detected after reflection from silicon-based test samples which feature periodic nanoscale deep trenches.

In a pulsed reflection acoustic microscope, sound waves are generated by an electromechanical transducer and brought to a focus at the surface of the sample which is immersed in a fluid coupling medium. Since the demonstration of a scanning acoustic microscope by Lemons and Quate [4], many researchers have explored the possibilities of acoustic microscopy. Using water as the coupling medium between the transducer and the sample, a resolution on the order of an optical wavelength has been demonstrated [5]. One of the appeals of using sound for imaging is the possibility of detecting subsurface defects in materials that are opaque optically, but relatively transparent (i.e., non-attenuating) acoustically. In addition, an acoustic microscope can image systems where the optical contrast is small but the acoustic contrast, or the difference in mechanical properties, is high [6]. The

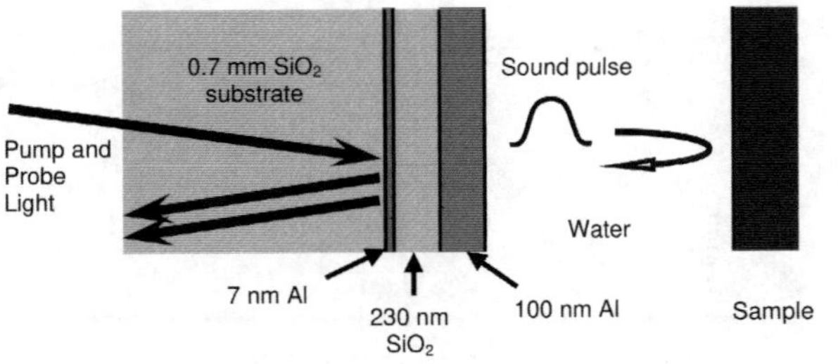


**Figure 1.** Schematic diagram of the optoacoustic transceiver. The Al/$SiO_2$/Al film stack acts as a Fabry-Perot optical cavity. The thickness of the middle $SiO_2$ film is chosen such that the optical reflectivity is sensitive to changes caused by the returning sound pulses.

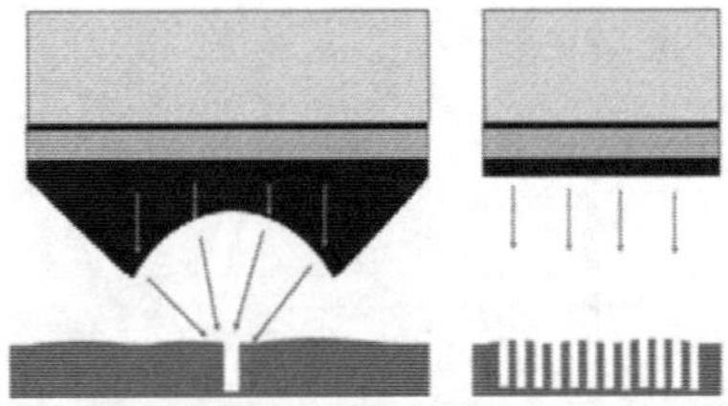

**Figure 2.** Left: Schematic of a scanning opto-acoustic microscope. Acoustic pulses generated by the transceiver structure are focused onto individual nanostructures by an acoustic lens. Right: Schematic of planar opto-acoustic microscope.

best resolution for an instrument using water as a coupling medium was reported by Hadimioglu and Quate [7] as 240 nm. In that work they used 3 ns rf pulses with a frequency of 4.4 GHz to excite a piezoelectric transducer. It is difficult to generate pulses much shorter than this electronically. In this work we examine the use of the picosecond ultrasonics technique to generate ultrashort (~1 ps) acoustic pulses and apply them to acoustic microscopy. In section 2 we address the problem of designing and testing an opto-acoustic transceiver, and in section 3 we present results from a first step implementation of this kind of acoustic microscope where the sound is not focused and a periodic array of nanostructures are probed.

## 2. Ultrafast Opto-Acoustic Transceiver Design and Testing

The picosecond ultrasonics technique provides a convenient means of generating ultrashort acoustic pulses, but detecting these pulses after they have propagated through a layer of coupling fluid, in this case water, presents some difficulties. There are several factors that reduce the amplitude of the reflected pulse. At an interface between aluminum and water, for example, the reflection coefficient for the strain amplitude is 0.84. Since the sound has to cross interfaces both into and out of the water, the returning signal will be significantly reduced. In addition, as the sound pulse propagates through the coupling fluid, the viscosity of the water makes the width of the pulse increase and the amplitude decrease. For example, when an acoustic pulse whose width is initially described by a delta function propagates through 1 micron of water, its profile will stretch out into a Gaussian shape with a FWHM of 117 nm.

In order to detect the returning acoustic pulses, we use an opto-acoustic transceiver structure that consists of films of aluminum, silicon dioxide, and aluminum with thicknesses of 7 nm, 230 nm, and 100 nm respectively as shown in figure 1.These layers were deposited onto a silicon dioxide substrate using a sputtering system with base pressure $1\times10^{-6}$ mbar. The transceiver structure acts as a Fabry-Perot optical resonator, and the optical reflectivity as a function of wavelength of one such structure is shown in figure 3. The pump and probe beams enter the cavity through the 7 nm Al film. The Al film on the far side of the cavity is thick enough to prevent any light reaching the water and sample. The effective quality factor Q of the transceiver is modest, by design, for optimizing its overall performance, since the reflectivity of the thinner Al film is approximately 0.5.

The thickness of the silicon dioxide layer was chosen so as to make the transceiver structure most sensitive to changes in the optical thickness of this layer. The calculated variation in the optical reflectivity of the transceiver structure with respect to the thickness of the silicon dioxide film for a wavelength of 800 nm is shown in figure 4. The returning sound will also perturb the optical

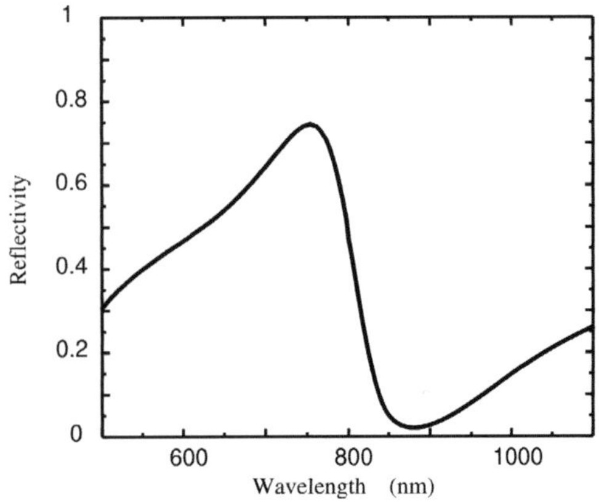


**Figure 3.** Measured reflectivity of the optical cavity as a function of wavelength.

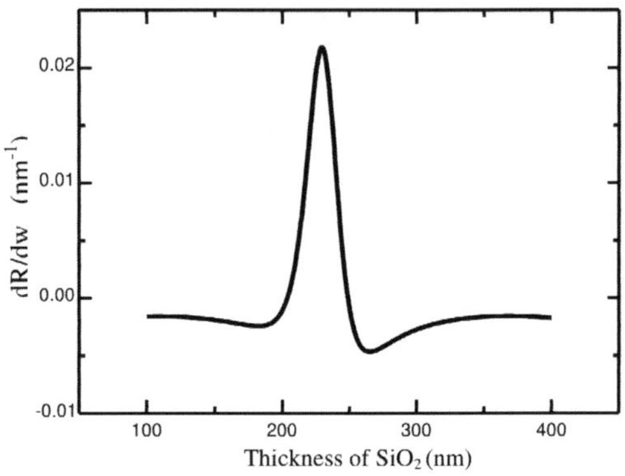


**Figure 4.** Calculated variation of the optical reflectivity of the transceiver structure with thickness change of the $SiO_2$ film, *w*, at 800 nm as a function of the film thickness.

properties of the aluminum films; the size of this additional perturbation is determined by the piezo-optic constants of Al. These coefficients have been measured by Jiles and Staines [8] but are found to vary rapidly for wavelengths around 800 nm. Their precise values are also likely to depend on the deposition conditions for the aluminum films, and consequently we will not comment on the size of this cavity-enhancement effect in detection between experiment and a model here.

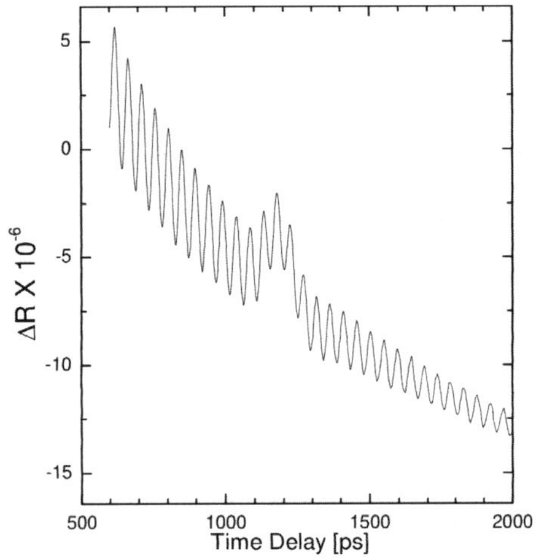


**Figure 5.** Measured change in optical reflectivity taken with a water layer of thickness 830 nm at a temperature of 23 °C. The sample is an unpatterned silicon wafer.

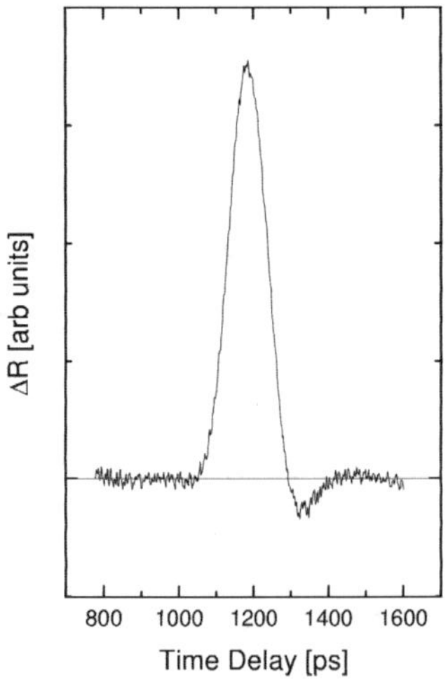


**Figure 6.** The first acoustic echo through a water layer of thickness 830 nm at 23 °C. This is obtained by subtraction of the background and the Brillouin oscillations from the data shown in fig. 5

To summarize the time sequence in the ultrafast optoacoustic experiments: the sound that is generated when light is absorbed in the Al films propagates across a thin water layer, scatters off of the surface of the sample, and then returns to be detected in the transceiver. Experimental results with a water thickness of 830 nm and using a flat surface of a silicon wafer as a sample are shown in figure 5. The slowly changing background is caused by the initial heating of the films in the transceiver structure; the temperature change perturbs the optical properties of the transceiver. The fast oscillations in the data [9], known as Brillouin oscillations, are caused by alternating constructive and destructive interference between the probe light that has reflected from the cavity and the probe light that has scattered from a sound pulse that propagates into the silicon dioxide substrate. These oscillations have no effect on the operation of the transceiver and they can be easily subtracted out numerically using a Fourier filter. In figure 6, we show the first echo from the silicon surface after subtracting out the slowly decaying background and the fast oscillations. By varying the thickness of the water and analyzing the Fourier components of the returning echoes, the velocity and attenuation of sound in water for frequencies up to 11 GHz can be measured [10]. The dip that occurs at around 1300 ps is attributed to sound which has been reflected at the transceiver-water interface with a sign change, reflected at the transceiver-substrate interface, transmitted into the coupling fluid, reflected at the sample and then returned to the transceiver.

## 3. Planar Opto-acoustic Microscopy

We are currently working on implementing an acoustic lens to focus the sound generated by the transceiver structure to develop a full scanning instrument. However, there is also considerable value in using an "unfocused" version of this technique in which planar sound pulses are scattered off of periodic nanostructures (see figure 2). While this technique cannot resolve individual nanostructures, experiments of this kind can provide information as to the average profile of the structures being examined.

As a first experimental test of these ideas, we have measured sound propagation within a number of trench structures and here report on two samples shown in figures 7 and 8. The samples consist of silicon dioxide lines on a silicon wafer with a 5 nm thick silicon nitride layer deposited on the surface of the lines and on the substrate. The depth of the trenches is measured by SEM to be 405 nm. The lines cover an area on the silicon wafer of about 500 by 100 μm, much larger than the area of the pump and probe light spots, which have a radius of approximately 10 μm. Sample A has a repeat distance of 240 nm and an average trench width half way down of 43±4 nm. Sample B has a repeat

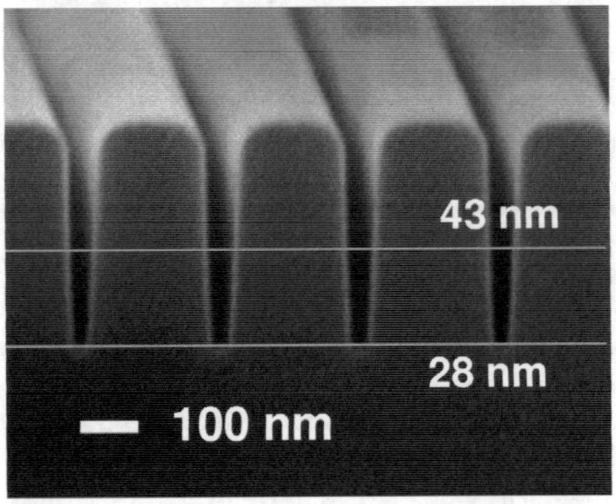


**Figure 7.** Cross section SEM image of sample A. The average width of the channels halfway down and at the bottom is 43 and 28 nm, respectively. The period is 250 nm, and the depth is 405 nm.

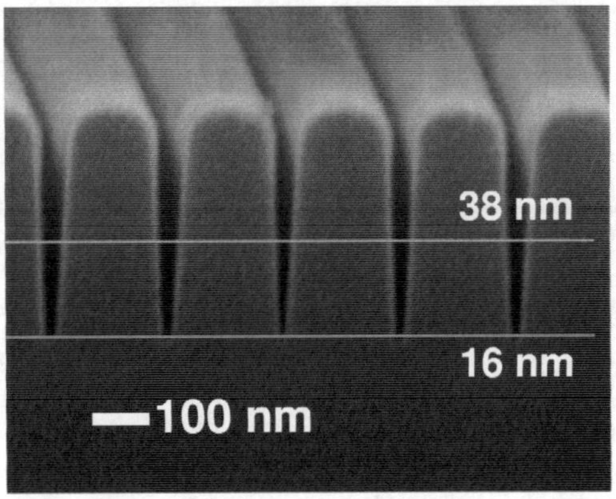


**Figure 8.** Cross section SEM image of sample B. The average width of the channels halfway down and at the bottom is 38 and 16 nm, respectively. The period is 200 nm, and the depth is 405 nm.

distance of 200 nm and an average trench width half way down of 38±6 nm. As can be seen from figures 7 and 8, there are appreciable fluctuations in the width of the trenches, particularly towards the bottom of the channels in sample B.

The experiment was performed using the opto-acoustic transceiver described in the previous section. The data set for sample A taken at a temperature of 23 °C is the black curve in figure 9. The first echo at 1147 ps is from sound reflected at the top of the structure, and the second smaller echo at 1730 ps from sound that has propagated into the trench and reflected at the bottom. The difference in the arrival times for the echoes in figure 9 is 583±6 ps and corresponds to the time required to traverse the channel twice. We have simulated the reflection of a sharp acoustic pulse from the structure shown in figure 10, using periodic boundary conditions in the lateral direction and taking into account both the shear and bulk viscosity of the water. The red curve in figure 9 shows the simulated results for the average pressure on the face of the transceiver versus time. The transceiver was taken to be 855 nm above the top surface of the sample. The difference between the arrival times for the echoes found in the simulation is 594 ps, close to the measured time of 583 ps. These values are significantly greater than the time of 541 ps required for sound to propagate through 810 nm (i.e., twice the depth of the channels) of bulk water; the velocity of the sound pulse down the narrow channel is decreased because of the viscous interaction with the wall.

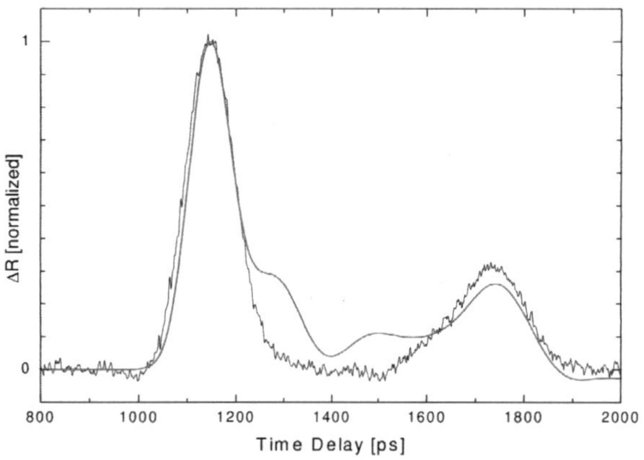


**Figure 9.** Experimental (black) and simulated (red) change in reflectivity for sample A, normalized so that the height of the first peak is unity.

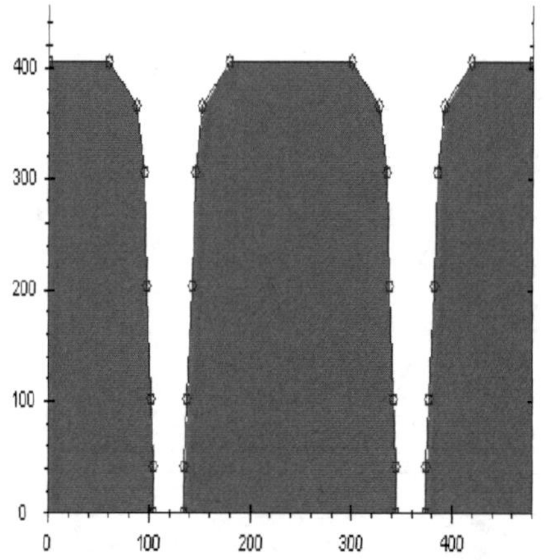


**Figure 10.** Profile used for the simulation for sample A. Dimensions taken from SEM profile measurements shown in figure 7. Units are nm.

The experimental data for sample B are shown by the black curve in figure 11. Narrowing the channels significantly reduces the magnitude of the echo from the bottom. The difference in the arrival times of the top and bottom echoes is now 563±12 ps, which again is slower than in a bulk fluid, but slightly larger than the velocity in the trench of sample A. This is surprising and is presumed to be a consequence of the different profiles of the trenches in the two samples (see figures 7 and 8). The simulation of the reflection of a sharp acoustic pulse from the structure shown in figure 12 is shown by the red curve in figure 11. The transceiver was taken to be 1010 nm above the top surface of the sample. The difference between the arrival times for the echoes in the simulation is 574 ps. To measure the geometry of structures of this type (ratio of depth to width $\geq 11$) is difficult using atomic force microscopy. Thus the ultrasonic technique described here may be a valuable new method for characterizing the geometry of this type of structure.

In the experimental and simulated results for both samples A (figure 9) and B (figure11), there are other features in the data whose arrival times fall between those of the echoes from the top and bottom. For example, there is a feature in both the experimental and simulated data slightly after 1600

Journal of Physics: Conference Series **278** (2011) 012037 doi:10.1088/1742-6596/278/1/012037

ps in figure 11. These features are presumed to be due periodic nature of the nanostructures, but at this time we cannot offer a more quantitative explanation as to their origin.

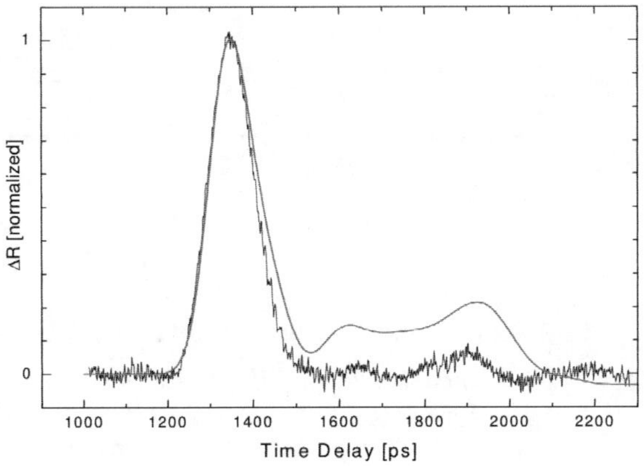


**Figure 11.** Experimental (black) and simulated (red) change in reflectivity for sample B, normalized so that the height of the first peak is unity.

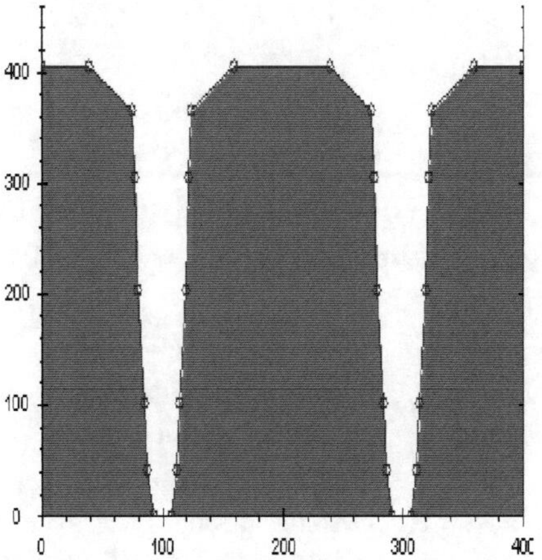


**Figure 12.** Profile used for the simulation for sample B. Dimensions taken from SEM profile measurements shown in figure 8.

## 4. Summary and Future Outlook

In this work we have presented preliminary steps towards the construction of a new type of acoustic microscope based on optical generation and detection of sound pulses. In future work, we plan to study a series of samples to explore the effect of the trench width and profile on the shape, amplitude and time of arrival of the acoustic echoes. By increasing the temperature of the water, we can improve the signal to noise ratio. We are also developing a focusing version which uses an acoustic lens to focus sound onto individual nanostructures.

This work was supported in part by the Department of Energy through grant DE-SC0001988.

## References

[1] Thomsen C, Grahn H T, Maris H J and Tauc J 1986 *Phys. Rev. B* **34** 4129-38
[2] See, for example, Antonelli G A, Malhotra S G, Harper J E M , and Maris H J 2002 *J. Appl. Phys.* **91** 3261-7; Crut A, Maioli P, Del Fatti N, and Vallée F 2009 *Phys. Chem. Chem. Phys.* **11** 5882
[3] Daly B C, Holme N C R, Buma T, Branciard C, Norris T B, Tennant D M, Taylor J A, Bower J E, and Pau S 2004 *Appl. Phys. Lett.* **84** 5180-2
[4] Lemons R A and Quate C F 1974 *Appl. Phys. Lett.* **24** 163-5
[5] Gibson V and Quate C F 1978 *Appl. Phys. Lett.* **32** 789-91
[6] Briggs A 1992 *Rep. Prog. Phys.* **55** 851-909
[7] Hadimioglu B and Quate C F 1983 *Appl. Phys. Lett.,* **43** 1006-7
[8] Jiles D and Staines M P 1983 *Sol. State Comm.* **47** 37
[9] Lin H N, Stoner R J, Maris H J and Tauc J 1991 *J. Appl. Phys.* **69** 3816-3822
[10] Yang F, Grimsley T J, Che S, Antonelli G A, Maris H J and Nurmikko A V 2010 *J. Appl. Phys.* **107** 103537

# Laser optoacoustic method of local porosity measurement of particles reinforced composites

**N B Podymova[1], A A Karabutov[2], L I Kobeleva[3] and T A Chernyshova[3]**

[1] Faculty of Physics, M.V. Lomonosov Moscow State University, Leninskie Gory, 119991, Moscow, Russia
[2] International Laser Center, M.V. Lomonosov Moscow State University, Leninskie Gory, 119991, Moscow, Russia
[3] A.A. Baikov Institute of Metallurgy and Materials Science, 49 Leninsky Prospect, 117334, Moscow, Russia

E-mail: npodymova@mail.ru

**Abstract.** In the present work we have realized experimentally the laser optoacoustic method for the measurements of local volume fraction of air pores (porosity *P*) of isotropic composite materials. It is based on measurements of phase velocities of thermooptically excited longitudinal ultrasonic waves in composite samples in the frequency range 0.5÷50 MHz. The local porosity (lateral resolution 1÷2 mm) is determined using the dependence of phase velocity of longitudinal acoustic wave on porosity for porous metals and theoretical calculation of phase velocity in a composite with the two-phase medium model. A number of aluminum alloy (silumin) matrix composite samples reinforced by SiC particles of a different mass concentration with the mean particle size of 14 μm was investigated. The samples were disks with the diameter $d$ = 40 mm and the porosity in the center and in the periphery area of each sample was determined. The increase of mass concentration of the filler SiC leads to the growth of $P$. The results coincide within relative inaccuracy of 2-3% with the gravimetrical measurements of average $\langle P \rangle$ value, the porosity in the center of each sample was slightly higher than in the periphery.

## 1. Introduction

The problem of nondestructive testing of the actual state of composite materials is of great importance because of essential decrease of their strength caused by defects and structural changes of a material that arise during exploitation [1]. In particular, fatigue changes of composite structure lead to the increase of the volume fraction of pores (porosity) that in turn leads to the decrease of material strength even by the absence of apparent defects. So the development of local porosity measurement methods is of great practical importance to provide valuable information for residual service life evaluation of composite products.

The structural fatigue alteration leads to the change of attenuation and velocity of ultrasonic waves in composite materials. So the most widely used technique of its nondestructive testing is the ultrasonic one [2-3]. Composites are acoustically inhomogeneous, so to provide the quantitative evaluation of structural features one needs to carry out measurements of attenuation coefficient and phase velocity in a wide frequency band. Besides the significant attenuation of ultrasound in composite materials requires the utilization of sources of powerful wideband acoustic signals.

Therefore it is difficult to carry out the testing of samples and items of several centimeters thickness with conventional piezoelectrical transducers because of low efficiency of wideband signals excitation and of technical problems to provide the uniform frequency response of an ultrasonic source and detector in a wide spectral band [4]. The latter feature could reduce the dynamic range by testing of materials with the strong frequency dependence of ultrasound attenuation.

To overcome these difficulties we propose to employ the laser thermooptical excitation of ultrasound – pulse optoacoustic effect [5]. The main goal of the present work is the development of laser optoacoustic method of local porosity measurements of isotropic composite materials.

## 2. Investigated composite samples and brief theoretical analysis

In the present work a number of aluminum alloy (silumin) matrix composite samples reinforced by SiC particles of a different mass concentration was investigated. The samples have been manufactured by casting of molten silumin mixed with the certain amount of the fine powder of SiC particles with the mean size of 14 μm.

The porosity $<P>$ averaged over the entire volume of the sample is expressed as:

$$<P> = (1 - \rho/\rho_0) \cdot 100\% \tag{1}$$

where the calculated density of the sample $\rho_0$ is determined with the known densities of the matrix – silumin and the filler SiC, and with its known mass concentrations; the actual density of the sample $\rho$ is measured gravimetrically.

The local porosity in different areas of a sample could be different because of nonuniform reinforcing fabrication method. To determine the local porosity $P$ we use the theoretical model of ultrasound propagation in a porous metal [6], where the phase velocity of longitudinal acoustic waves in a material $c_l$ depends on $P$ as follows (at the values of $P < 20\%$):

$$c_l^2 = c_{l_0}^2 \left(1 - P^{2/3}\right) \tag{2}$$

where $c_{l_0}$ is the calculated with the two-phase medium model [7] phase velocity in a sample:

$$c_{l_0}^2 = \frac{1}{\rho_0^2} \left[ \frac{n_{SiC}}{(\rho_{SiC} c_{l_{SiC}})^2} + \frac{n_m}{(\rho_m c_{l_m})^2} \right]^{-1} . \tag{3}$$

In the equation 3 $n_m$ and $n_{SiC}$ are the mass concentrations of silumin and SiC in the sample under study, $\rho_m = 2.735 \times 10^3$ kg/m³, $\rho_{SiC} = 3.2 \times 10^3$ kg/m³, $c_{l_m} = 6.86 \times 10^3$ m/s, $c_{l_{SiC}} = 11.82 \times 10^3$ m/s are the known densities and phase velocities of longitudinal acoustic waves for the matrix and the filler [8, 9]. So to determine the local porosity $P$ from the equation 2 we have to measure the phase velocity of longitudinal acoustic waves $c_l$ in the corresponding area of the sample under study.

## 3. Experimental technique of phase velocity measurement of longitudinal acoustic waves

We used the experimental setup (figure 1) for optoacoustic investigations in the transmission mode [7, 9]. A pulse of Q-switched Nd:YAG laser at the fundamental harmonic ($\lambda$ = 1.064 μm) is absorbed in the specially designed optoacoustic (OA) source – the plate of optical filter blue-green glass, that leads to excitation of the pulse of longitudinal acoustic waves (OA signal). The acoustic signal excited in the OA source – the reference ultrasonic pulse – passes through an investigated sample and is detected with the specially designed wideband piezoelectric detector being in the acoustic contact with the sample through distilled water. The OA source, the sample and the detector are mounted in the special cell. The small duration ($\approx$200 ns) and the large amplitude (up to several of MPa) of reference ultrasonic pulses allow to test both thin composite samples (the thickness from hundreds of microns) and samples of the thickness up to several centimeters of strong ultrasound scattering and absorptive materials. The cross-section of the reference ultrasonic beam irradiated into the sample coincides with the mean diameter of the laser beam and is approximately 1.5 mm. This dimension determines the locality of testing in the lateral plane.

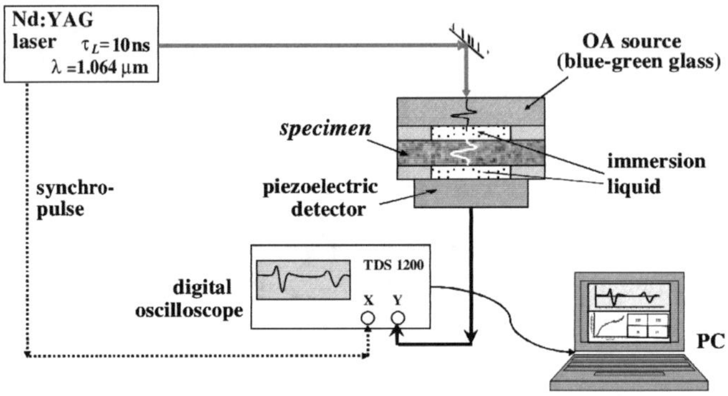


**Figure 1.** Experimental laser optoacoustic setup

Electrical signals from the piezoelectric detector come into the two-channel digital oscilloscope (analog bandwidth 100 MHz), the signal-to-noise ratio for the registration system is 50÷60 dB. The dispersion of the phase velocity of longitudinal acoustic waves is calculated in real time due to high signal-to-noise ratio and high stability of temporal profile of reference ultrasonic pulses.

The absolute value of the phase velocity of longitudinal acoustic waves is determined as follows in the absence of the significant dispersion (the relative change of velocity doesn't exceed 5÷7% in the investigated frequency band 0.5÷50 MHz):

$$c_l = \frac{2H}{\Delta T_l} \ . \tag{4}$$

where $H$ is the known thickness of the sample, $\Delta T_l$ is the difference of arrival times to the receiver of the ultrasonic pulse once passed through the sample and of the pulse after the "tripple pass" – passed through the sample and reflected at the sample - immersed liquid interfaces. So the uncontrolled influence of difference in the liquid layers thickness could be eliminated by the mounting of OA cell for detection of the reference signal without the sample and the OA cell with the sample.

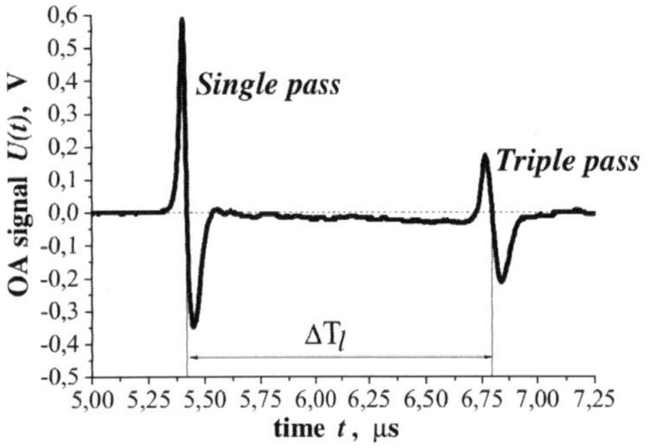


**Figure 2.** Typical temporal profile of the OA pulse of longitudinal acoustic waves in a silumin-matrix composite sample [9].

As an example the typical temporal profile of these two pulses in a silumin matrix composite sample is shown in figure 2 [9]. The time interval $\Delta T_l$ is measured between the instants of transition through the "zero" from the compression to the rarefaction phases of the ultrasonic pulses. As it follows from the theoretical model of the wideband OA signal propagation in an absorbing medium [5], just the propagation velocity of the "zero" point of the bipolar OA pulse is the most close to the acoustic wave phase velocity in the absence of the significant dispersion. The small duration of the longitudinal acoustic wave pulses provides sufficiently low relative inaccuracy of the phase velocity measurement: $\delta(c_l) \approx 0.5\%$. This value is governed basically by the relative inaccuracy of the sample thickness measurement: $\delta H \approx 0.5\%$, because of the value of $\Delta T_l$ (see figure 2) is determined accurate within 2÷3 ns, that is of the order of $10^{-3}\,\Delta T_l$ for the sample thickness $H \sim 10$ mm.

## 4. Experimental results and discussion

As the test samples a number of aluminum and copper samples have been studied. The experimental results: $c_{l_{Al}} = (6.28\pm0.03)\times10^3$ m/s, $c_{l_{Cu}} = (4.72\pm0.02)\times10^3$ m/s coincide within the inaccuracy limits with the reference data [8]. This fact confirms the reliability of the laser optoacoustic method for the measurement of the phase velocity of longitudinal acoustic waves in isotropic solids.

All composite samples were the discs of the diameter $d$ = 40 mm. We determined the porosity in the center ($P_1$) and in the periphery ($P_2$) areas of each sample using the equation 2 with the measured values of the phase velocities in these areas $c_{l1}$ and $c_{l2}$ and the calculated value of $c_{l_0}$ (the equation 3) for the given sample (see table 1). The relative inaccuracy of 0.5% of the phase velocity measurements leads to the relative inaccuracy of 2÷3% of porosity determination. The experimental results show, that the increase of the mass concentration of the filler $n_{SiC}$ leads to the growth of the sample porosity *P*. All investigated samples have quite uniform distribution of pores, the local porosity values coincide within the inaccuracy limits with the gravimetrical measurements of the average <*P*> value. The porosity in the center of each sample is slightly higher than in the periphery.

**Table 1.** Results of optoacoustic measurements

| sample # | $n_{SiC}$ | $c_{l_0}$, m/s | $c_{l1}$, m/s | $c_{l2}$, m/s | $P_1$, % | $P_2$, % | <$P$>, % |
|---|---|---|---|---|---|---|---|
| *1* | 0.00 | 6860 | 6670 | 6730 | 1.30±0.03 | 0.77±0.02 | 0.77 |
| 2 | 0.038 | 6920 | 6720 | 6740 | 1.43±0.03 | 1.18±0.02 | 1.45 |
| *3* | 0.077 | 6990 | 6550 | 6590 | 4.24±0.09 | 3.69±0.07 | 3.65 |
| *4* | 0.155 | 7140 | 6660 | 6680 | 4.69±0.09 | 4.41±0.09 | 4.93 |

The proposed laser optoacoustic method may be useful by technological development and improvement of the materials fabrication process and to reveal the areas in a material with the less strength before manufacturing of items and products.

## References

[1] *Handbook of Composites* 1988 ed. C Herakovich and Yu Tarnopol'skii (Amsterdam: Elsevier)
[2] Vary A 1980 *Research Techniques in NDT* **4** 160
[3] Fitting D W and Adler L 1981 *Ultrasonic spectral analysis for nondestructive evaluation* (New York: Plenum)
[4] Truell R, Elbaum C and Chick B 1969 *Ultrasonic methods in solid state physics* (New York: Academic)
[5] Gusev V E and Karabutov A A 1993 *Laser Optoacoustics* (New York: AIP)
[6] Polyakov V V and Golovin V A 1994 *Technical Physics Lett.* **20** 54
[7] Zharkii S M, Karabutov A A, Pelivanov I M, Podymova N B and Timoshenko V Yu 2003 *Semiconductors* **37** 468
[8] *Handbook of Physical Quantities* 1997 ed. I S Grigoriev, E Z Meilikhov and A A Radzig (Boca Raton: CRC-Press)
[9] Karabutov A A, Kobeleva L I, Podymova N B and Chernyshova T A 2009 *Zavodskaya Laboratoriya - Diagnostika Materialov* **75** 27 (in Russian)

Journal of Physics: Conference Series **278** (2011) 012039 doi:10.1088/1742-6596/278/1/012039

# Rejection of crosstalk and noise by a quasi balanced CFPI for remote ultrasound detection

**B Reitinger, T Berer, C Hornhuber and P Burgholzer**

Research Center for Non Destructive Testing GmbH (Recendt), Hafenstrasse 47-51, 4020 Linz, Austria

E-mail: bernhard.reitinger@recendt.at

**Abstract**. In this paper we show the benefits of a quasi balanced fringe hopping CFPI (confocal Fabry-Pérot interferometer) with broadband CMRR (common mode rejection ratio) for remote ultrasound detection. Ultrasonic information in general lies in the phase modulation of laser light which in this case is demodulated by using the CFPI at a certain working point on a fringe. By hopping from the positive to the negative slope on the same fringe the detected ultrasonic signals are inverted. In contrary interference signals like crosstalk from the generation, ghosts, or noise correlated to pulse laser excitation are not influenced and hence get rejected by subtracting the signals from both slopes. Hence, a minimum of two measurements is needed for common mode rejection. The fringe hopping from the positive to the negative slope is done by changing the distance of the CFPI mirrors with a precise piezoelectric-stack and a fast high resolution digital controller. As only one photo-detector with a transimpedance-amplifier is needed a high CMRR can be accomplished which is not affected by the symmetry of the fringe but only by pulse to pulse energy fluctuations of the generation laser. We show that with fringe hopping and averaging the signal to noise ratio increases much faster than with averaging without fringe hopping. This is due to the correlation of the quasi-noise with the generation cycle.

## 1. Introduction

The CFPI (confocal Fabry Pérot interferometer) is the workhorse for the contactless detection of laser ultrasound in many industrial applications. In laser ultrasound an ultrasonic wave is generated by a short laser pulse and further contactless detected with an interferometer setup. Unfortunately the sensitivity of contactless interferometer setups is much smaller compared to contacting ultrasonic detectors like piezo transducers. Therefore improvements to the signal to noise ratio are very important. In this paper we present a simple method to improve SNR (signal to noise ratio) by rejecting common mode noise like generation crosstalk and high frequency amplitude fluctuations with a fringe hopping technique. With the proposed technique no additional costs compared to a standard CFPI setup are incurred, but it needs a minimum of two measurements to achieve common noise rejection. Dubois et al [1] demonstrated a more cost intensive way to improve SNR with two CFPI but with the advantage of common mode rejection with only one measurement.

## 2. Principle of common mode rejection by differential signals and balance lines

In electronics often very small signals have to be detected, amplified or transmitted. The signal to noise ratio of these signals suffers from their small amplitude and often also from the high input impedance of the amplification circuit. Both lead to a poor SNR ratio. To enhance the signal to noise

Journal of Physics: Conference Series **278** (2011) 012039 doi:10.1088/1742-6596/278/1/012039

ratio of signals the balanced circuit principle with differential source was implemented (figure 1a). In principle the signal is separated into two differential signals (lines) which are symmetrical in respect to a defined amplitude level. At a balanced detector the signals are subtracted and the noise (figure 1b) which is common on both signals is rejected (figure 1c and 1d).

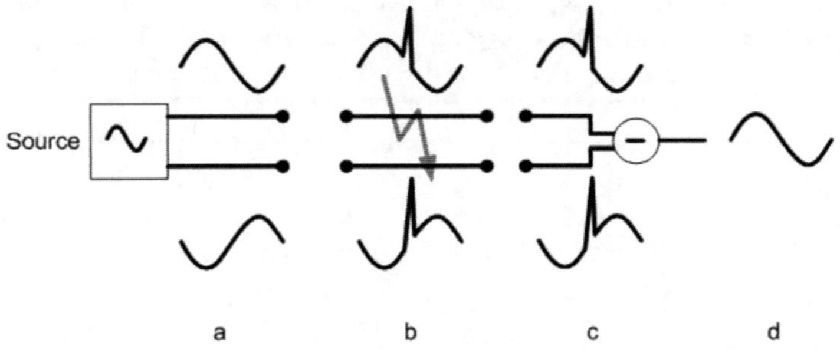


**Figure 1.** Principle of ambient noise cancellation by differential signals and balanced lines.

This principle can be found on many devices in daily life, for example in audio (differential signal from microphone capsule), balanced photo-detectors (two photo-detectors in complimentary), computer data transmission (twisted pair lines) et cetera.

## 3. The confocal Fabry Perot detector and implementation of quasi balanced principle

The proposed setup is based on the well known CFPI detector setup [2, 3, 4] for remote laser ultrasound. An ultrasonic wave is generated by a short laser pulse. The generation process can be either in thermoelastic or ablative regime which affects both the amplitudes and directions of the ultrasonic waves [5, 6]. Detection is done with a continuous wave (or long pulse [7] laser) which is back reflected from the surface and gets phase modulated by the surface vibrations caused by the ultrasonic waves. Subsequently the phase modulation is demodulated by the CFPI to an amplitude modulation which is measured by a fast optical detector. To achieve high bandwidth detection a Si-detector was utilized to measure ultrasonic signals in reflection scheme of the CFPI.

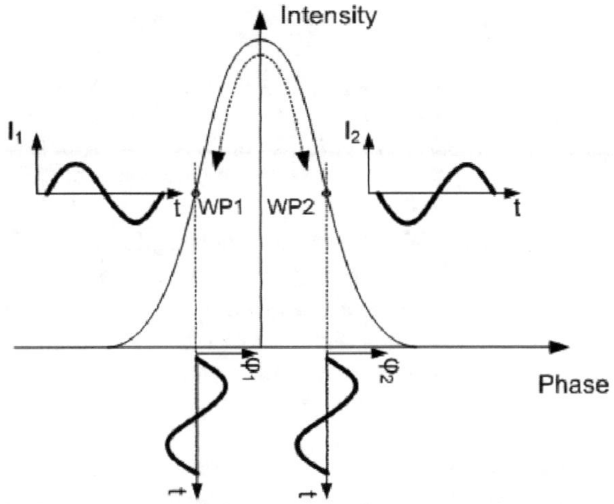


**Figure 2.** Principle of differential detection with fringe hopping technique

Demodulation is done at a certain working point (WP). The working point can be set by controlling the distance of the CFPIs mirrors. The WP which regularly is used is at half the height of the fringe and on the positive slope (figure 2 WP1).

The principle of differential and balanced detection can be applied to the CFPI detector by using quasi differential signals and signal averaging. By measuring first at the positive slope (figure 2 WP1), hopping to the negative slope, and following another measurement at the negative slope (figure 2 WP2) quasi-differential signals are acquired. By subsequently subtracting both signals common mode noise is rejected.

Stabilization of the working points and jumps between the slopes to the desired WPs were realized by a 12 bit A/D, ARM7 digital control and a precise piezoelectric stack. Before the laser ultrasound measurement is started FSR (free spectral range) and finesse of the CFPI are analyzed on. These variables are used to calculate a trajectory to hop from one WP to the other. The hopping including settling time was achieved within 3 ms.

## 4. Measurements

Measurements were carried out on a 1 mm thick steel sample. Ultrasonic waves were generated on one side of the sample by short laser pulses with a pulse duration of 20 ps and energy of 2 mJ. The generation laser was focused to a spot with a diameter of around 100 $\mu$m. Detection was done on the opposite side of the sample (measurement in transmission) with the proposed CFPI detector.

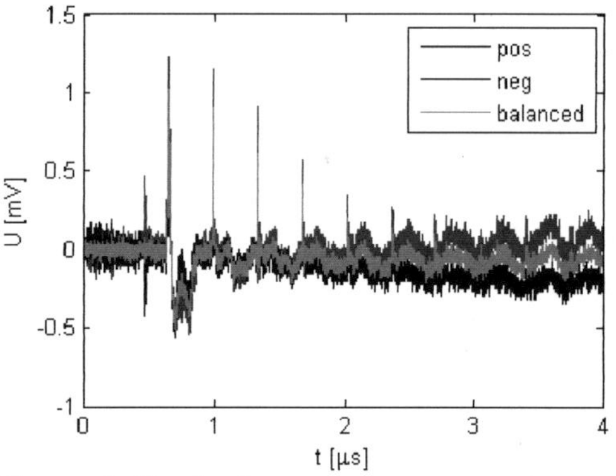


**Figure 3.** Single shot measurement on 1mm a thick steel sheet. The red trace shows the balanced detection. A direct signal is eliminated at about 0.5 $\mu$s.

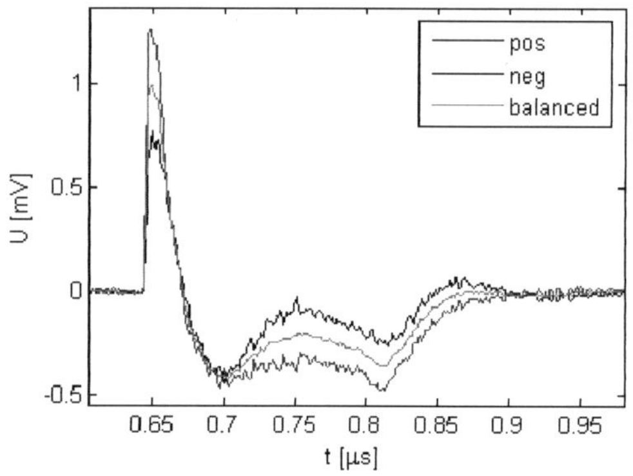


**Figure 4.** LUS measurement on 1 mm thick steel sheet. Signals were averaged 180 times. The traces show zoom-in of first echo. Red trace shows elimination of quasi-noise.

Figure 3 shows a typical measurement. Several echoes can be identified. The blue trace shows the detected signal on the positive slope, the black on the negative slope, and the red curve shows the subtraction of both signals. As can be easily seen the first direct signal (figure 3 at 0.5 $\mu$s) from the generation laser is rejected. Figure 4 shows the averaging results over 180 signals. A zoom on the first echo is depicted to show the rejection of ripples on the subtracted signal. The ripple is apparent on both negative and positive slopes and cannot be eliminated by averaging, even by a multitude of measurements.

By looking at the frequency spectra (see figure 5) of 180 averaged signals (see figure 4 for time traces) the frequency bands of the ripples are apparent. These frequency bands get almost rejected by subtraction of the differential signals in the time domain.

As this noise is not eliminated by averaging it must be time correlated to the cycle of the generation laser pulse and is further called quasi noise.

Most prominent quasi noises which apply to that symptom are:

- Generation crosstalk – electric field: The first eliminated signal (figure 3 at 0.5 $\mu$s) is attributed to stray pickup of the high electric field of the flashlamps of the ps-generation laser.
- Generation crosstalk – ablative generation: At the same time as the stray pickup of electric field takes place a small amount of plasma radiation generated by the focused generation laser reaches the detector. The polarity of this crosstalk is not affected of the chosen slope.
- Amplitude ghosts [1]: Amplitude ghosts are high frequency amplitude fluctuations which travel inside the cavity and therefore have a period time of the cavity length. Also these amplitude fluctuations are not affected by the chosen WP.

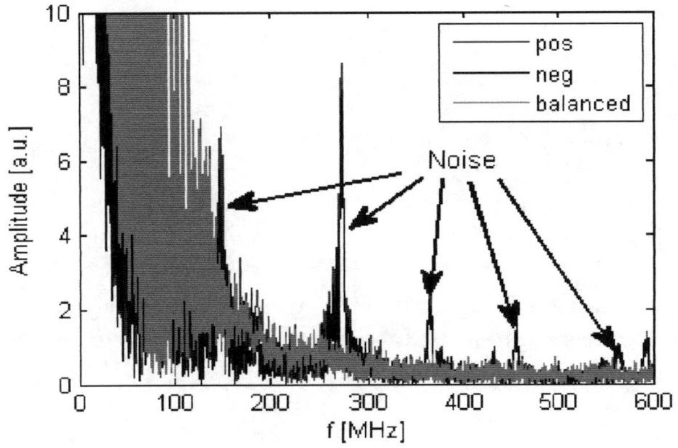


**Figure 5.** Signal and noise spectra of averaged measurements.

## 5. Conclusion

The benefits of a quasi balanced fringe hopping CFPI for ultrasound detection was shown. It is evident that only with a differential type CFPI quasi noise can be rejected. Dubois et al [1] demonstrated a more cost intensive way of implementing differential signals with CFPI: a detector with two CFPIs and working points on different sides of the fringes was used. Therefore, with a single shot measurement quasi noise signals can be rejected. The setup shown in this work needs two measurements to achieve common noise rejection. However, as measurements are usually averaged to achieve a good SNR this is no disadvantage for most applications. The proposed technique offers a method to reject quasi noise at lower material cost and low complexity. The work also shows that SNR is enhanced significantly compared to standard measurements without fringe hopping.

## 6. Acknowledgments

This work has been supported by the European Regional Development Fund (EFRE) in the framework of the EU-program Regio 13, and the federal state Upper Austria

## References

[1] Dubois M, Burr KC and Drake T E 2004 *Applied Optics* **43** 22 4399
[2] Monchalin J P and Heon R 1986 *Materials Evaluation* **44** 1231
[3] Shan Q, Bradford A S and Dewhurst R J 1998 *Meas. Sci. Technol.* **9** 24
[4] Dewhurst R J and Shan Q 1999 *Meas. Sci. Technol.* **10** 139
[5] Scruby C B and Drain L E 1990 *Laser ultrasonics: techniques and applications* Adam Hilger
[6] Wang X and Xu X 2001 *Applied Physics* A **73** 107
[7] Carrion L, Blouin A, Padioleau C, Bouchard P and Monchalin J P 2004 *Measurement science and technology* **15** 1939
[8] Nadeau A, Martin F, Blouin A, Nadeau F, Choquet M and Lord M, 2007 *AIP Conference Proceedings* **894** 225

Journal of Physics: Conference Series **278** (2011) 012040 doi:10.1088/1742-6596/278/1/012040

# Rubidium vapour based adaptive interferometer for laser ultrasound detection

**N Korneev, P Rodriguez Montero**

Instituto Nacional de Astrofísica, Óptica y Electrónica, Apt. Postal 51 y 216 CP72000, Puebla, Pue., Mexico

korneev@inaoep.mx

**Abstract**. We describe an application of dynamic holography in rubidium vapour for laser ultrasound detection. This medium combines good signal to noise ratio with high cut-off frequency of approximately 1MHz.

## 1. Introduction

The generic scheme of adaptive interferometry is presented in figure 1. The dynamic holographic medium with a response time $\tau$ produces a thick phase hologram, which can be regarded as a beamsplitter reflecting part of the beam 1 in the direction of the beam 2 and vice versa. In a steady state there is no intensity transfer between writing beams, but instantaneous phase shift in a signal beam results in a transient energy transfer between the signal and the pump. The magnitude of this transfer is proportional to the phase shift. The duration of the transition process is determined by the hologram writing /erasing time $\tau$. The adaptive interferometer automatically maintains an optimal quadrature condition and can compensate for speckles and slow environment - induced phase shifts. The use of photorefractive crystals in adaptive interferometers is described in [1]. The conceptually similar device based on holographic currents in photorefractive crystals (non-steady state photo-EMF effect) was also studied extensively [2].

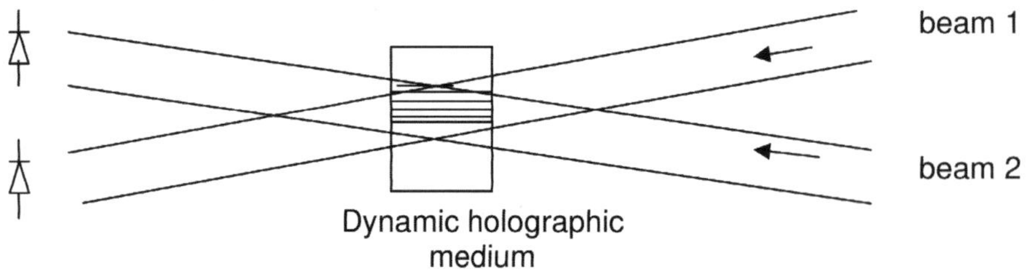


**Figure 1.** The principle of adaptive interferometry. Two beams write a hologram in a medium and diffract on it. The photodetector pair detects rapid phase shifts, which are not compensated by a medium.

Journal of Physics: Conference Series **278** (2011) 012040 doi:10.1088/1742-6596/278/1/012040

The adaptive interferometer ability to compensate for environment-induced phase drifts and beam distortions is in direct proportion to the material cut-off frequency $\nu = (2\pi\tau)^{-1}$. For semiconductor photorefractives, such as GaAs, and laser power in 100 mW range, this frequency is 0.1-1 kHz , which can make difficult device operation in noisy conditions . For the non-steady-state photo EMF detector, the cut-off frequency of a few MHz is possible, but the signal to noise ratio is typically worse than for holographic devices [2], which limits its performance in laser ultrasound detection. For laser ultrasound applications, where spikes typically have submicrosecond duration, the optimal cut-off frequency has to be in the MHz range.

Rubidium vapor has a number of attractive properties as a dynamic holographic medium. For power range of tunable semiconductor lasers, its diffractive efficiency is high (>10%) , and the response time is 20-200 ns [3 ], which results in a sensitivity (defined as the energy per $cm^2$ necessary to write a 1% efficient hologram ) approximately four orders of magnitude better, than that one of the most sensitive photorefractives. The rubidium cells are cheap, they have excellent optical quality, and holographic properties of all of them are identical. The vapor concentration is easily controlled by temperature. The difficulties of work with rubidium include the necessity of fine laser frequency tuning, the geomagnetic field protection, and the polarization control. Additionally, optimal conditions for hologram writing can involve nontrivial combinations of writing beam polarizations, laser frequency and external magnetic field. The nonlinear properties of rubidium, though quite complicated, can be derived using a density matrix formalism with a computationally intensive numerical procedure [4].

## 2. Experimental results

The setup we used to demonstrate ultrasound detection with rubidium based adaptive interferometer is shown in figure 2 (see also [5]). The beam of 780 nm <50 mW produced by a tunable semiconductor laser with an external cavity is split. The signal beam is reflected by the ultrasound-driven mirror, and crosses a pump beam at a small angle within a 75 mm long natural rubidium cell. The polarization of the reference beam is rotated 90 degrees with a half wave plate. The rubidium cell is protected from the geomagnetic field by a mu-metal chamber with incorporated electric heater.

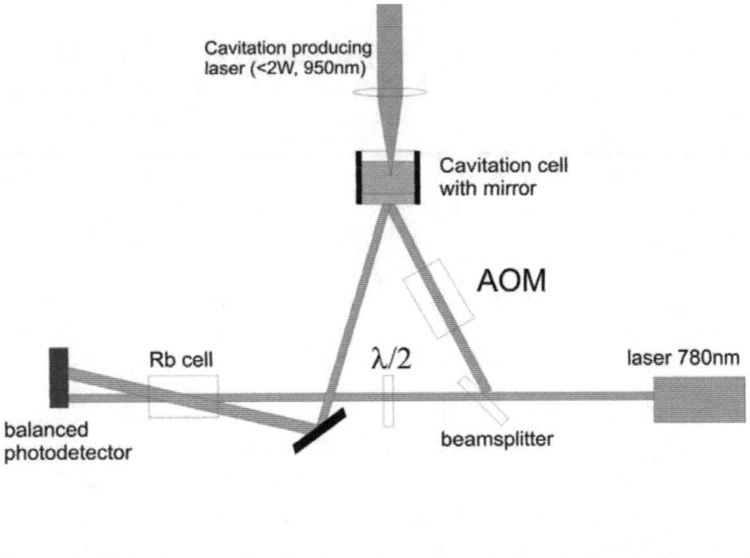


**Figure 2.** Optical setup for the detection of cavitation-induced ultrasound.

The optimal operation conditions for beams of nearly equal intensities are obtained for the magnetic field close to zero. The electro-optical modulator giving the sinusoidal phase modulation of 150 mrad at 8 MHz is included for calibration purposes. For operation, the cell is heated to 80-100 °C.

The best results for our laser power are obtained in $^{87}$Rb D2 Fg=2 transition. For better signal to noise ratio, the signal and pump must have mutually orthogonal nearly linear polarizations. This is related to the complicated transition structure, and this feature is confirmed by the numerical analysis, similar to reported in [4]. The atomic coherent states are sensitive to the polarization ellipticity, thus the optical properties of the illuminated atoms depend on the local polarization state, which changes, and not on the total local intensity, which is constant. This permits to write a hologram with a complicated polarization properties.

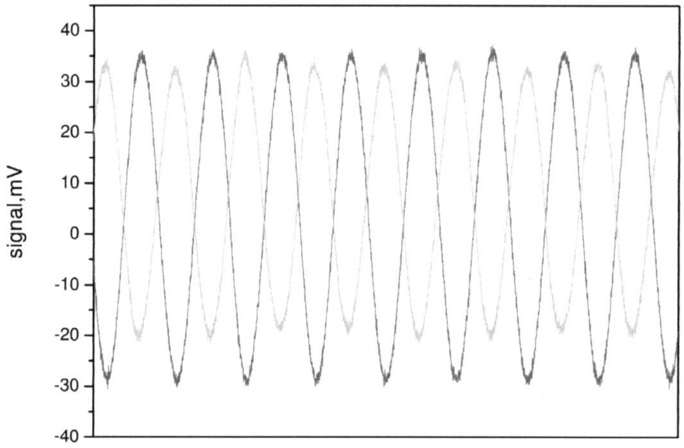


**Figure 3.** The photodetector pair response to 8.3 MHz phase modulation with 150mrad modulation amplitude.

In figure 3 we present the photodetector pair response to the sinusoidal phase modulation introduced in one arm of the interferometer with an electro-optic modulator. It is seen, that the signals have nearly equal amplitudes and 180 degrees phase difference. The balanced photodetector is used to suppress the laser intensity noise. The noise equivalent phase modulation in 20 MHz frequency band is 1-2 mrad, which corresponds to ~0.1 nm noise equivalent surface displacement. Since the cut-off frequency in rubidium is ~1.0 MHz, even the additional 1000 rad phase modulation at 200 Hz introduced with a loudspeaker driven mirror does not diminish the signal more than 1% [5].

The optical detection of the ultrasound induced in the aluminum plate by the Nd:YAG laser pulse, and by the piezoelectric transducer was reported in [5]. Here we demonstrate the adaptive detection of ultrasound generated by a CW-laser induced cavitation. The 5 mm thick cell is filled with a liquid strongly absorbing in infrared (water solution of cupric nitrate ). The front window is made with 1.7 mm thick glass, the back window is 2.7 mm thick and coated with aluminum. By loosely focusing 0.1-2 W laser beam inside a liquid, the single cavitation bubbles are created. On collapse, the strong ultrasound shock wave is generated. The details are reported elsewhere [6]. The obtained signals have good SNR, and the surface displacement is comparable to that one produced by a Nd:YAG laser pulse.

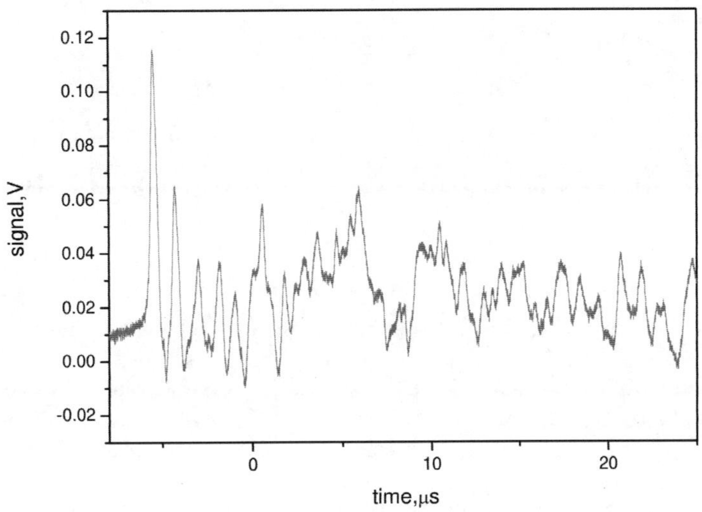


**Figure 4.** The differential photodetector signal produced by the mirror displacement resulting from the cavitation shock wave in a cell. Multiple wave reflections on boundaries are observed.

## 3. Conclusions

In conclusion, the use of rubidium as a dynamic holographic medium permits to detect the laser - induced ultrasound with a good signal-to noise ratio, comparable to that of the photorefractve crystals based adaptive interferometers. Rubidium demonstrates excellent adaptive properties because the hologram writing time in it is less than 1 microsecond, which is 3-4 orders of magnitude smaller, than for the fastest photorefractive crystals.

## References

[1] Solymar L, Webb D J and Grunnet-Jepsen A 1996 *The Physics and Applications of Photorefrctive Materials* (Oxford: U.Press,) chapter 13.

[2] Stepanov S 2001 Photo-electromotive force effect in semiconductors *Handbook of Advanced Electronic and Photonic Materials and Devices* (*Semicondutor Devices* vol.2), ed H S Nalwa (Academic)chapter 6, pp.205-272.

[3] Korneev N and Benavides O 2008 *JOSA B* **25** 899

[4] Korneev N and Benavides O 2009 *J. Mod. Opt.* **56** 1194

[5] Korneev N, Rodriguez-Montero P and Benavides O 2009 *Opt.Lett.* **34** 1964

[6] Ramirez-San Juan J C, Rodriguez-Aboyetes E, Martinez-Canton A, Baldovino-Pantaleon O, Robledo-Martinez A, Korneev N and Ramos-Garcia R 2010 *Opt. Express* **18** 8735

Journal of Physics: Conference Series **278** (2011) 012041 doi:10.1088/1742-6596/278/1/012041

# Laser Ultrasonic Technique for Evaluating Human Dental Enamel

**D H-C Wang[1], S Fleming[1], Y-C Lee[2], M Swain[3], S Law[1], J Xue[3]**

[1]Insititue of Photonics and Optical Science, School of Physics, University of Sydney, NSW 2006, Australia
[2]Department of Mechanical Engineering, National Cheng Kung University, Tainan City, Taiwan
[3]Faculty of Dentistry, University of Sydney, NSW 2006, Australia

E-mail: hsiao-chuan.wang@sydney.edu.au

**Abstract**. A non-destructive laser ultrasonic surface acoustic wave technique has been demonstrated to quantitatively evaluate the elastic response of human dental enamel. We demonstrate the system performance by measuring surface acoustic wave velocity in sound and demineralised enamel. In addition, progressive measurements were made to monitor the change in the enamel elasticity during a two week remineralisation process. The results are presented and they confirm the efficacy, as well as illuminating the progress, of the treatment.

## 1. Introduction

Dental caries, one of the most common diseases, is caused by the excessive mineral dissolution of tooth enamel [1]. The level of enamel mineralisation, which correlates closely to its elastic response [2-4], is hence an indicator for the healthiness of the tooth. However, conventional diagnostic methods, such as tactile and X-ray [5], are relatively ineffective for providing quantitative assessments. In this regard, a non-contact and non-destructive technique for localised quantitative evaluation of the elastic response of human dental enamel is desirable.

A surface acoustic wave (SAW) is a type of elastic ultrasound which propagates across the material surface with a wavelength dependent probing depth [6]. Its propagation velocity is directly related to the elastic constants of all the material layers it probes [7]. Determining the SAW velocity of different frequency components (different penetration depths) gives a dispersion spectrum whose slope and curvature reveal information about the elastic and geometric parameters of the specimen [8-10].

In this paper we present a validated laser SAW method [11] and its novel application of quantitative evaluation of the elastic response of human dental enamel. The measurement technique is briefly discussed. The elastic properties of human enamel are then considered and the measurement results of sound and demineralised enamel, as well as progressively remineralised enamel, are presented and discussed.

## 2. Laser Ultrasonic Technique

Detailed discussion and extensive validation testing results of this laser ultrasonic SAW dispersion technique can be found in a previous publication [11]. Key details only are presented here. The schematic of the measurement system is illustrated in figure 1.

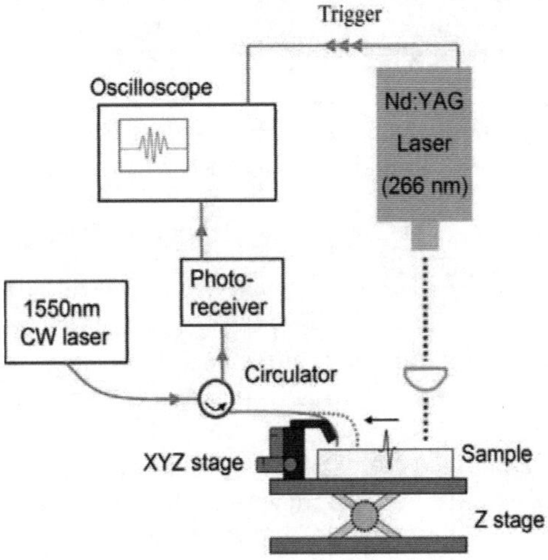


**Figure 1**. Arrangement of the laser ultrasonic system for the SAW dispersion measurement.

A quadrupled Nd:YAG laser operating at 266 nm (~5 ns pulse duration) was used to excite surface waves on the samples. The source was focused into a thin line by a cylindrical lens onto the sample surface, as shown on the right of figure 1. The absorbed laser energy produces localised heating and causes rapid thermo-elastic expansion around the illuminated region, hence generating ultrasound [12, 13]. In this setup, test samples were placed on a manual Z-stage such that the irradiated line-source width can be tuned sufficiently narrow to generate broadband SAW, which is desirable for dispersion measurement. In the current study, the bandwidth of the generated SAW typically spanned from 1 ~ 25 MHz [14]. During measurements the laser pulse energy and the sample surface were constantly monitored to ensure non-destructive operation.

The SAW impulse was detected by an Optical Fibre Interferometer (OFI). The OFI consisted of a 1550 nm continuous-wave laser source, a 3-port optical fibre circulator and a photo-receiver. The OFI operation is similar to a reference beam interferometer [11]. The coherent laser light from the source was coupled into the circulator. As the light arrives at the end of the cleaved middle port (the sensing tip), it is split into the reference and the measurement beams by internal reflection and external reflection, respectively. The path difference of the two beams, and hence their relative phase, depends only on the separation between fibre tip and sample surface. In a typical ultrasonic measurement the surface wave modulates the measurement beam path-length and the interference pattern becomes a good measure of the absolute acoustic waveform. The sensing port of the circulator was placed on a micron-precision three axis (XYZ) positioning-stage to adjust the measurement position. All measurements were made with a probe-to-sample separation of 5 mm or smaller and between 1 mm and 12 mm from the line-source.

The propagation velocity of the SAW, $c_R$, is governed by the elastic properties of the material layers within the frequency dependent penetration depth. The SAW penetration depth, $z$, can be estimated from the relation [7, 15]:

$$z \approx c_R / f \, , \tag{1}$$

where $f$ is the SAW frequency. As a broadband SAW propagates through a material surface, the higher frequency components have shallow penetration depth and are more influenced by the elastic parameters of the surface layer. On the other hand, the lower frequency components penetrate deeper and are influenced more by the bulk elastic parameters. Typically $c_R$ is proportional to the elastic modulus but inversely proportional to the density [7]. In an isotropic homogeneous medium $c_R$ is independent of the penetration depth (having a constant value for all frequency components) as well as propagation direction. In a complex inhomogeneous medium $c_R$ becomes a function of signal frequency and the SAW is dispersed [15]. The characteristics of the dispersion curves depend on the nature of the specimen.

The experimental frequency dependent $c_R$ and hence the dispersion curve was determined from two of the measured SAW signals, at locations $x_1$ and $x_2$, using the following equation [11]:

$$c_R(f) = 2\pi f \frac{x_2 - x_1}{\varphi(f)}, \tag{2}$$

where $f$ is the SAW frequency and $\varphi(f)$ is the phase difference between the two signals derived from their Fourier spectra.

## 3. Evaluation of Human Dental Enamel

Kushibiki et al. [16] used line-focused-beam scanning acoustic microscopy and measured the SAW velocity as a function of propagation direction on extracted human incisors and reported that the velocity value varied between 3105 $m.s^{-1}$ to 3155 $m.s^{-1}$. Peck and colleagues [17] repeated the investigation on the surface of human permanent molars and reported that the SAW velocity varied between 3075 $m.s^{-1}$ to 3142 $m.s^{-1}$.

A recently extracted sound human incisor was selected for this initial study because this type of tooth has a large flat area of enamel on its front surface. In addition, the enamel thickness is relatively constant (~1 mm) over a wide region near the centre of the incisor front surface. This is desirable because the measured SAW dispersion will not depend significantly on the uneven thickness of the enamel layer but rather on the elasticity variation.

Artificial WSL, which is similar to a natural WSL and widely used for dental research [18], was created on the sample. About ~70% of the front surface of the incisor was first abraded and polished and a window of about 3 mm by 3 mm was left exposed while the rest of the tooth was coated and protected with nail varnish. The sample was then placed in demineralisation solution (pH = 4.5) [19] for five days. After the treatment, the nail varnish was removed. The WSL enamel is low in mineral (10 ~ 70% of the value for sound enamel) and thus has lower elastic modulus and consequently lower SAW velocity. A WSL layer developed on top of sound enamel is similar to the two-layer system and dispersion is expected.

The healthy region of the sample was first measured, as depicted in figure 2(a). The line-source (illustrated as the line) was irradiated near the edge of the enamel surface and the SAW propagated in the direction horizontal to the tooth axis. The recording of the surface waves was made at several positions along the propagation path (illustrated as the dots). The measurement was repeated ten times.

**Figure 2.** Illustrations showing the positions of the SAW generation and detection on (a) the healthy region and (b) the WSL region of the tooth sample.

For the measurements on the WSL region, as illustrated in figure 2(b), we irradiated the line-source near the top edge of the tooth and allowed propagation across the lesion in the direction parallel to the tooth axis. Surface waves were recorded at various locations especially near the junctions of the sound enamel and the WSL (illustrated by the dots). Again ten measurements were repeated. Dispersion spectra of sound enamel and WSL measurements were calculated and the ten-times-averaged values were used as the final results and are presented in figure 3 with standard deviation error-bars. It is important to note here that we only investigate frequency components between 7 ~ 25 MHz, corresponding to a penetration depth of 0.5 ~ 0.1 mm (using Eq. 1 and $c_R = 3100$ m.s$^{-1}$). Frequency components of 1 ~ 7 MHz have deep penetration depth and were significantly influenced by the underlying dentin is not important for this study.

In the sound enamel the SAW propagates as in a relatively homogeneous single layer system with ~3150 m.s$^{-1}$. This value compares well with previously reported values [16, 17]. In the WSL region, the influence of the lower elastic lesion layer results in a slower initial SAW velocity value which continues to drop as the influence of the lesion layer became more dominant for higher frequency components. This dispersion profile fits well with expectation and the differences between the two dispersion spectra are significant (about 10 % drop in velocity value).

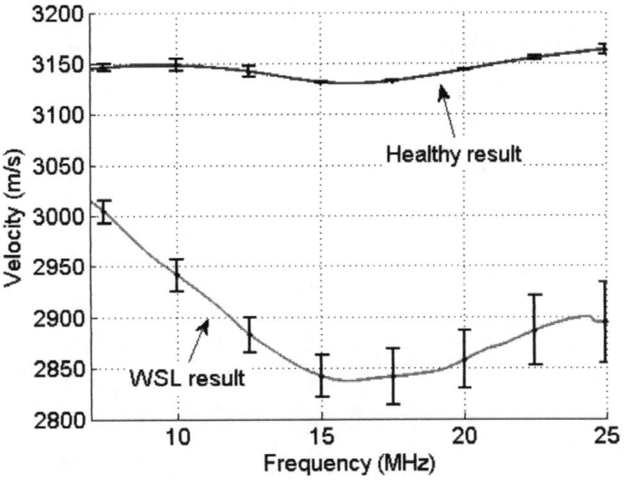


**Figure 3.** Dispersion curves of the final averaged results from the healthy region (blue) and the WSL region (red) of the sample enamel.

Having demonstrated the ability of the laser ultrasonic technique in evaluating and differentiating sound and WSL enamels, we extended the study to measure the progressive elasticity variation of artificially demineralised human enamel during remineralisation treatment. The same incisor sample was again used in this study. It was soaked entirely in the remineralisation solution (pH = 7.0) [20] for two weeks, the solution being renewed after the first week. Four sets of ten measurements were taken during the two week period after 4, 7, 10 and 14 days of treatment. The measurement procedure and the measurement regions were identical as before.

The final averaged dispersion curves (ten measurements for each set) of the sound enamel and the WSL measurements after 4, 7, 10 and 14 days of remineralisation are shown in figures 4 and 5, respectively (with different vertical scales). All of the results have less than 3 % standard deviation and the error-bars are not shown for the clarity of the figures.

For the sound enamel the SAW velocities are almost unchanged at ~3150 m.s$^{-1}$ (less than 1% variation) during the two weeks of remineralisation treatment, which indicates that the treatment causes no change in the sound enamel. On the other hand, significant changes can be observed in the

WSL results with a trend towards increasing velocity values in the dispersion curves as the number of days of remineralisation increases. This observation suggests that the elastic modulus of the lesion layer has increased in value during the first 7 days of remineralisation. In the second week of the treatment, the increase in SAW velocity is minimal, which could be explained by a saturation of the remineralisation process.

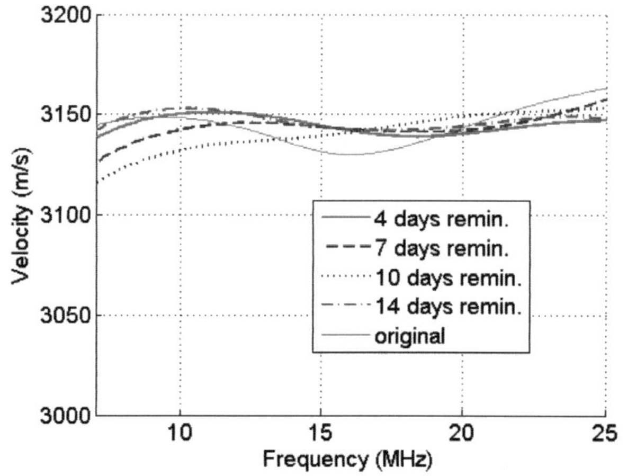


**Figure 4.** Comparison diagram of the sound enamel results during the two weeks remineralisation treatment.

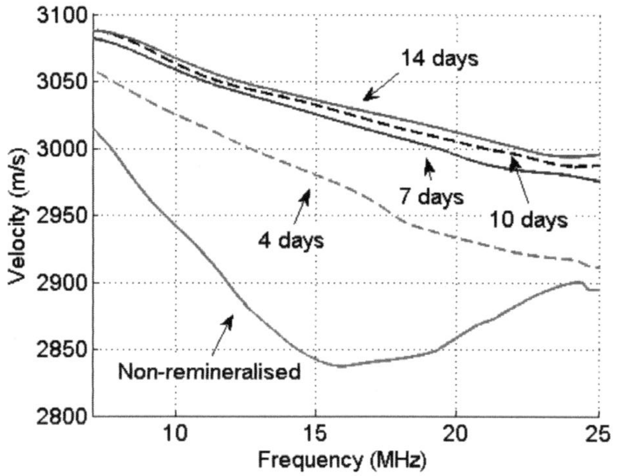


**Figure 5**. Comparison diagram of the WSL results from the two weeks remineralisation treatment.

From these results it is clear that we have successfully evaluated the change in the SAW dispersion curves, which directly correlate to the elastic properties, of a human tooth under remineralisation treatment.

## 4. Conclusion

In this paper we presented a laser ultrasonic technique and applied it to evaluate the elastic response of extracted human dental enamel. In sound enamel, the measured SAW velocity matched well with previously reported values. In the artificial lesioned region, the SAW velocity value was significantly lower and displayed significant dispersion. We have also presented the results of progressive monitoring of a remineralisation process and shown that the SAW velocity increases in the lesion region only.

Although *in-vivo* tests have not yet been performed, the laser ultrasonic technique is a good candidate as a new examination tool for dental research purposes. Contrasted with the conventional destructive nano-indentation method and the limited discrimination provided by X-ray procedures, laser ultrasound provides repeatable non-destructive evaluation which can be used to study lesions of different severity and depth as well as verify the efficacy of different remineralisation methods.

**Acknowledgement**

This work was funded by the Australian Government and Bio-Dental Technology Pty. Ltd. under an ARC Linkage Grant, No. LP0561184.

**Reference**

[1] Krasse B 1988 Biological factors as indicators of future caries *International Dental Journal* **38** 219-225
[2] Feagin F, Koulourides T, Pigman W 1969 The characterization of enamel surface demineralization, remineralization, and associated hardness changes in human and bovine material *Archives of Oral Biology* **14** 1407-1417
[3] Habelitz S, Marshall S J, Marshall Jr G W, Balooch M 2001 Mechanical properties of human dental enamel on the nanometre scale *Archives of Oral Biology* **46** 173-183
[4] Cuy J L, Mann A B, Livi K J, Teaford M F, Weihs T P 2002 Nanoindentation mapping of the mechanical properties of human molar tooth enamel *Archives of Oral Biology* **47** 281-291
[5] Ismail A I 2004 Visual and visuo-tactile detection of dental caries *Journal of Dental Research* **83** C56-C66
[6] Soczkiewicz E 1997 The Penetration Depth of the Rayleigh Surface Waves *Nondestructive Testing and Evaluation* **13** 113-119
[7] Achenbach J D 1984 *Wave Propagation in Elastic Solids (*Elsevier Science Ltd)
[8] 2004 *Ultrasonic nondestructive evaluation: engineering and biological material characterization* ed Kundu T (CRC Press)
[9] Schneider D, Schultrich B, Scheibe H J, Ziegele H, Griepentrog M 1998 A laser-acoustic method for testing and classifying hard surface layers *Thin Solid Films* **332** 157-163
[10] Glorieux C, Gao W, Kruger S E, Van de Rostyne K, Lauriks W, Thoen J 2000 Surface acoustic wave depth profiling of elastically inhomogeneous materials *Journal of Applied Physics* **88** 4394-4400
[11] Wang H C, Fleming S, Lee Y C 2009 A remote, non-destructive laser ultrasonic material evaluation system with simplified optical fibre interferometer detection *Journal of Nondestructive Evaluation* **28** 75-83
[12] Davies S J, Edwards C, Taylor G S, Palmer S B 1993 Laser-generated ultrasound: its properties, mechanisms and multifarious applications *Journal of Physics D: Applied Physics* **26** 329-348
[13] Achenbach J D 2003 Laser excitation of surface wave motion *Journal of the Mechanics and Physics of Solids* **51** 1885-1902
[14] Wang H C, Fleming S, Lee Y C 2009 Non-contact, non-destructive elasticity evaluation of sound and demineralised human dental enamel using laser ultrasonic surface wave dispersion technique *Journal of Biomedical Optics* **14** 054046
[15] Rose J L 1999 *Ultrasonic Waves in Solid Media* (Cambridge University Press)
[16] Kushibiki J, Ha K L, Kato H, Chubachi N, Dunn F 1987 Application of Acoustic Microscopy to Dental Material Characterization *IEEE 1987 Ultrasonics Symposium* 837-842
[17] Peck S D, Rowe J M, Briggs G A 1989 Studies on sound and carious enamel with the quantitative acoustic microscope *Journal of Dental Research* **68** 107-112
[18] Arends J, Christoffersen J 1986 The Nature of Early Caries Lesions in Enamel *Journal of Dental Research* **65** 2-11
[19] ten Cate J M, Duijsters P P E 1983 Influence of fluoride in solution on tooth demineralization. I. Chemical data *Caries Research* **17** 193-199
[20] Iijima Y, Takagi O, Ruben J, Arends J 1999 In vitro remineralization of in vivo and in vitro formed enamel lesions *Caries Research* **33** 206-213

# Two color probing of the ultrafast photo-acoustic response in a single biological cell

**M Ducousso[1], T Dehoux[1], C Rossignol[1], B Audoin[1], O Zouani[2], C Chollet[2], C Chanseau[2] and M C Durrieu[2]**

[1] Université de Bordeaux, CNRS, UMR 5469, Talence, F-33405, France
[2] Université de Bordeaux, INSERM, U 577, Bordeaux, F-33000, France

mathieu.ducousso@u-bordeaux1.fr

**Abstract** The measurement of the mechanical properties of single biological cells with a nanometer depth resolution using only coherent light is proposed. A pump-probe set-up based on an ultrafast laser (100 fs pulses) is used to excite and detect acoustic frequencies in the GHz range. Experiments are performed on single fixed mouse MC3T3 cells adhering on titanium alloy substrate. Using two different probe wavelengths, the contributions to the optical detection resulting from the cell interface displacements and from interactions between acoustic waves and the laser light are identified. Semi-analytical calculations allow the determination of acoustic celerities and thicknesses in cells thinner than 150 nm.

## 1. Introduction

Since the 80's, [1] interest in the picosecond acoustics technique has been growing owing to its wide range of applications in non-destructive control and solid-state physics. By this technique high frequency acoustic waves are both generated and detected with laser light pulses (duration<1 ps), allowing the measurement of the optical, thermal, or mechanical properties of submicrometric materials. [2] Some applications of picosecond acoustics are thickness measurement and bonding control at the nanometer scale, or studies of material microstructures. Experiments can be performed in opaque or transparent media.

In this paper, the picosecond acoustics technique is used in biological cells, the smallest unit of life that is classified as a living thing. Innovative experiments have already been performed in vegetal *allium cepa* cells and the potentialities of the picosecond acoustics technique to improve current cell imaging resolution have been demonstrated. [3-4] The method is now applied to single mouse MC3T3 fixed cells adhering on a titanium alloy substrate. Acoustic celerities are measured in cell thinner than 150 nm. In the second section of the paper, the optical detection in a cell of finite thickness adhering on an opaque substrate is proposed. We use two different probe wavelengths at the same point in the cell in order to identify the different contributions to the opto-acoustic detection. Simulation results and measurements are compared in the third section to determine the acoustic celerities and thicknesses of the cells.

## 2. Picosecond acoustics detection in a thin viscoelastic film adhering on a half-space substrate

The picosecond acoustics technique has already been used to study thin transparent films, from nanometric to micrometric thicknesses. [5-7] The cell is modeled as a thin transparent elastic film of thickness $d$ adhering on an opaque substrate. A short light pulse is absorbed in the vicinity of the substrate and the subsequent thermal expansion generates a short strain pulse. The continuity of displacement and stress at the cell/substrate interface launches two acoustic pulses, one propagating in the cell and one propagating in the substrate. The acoustic wavelength of the acoustic pulse propagating in the cell is dictated by the acoustic celerity and the optical penetration depth in the titanium alloy substrate, [1] around 5.8 nm/ps and 20 nm, respectively. This acoustic wavelength is about 5 nm in the present study. We impose a zero stress boundary condition at the cell/air interface. The acoustic pulse propagates in the cell and is reflected at the cell interfaces. The period of the first harmonic of the acoustic resonances in the cell is:

$$T_A = \frac{4d}{v}, \tag{1}$$

where $v$ and $d$ are the acoustic celerity and the cell thickness of the cell, respectively.

The optical detection of the acoustic response is the sum of the interfaces displacements and the acousto-optic interaction between the acoustic pulse and the probe laser radiation. [6] The acousto-optic interaction leads to the so-called Brillouin oscillation. [5] Their period for a normal probe incidence is:

$$T_B = \frac{\lambda}{2nv}, \tag{2}$$

where $\lambda$ and $n$ are the laser wavelength and the optical index of the medium. The amplitude of these oscillations is proportional to the piezo-optic coefficient. [6]

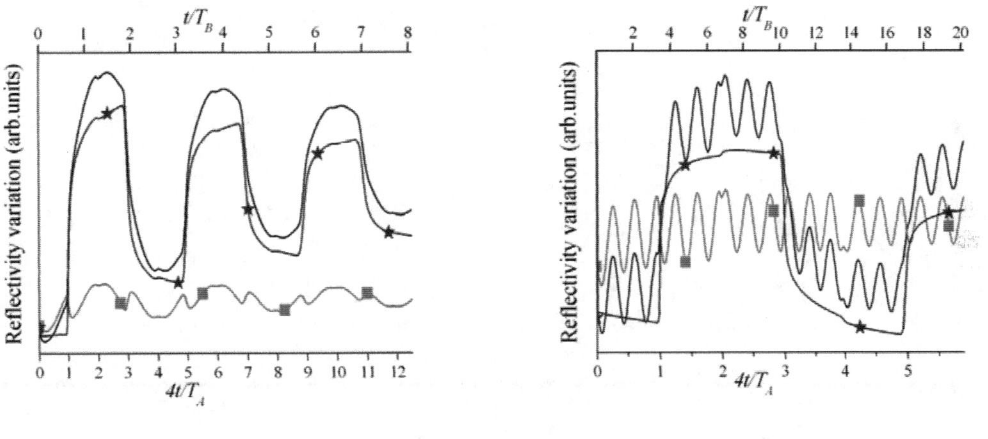


(a) (b)

**Figure 1.** Picosecond acoustics of a thin biological cell. For each plot, calculated signals labelled with squares and stars markers represent the acousto-optic contribution and the interface displacement contribution, respectively. We compare cells where (a) the acoustic frequency is higher than acousto-optic frequency and (b) the inverse situation. Unlabelled plots are the sum of the acousto-optic and interface displacement contributions.

Figure 1 presents the relative reflectivity changes calculated for different cell thicknesses compared to the acousto-optic interaction period $T_B$. Plots are function of the dimensionless times $4t/T_A$ (down abscissa) and $t/T_B$ (up abscissa). Plots with squares and stars represent the acousto-optic and the interface displacement contributions to the detection, respectively. The square-like modulation is caused by the step-like acoustic displacement of the interface. Frequencies of interface displacement and acousto-optic interactions are dictated by equation 1 and equation 2, respectively. On figure 1(b),

$\lambda << d$, that is $T_B << T_A$: Brillouin oscillation arise in the cell. On figure 1(a), $\lambda >> d$ and Brillouin oscillations cannot arise.

Let us now compare experimental data obtained in submicrometric cells with the modeling presented above.

## 3. Experimental evaluation of the acoustic celerity and thickness in submicrometric cells

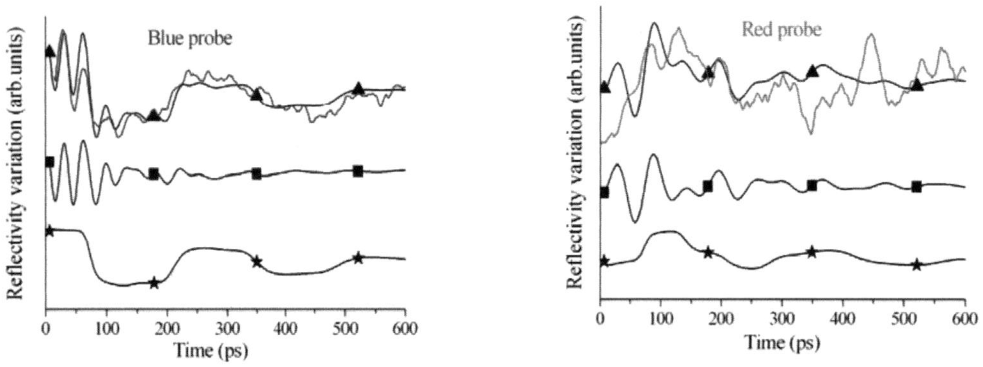


Cell #1

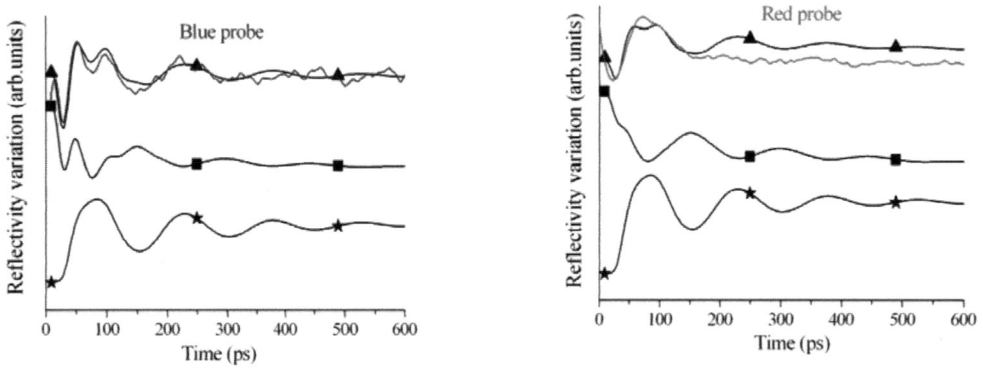


Cell #2

**Figure 2.** Experimental data (unlabelled plots) obtained in two different cells. On the left-hand side plots, detection is done using blue probe, and on right-hand side plots, detection in performed at the same point in the cell with a red probe. Triangles are corresponding calculated signals, sum of the acousto-optic detection (squares) and the interface displacement detection (stars).

The experimental set-up is a classical pump-probe setup used to generate and to detect acoustic waves in a single MC3T3 fixed cell adhering on a Ti6Al4V substrate. [8] Red pulses (800 nm) of 100 fs duration are generated by a mode-locked Ti:sapphire laser, of repetition rate 82 MHz. A polarizing beam splitter divides the laser beam in pump and probe beams. The pump beam passes through an acousto-optic modulator (330 kHz) to provide the reference signal for lock-in amplification. One of the laser beams is converted into blue light using a BBO crystal and the other beam passes through a delay line to provide a tunable time delay between the pump and the probe. Both beams propagate through the transparent cell and are focused at a normal incidence with a X20 microscope objective at the Ti6Al4V surface and can be used either as pump or probe. The width at half-height of the space cross-correlation of the pump and probe beams is approximately 5 μm. Reflectometric measurements are presented.

Four experimental results obtained on two different cells, one cell per line, and corresponding simulations are presented on figure 2. Unlabelled plots are experimental detection. On the left plots, signals are detected using blue probe while on right plots the detection is done using red probe at the same point of the cell. The experiments are led on the vacuole across the nucleus, the cytoplasm and the cytoskeleton of the cell. Lines with triangles are the corresponding simulation. Physical parameters are identical for each cell. The piezo-optic coefficients are adjusted to match the ratio of the acousto-optic to interface displacement contributions to the change of reflectivity. For each cell, the signal detected using the blue probe allows the measurements of the mechanical properties of the inspected cell. Assuming a cell optical index equal to 1.4, [9] the acoustic celerity is measured using the acousto-optic detection (square plots) equal to 4.4 and 3.7 nm/ps in cells 1 and 2, respectively, corresponding to 33 and 26 GHz acousto-optic frequency signal, respectively. Taking a cell density equal to 1100 kg.m$^{-3}$, the measured stiffness of the cell nucleus is evaluated between 15 to 21 GPa. Comparable rigidity has already been measured in osteoblasts cells on titanium alloy substrate. [10] The interface displacement contribution (star plots) allows the measurements of the cell thicknesses, 300 and 140 nm for cell 1 and 2, respectively. Data obtained using the red probe do no permit the mechanical evaluation of the cell for two reasons. For cell 1, the signal is very low. This may be attributed to a weak piezo-optic coefficient in the MC3T3 cell at the probe wavelength. The cell 2 is very thin compared to the red probe optical wavelength and the probe is almost not sensitive to the acousto-optic effect. It is not possible to distinguish the acoustic celerity from the cell thickness value in the detection.

## 4. Conclusion

Optical detection of acoustic frequencies higher than 30 GHz has been successfully performed in MC3T3 cells allowing the determination of acoustic celerity and thickness in single cells thinner than 150 nm. This technique suggests promising perspectives in single cells imaging and biomedical applications, as various as cancer studies or cell adhesion on biomaterials.

## References

[1] Thomsen C, Grahn T C, Maris H J and Tauc J 1986, *Phys. Rev. B* **34** 4129
[2] Antonelli G, Perrin B, Daly C D, Cahill D G 2006, *Mater. Res. Bull.* **31** 607
[3] Rossignol C, Chigarev N, Ducousso M, Audoin B, Forget G, Guillemot F and Durrieu M C 2008, *Appl. Phys. Lett.* **93** 123901
[4] Audoin B, Rossignol C, Chigarev N, Ducousso M, Forget M, Guillemot F and Durrieu M C 2010, *Ultrasonics* **50** 202
[5] Rossignol C, Perrin B, Laborde S, Vandenbulcke L, De Barros M I and Djemia P 2004, *J. Appl. Phys.* **95** 4157
[6] Wright O B 1992, *J. Appl. Phys.* **71** 1617
[7] Côte R and Devos A 2005, *Rev. Sci. Instrum.* **76** 053906
[8] Chollet C, Lazare S, Guillemot F and Durrieu M C 2010, *Colloid. Surface. B* **75** 107
[9] Beuthan J, Minet O, Helfman J, Herrig M and Müller G 1995, *Phys. Med. Biol.* **41** 369
[10] Saruwatari L, Aita H, Butz F, Nakamura H, Ouyang J, Yang Y, Chiou W A and Ogawa T 2005 *J. Bone Miner. Res.* **20** 2002

Journal of Physics: Conference Series **278** (2011) 012043 doi:10.1088/1742-6596/278/1/012043

# Mechanical characterization of temperature-sensitive objects using picosecond ultrasonics

**T Dehoux**[1], **B Audoin**[1], **O Zouani**[2] **and M C Durrieu**[2]

[1] Université de Bordeaux, CNRS, Laboratoire de Mécanique Physique, UMR 5469, F-33405 France

[2] Université de Bordeaux, INSERM, Laboratoire Biomatériaux et Réparation Tissulaire, U577, F-33076 France

E-mail: thomas.dehoux@u-bordeaux1.fr

**Abstract.** Biological objects are exquisitely sensitive to temperature variations and their mechanical characterization is often a challenge when using the picosecond ultrasonics technique. To reduce the laser-induced temperature rise, we place single biological cells on a thin metal transducer and we focus the laser beam that generates the acoustic waves at frequencies $\leq$ 150 GHz on the rear side of the transducer. The acoustic waves propagate through the transducer and are partially transmitted to the cell to create the so-called Brillouin oscillations. The frequency of these oscillations provides a direct measurement of the sound velocity. The simultaneous measurement of the acoustic reflection coefficient at the transducer/cell interface allows the determination of both the density and the compressibility of the cell.

## 1. Introduction

Temperature-sensitive materials (TSM) include materials with a phase transition close to the room temperature, or objects that can be damaged by temperature variations, such as water-based gels or biological media. The mechanical characterization of TSM is often a technical challenge when using a laser-based setup such as the picosecond ultrasonics technique (PU).

PU is an optical pump-probe technique [1] which has proven useful for the non-invasive mechanical characterization of viscous liquids,[2] Langmuir-Blodgett films,[3] or liquid mercury.[4] The measurement of the acoustic reflection coefficient from a Al/water interface has been used to design an acoustic microscope with nanometer resolution.[5] Sound velocity $v$ has also been determined in water by measuring the frequency of Brillouin oscillations.[6]

PU has recently been applied to the mechanical characterization of biological cells deposited on a metallic half-space. [7] The approach used by the researchers has been to generate acoustic waves through the absorption of laser pulses at the cell-substrate interface. The first drawback of this configuration is that the temperature rise is maximal at the interface, thus imposing to limit the fluence of the laser to avoid damage to the cell. Obviously, the same problem would arise in any TSM. The second drawback is that the presence of organelles, diffusing the pump light, and the strong acoustic absorption require high pumping fluence. These irreconcilable hindrances hamper the optimization of the signal-to-noise ratio (SNR) in cells.

The acoustic reflection coefficient, which can yield the acoustic impedance $Z$ of the TSM, and $v$ have only been measured in separate configurations. However, to fully characterize the mechanical properties of a soft material one needs to measure simultaneously $Z$ and $v$ to obtain

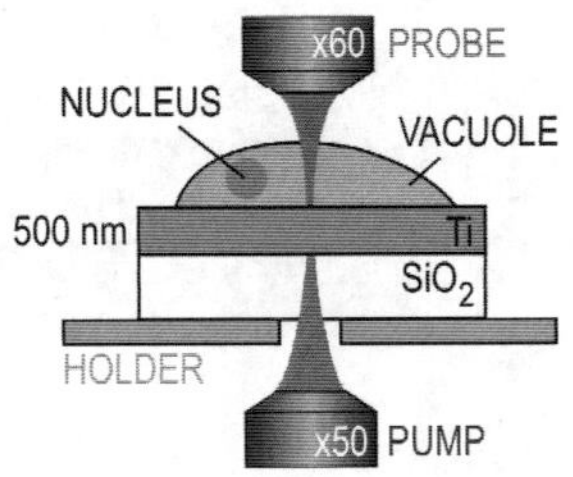


**Figure 1.** Transducer geometry. The pump is focused on the rear-side of the Ti film to reduce thermal stress in the cell.

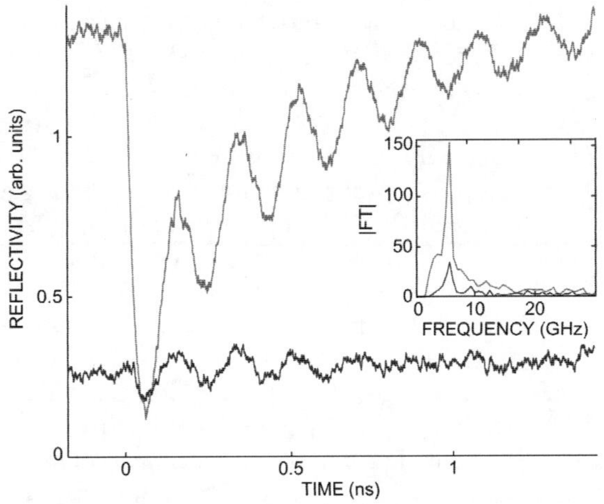


**Figure 2.** $|\Delta R/R(t)|$ in an *allium cepa* cell when the pump beam is focused on the top (top curve) or the rear side (bottom curve) of the Ti film.

the density $\rho = Z/v$ and the compressibility $\chi = 1/Zv$. In the present paper we place single *allium cepa* cells on a thin metal transducer. We focus the pump beam on the rear side of the transducer to reduce the laser-induced temperature rise in the cell. We measure simultaneously $\rho$ and $\chi$ of the cell using acoustic echoes in the film and Brillouin oscillations in the cell.

## 2. Transducer design: Reduction of thermal stress in the cell and increase in SNR

Titanium-based alloys are commonly used as structural biomaterials for the replacement of hard tissues in artificial joints and they display the most suitable biomedical properties. For this reason we sputtered a polycrystalline Ti film on a silica supporting layer transparent at both pump and probe wavelengths (figure 1). To define the optimum Ti thickness, consider the diffusion length $\mu = (\kappa/\rho C_p \pi f_m)^{1/2} = 660$ nm at the modulation frequency $f_m = 1$ MHz of the laser, where $\rho = 4500$ kg m$^{-3}$ and $C_p = 500$ J kg$^{-1}$ K$^{-1}$ are the density and specific heat capacity of the Ti film, respectively.[7] The thermal conductivity of the film $\kappa = 3.1$ W m$^{-3}$ K$^{-1}$ is assumed to be 1/7 of the single-crystal value.[8] In order to avoid thermal confinement at $f_m$ and to keep a thickness within rf sputtering capabilities, we chose a thickness $d = 500$ nm.

We use an optical pump-probe technique described in details in Ref. [1]. Longitudinal acoustic pulses are thermoelastically generated at the Ti-$SiO_2$ interface with optical pump pulses of duration 100 fs, repetition frequency $f_r = 80$ MHz, wavelength $\lambda_{pp} = 790$ nm from a Ti:Sapphire mode-locked laser. This light is chopped at $f_m = 1$ MHz for lock-in detection. The reflectivity change is measured with frequency-doubled probe pulses of wavelength $\lambda_{pb} = 395$ nm as a function of the pump-probe time delay $t$. The pump beam focus is focused either on the top or rear side of the Ti film, as discussed later. Each pump pulse is initially absorbed in the metal layer over a depth comparable to the optical skin depth of $\sim 15$ nm.[7] Conduction band electrons are excited and diffuse over a depth of $(\kappa/g)^{1/2} = 1$ nm during their thermalization with the lattice, where $g = 35 \times 10^{17}$ W m$^{-3}$ K$^{-1}$ is the electron-phonon coupling constant in Ti.[9] Therefore, when the pump is focused on the rear-side of the Ti film, neither the pump radiation nor the overheated electrons reach the top Ti surface where the cell is deposited.

We place an *allium cepa* cell on top of the transducer to characterize the efficiency of acoustic generation. The probe beam is focused at the Ti/cell interface (top curve) using a $\times 60$ objective lens (NA 0.7). The measured change of reflectivity is plotted in figure 2. The pump beam is alternatively focused at the Ti/cell interface (top curve) through the same objective lens as the probe, or at the rear-side of the Ti film (bottom curve) through a $\times 50$ objective lens (NA 0.8). The latter configuration is depicted in figure 1. The probe beam is partially reflected from the Ti surface and partially scattered by the acoustic wavefront propagating in the cell. Interferences arise from these two probe contributions causing the so-called Brillouin oscillations visible on both curves in figure 2.[7] Their frequency is $f_b = 2n_c^r v/\lambda_{pb}$, where $\lambda_{pb}$, $n_c^r$ and $v_c$ are the probe wavelength, the real part of the refractive index and the sound velocity in the cell, respectively. The amplitude of the Fourier spectrum is shown in the inset of figure 2. The ratio of the peaks observed at $f_b = 5.6$ GHz in the two configurations is $H = 0.25$.

A slow decay is observed on the top curve due to thermal diffusion and overheated electrons relaxation. As $d$ is larger than the diffusion length at $f_r$, these effects do not show on the bottom curve. The thermal stress in the cell is thus reduced importantly when the pump is focused on the rear-side of the Ti film. An offset is also observed at $t < 0$ resulting from the temperature rise at the modulation frequency. This temperature is five times lower when the pump beam is focused on the rear-side of the Ti film. The pump fluence could therefore be increased by a factor of 5 to obtain the same temperature rise in both configurations. The acoustic amplitude would also be increased by the same factor, resulting in an increase in SNR by a factor $5H = 1.25$ when changing the position of the pump from the top to the rear-side of the Ti film.

## 3. Simultaneous measurement of the density and compressibility of the TSM

We present in figure 3 the amplitude of the reflectivity change $|\Delta R/R(t)|$ for bare Ti (a), Ti/ethanol (b) and Ti/cell (c). $|\Delta R/R(t)|$ measured in Ti shows three acoustic echoes arising from successive reflections in the Ti film. The amplitude of the Fourier spectra of the echoes plotted in the inset of figure 3 is centered around 50 GHz. In the case of Ti/ethanol, Brillouin oscillations measured at 3.8 GHz from Fourier analysis allows the determination of the sound velocity $v_e = 1120$ m/s in ethanol assuming a refractive index of $n_e = 1.36$.[10] The value of $v_e$ agrees closely with literature.[11] In the case of Ti/cell, we measured a frequency of 5.6 GHz which gives a sound velocity in the cell of $v_c = 1610$ m/s, the expected value for such a cell.[7]

A closer look at $|\Delta R/R(t)|$ measured in ethanol shows at least two echoes with the same arrival time as in the case of bare Ti. In the case of the cell, there is apparently no echo. Consider a Morlet wavelet centered on the center frequency of the acoustic echoes in Ti defined by $W_m(t) = \cos(4\alpha t)\exp[-(\alpha t)^2/2]$, with $\alpha = 0.04$. The amplitude of the Fourier spectra of the wavelet is plotted in the inset of figure 3(a) (dash). The convolutions of $|\Delta R/R(t)|$ with $W_m(t)$ are plotted on figure 4. The density of acoustic energy resulting from the convolution reveals a first echo at around 137 ps and a second echo at around 288 ps in all three cases.

The ratio of the second to the first echo yields the acoustic reflection coefficient $R_{ts} = (Z_t - Z_s)/(Z_t + Z_s) = 0.22$ at the Ti/silica interface in the case of bare Ti, where $Z_t$ and $Z_s$ are the acoustic impedances of Ti and silica, respectively. The ratio of the second echo measured in the case of Ti/ethanol (and in the case of Ti/cell, respectively) to the second echo measured in Ti yields the acoustic reflection coefficients $R_{te} = 0.91$ and $R_{tc} = 0.77$ at the Ti/ethanol and Ti/cell interfaces, respectively. $Z_s = 12.7$ MPa s m$^{-1}$ has been obtained from a previous measurement, so $Z_t = 19.5$ MPa s m$^{-1}$ can be deduced from $R_{ts}$. We obtain the impedances of ethanol $Z_e = 0.9$ MPa s m$^{-1}$ and of the cell $Z_c = 2.5$ MPa s m$^{-1}$ similarly from $Z_t$, $R_{te}$ and $R_{tc}$.

The particularity of the present experiment is that we measure simultaneously $v$ and Z. So we can obtain the compressibility $\chi_e = 1/Z_e v_e = 1.0$ GPa$^{-1}$ and the density $\rho_e = Z_e/v_e = 805$ kg m$^{-3}$ for ethanol, values in close agreement with literature. Similarly we obtain $\chi_c = 1.0$ GPa$^{-1}$ and $\rho_c = 1550$ kg m$^{-3}$ for the cell. The value of the $\rho_c$ is larger than that commonly found in the literature $\sim$1100 kg m$^{-3}$. The cumulated errors on $v_c$ and on $R_{tc}$ are not likely to explain this discrepancy since we measured reasonable values for ethanol. However,

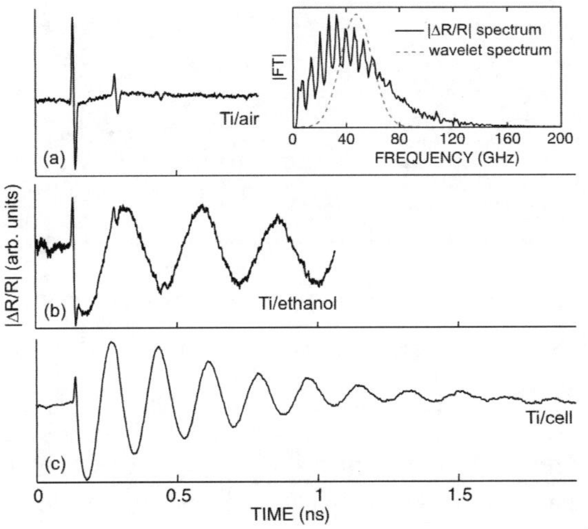


**Figure 3.** $|\Delta R/R|$ for (a) bare Ti, (b) ethanol on Ti and (c) cell on Ti. The amplitude of the Fourier spectra of $|\Delta R/R|$ in the case of bare Ti and of the Morlet wavelet (dash) used to extract the acoustic echoes in figure 4 are plotted as an inset.

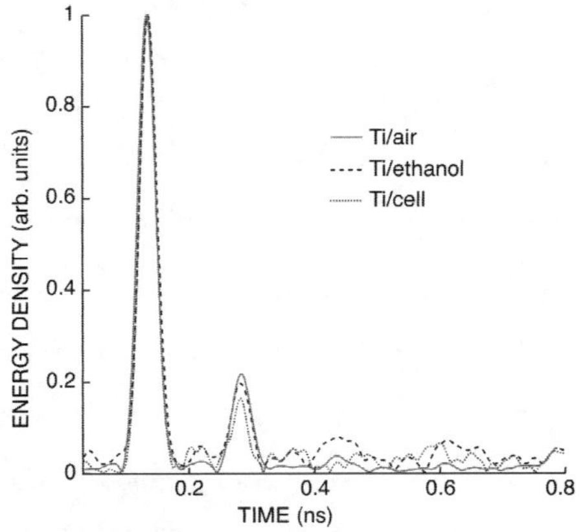


**Figure 4.** Convolution of $|\Delta R/R(t)|$ with a single Morlet wavelet of center frequency 50 GHz for bare Ti (plain), Ti/ethanol (dash) and Ti/cell (dotted).

there might be an influence of the frequency on the determination of $R_{tc}$ due to the imperfect contact at the Ti/cell interface and to the viscosity of the cell. It should also be noted that the value given in literature is an average value since the cell is composed of different compartments, notably the vacuole and the first and secondary cell walls. Indeed $R_{tc}$ might be sensitive to the wall, which exhibits a higher density, rather than to the vacuole.

## 4. Conclusion

Using a transducer composed of a thin Ti film we have measured simultaneously the density and the compressibility of a single biological cell by means of the PU technique. Focusing the pump beam on the rear side of the Ti film we have removed completely the pump radiation, the diffusion of overheated electrons and the thermal diffusion on a nanosecond time scale, thereby reducing the cell stress and improving the SNR. Further investigation of the sensitivity of our measurements to cell wall and contact quality should pave the way towards mechanical characterization of subcellular components and should enlighten the mechanics of cell adhesion.

## References

[1] Thomsen C, Strait J, Vardeny Z, Maris H J, Tauc J and Hauser J J 1984 *Phys. Rev. Lett.* **53** 989
[2] Pezeril T, Klieber C, Andrieu S and Nelson K A 2009 *Phys. Rev. Lett.* **102** 107402
[3] Rouai L, Bonello B, Louis G, Perrin B and Peretti P 1999 *J. Appl. Phys.* **85** 8155–8159
[4] Wright O B, Perrin B, Matsuda O and Gusev V E 2008 *Phys. Rev. B* **78** 024303
[5] Yang F, Grimsley T J, Che S, Antonelli G A, Maris H J and Nurmikko A V 2010 *J. Appl. Phys.* **107** 103537
[6] Shelton L, Yang F, Ford W and Maris H 2005 *Phys. Status Solidi B* **242** 1379–1382
[7] Audoin B, Rossignol C, Chigarev N, Ducousso M, Forget G, Guillemot F and Durrieu M C 2010 *Ultrasonics* **50** 202–207
[8] Dehoux T, Wright O B, Voti R L and Gusev V E 2009 *Phys. Rev. B* **80** 235409
[9] Lin Z, Zhigilei L V and Celli V 2008 *Phys. Rev. B* **77** 075133
[10] Rheims J, Köser J and Wriedt T 1997 *Meas. Sci. Technol.* **8** 601–605
[11] Zapka W and Tam A C 1982 *Appl. Phys. Lett.* **40** 310–312

# Adhesive bond strength evaluation in composite materials by laser-generated high amplitude ultrasound

**M. Perton, A. Blouin, J.-P. Monchalin**

Industrial Materials Institute, National Research Council of Canada, 75 de Mortagne Boucherville, Québec, Canada, J4B 6Y4

Email: Mathieu.Perton@imi.cnrc-nrc.gc.ca

**Abstract**. Adhesive bonding of composites laminates is highly efficient but is not used for joining primary aircraft structures, since there is presently no nondestructive inspection technique to ensure the quality of the bond. We are developing a technique based on the propagation of high amplitude ultrasonic waves to evaluate the adhesive bond strength. Large amplitude compression waves are generated by a short pulse powerful laser under water confinement and are converted after reflection by the assembly back surface into tensile waves. The resulting tensile stresses can cause a delamination inside the laminates or at the bond interfaces. The adhesion strength is evaluated by increasing the laser pulse energy until disbond. A good bond is unaffected by a certain level of stress whereas a weaker one is damaged. The method is shown completely non invasive throughout the whole composite assembly. The sample back surface velocity is measured by an optical interferometer and used to estimate stress history inside the sample. The depth and size of the disbonds are revealed by a post-test inspection by the well established laser-ultrasonic technique. Experimental results show that the proposed method is able to differentiate weak bond from strong bonds and to estimate quantitatively their bond strength.

## 1. Introduction

In aerospace, adhesive bonding of structural components presents many practical advantages when compared to other joining methods. However, its use for primary structures is impeded by the absence of reliable nondestructive methods that can ensure the integrity of the joint.

Laser shock generation and spallation have been studied for the measurement of the bond strength between a thin planar coating on a substrate [1], [2]. Recent work addresses the problem of adhesive bonding of structures made of carbon-epoxy composite [3].

In this paper, the laser shock wave technique is further explored for evaluating the bond strength of composite laminates joined by an adhesive layer. The principles of bond strength evaluation will be first detailed and then applied to bonded composite plates made of carbon fibers embedded in epoxy.

## 2. Composite laminate properties

The composite laminates were obtained by curing a stack of 4 carbon fibre plies pre-impregnated with epoxy (Cytec 5276-1). In each ply, the fibres (G40-800 – 24K) are unidirectional, and the orientation of the plies is $[0/90]_S$. The total thicknesses of the 4-ply is about 0.72 mm. Given that the plies are co-cured, the epoxy matrix has a quite homogeneous high strength, even between the plies. The composite sample shows strong anisotropy indicated by a longitudinal velocity of 3100 $m.s^{-1}$ in the $z$

direction, normal to the plies, and of 8300 $m.s^{-1}$ along the fibre direction of a single ply. Adhesive paste Hysol® EA9394 was also used to bond the composite laminates together.

## 3. Principle and experimental approach

A powerful Q-switched Nd:YAG laser which delivers optical pulses of 8 ns duration and up to 2 J energy at 1064 nm wavelength is used to induce shock waves or very high amplitude ultrasonic waves in the sample. The laser beam is focused to a spot diameter of about 4 mm. To avoid surface damage and to increase the ultrasonic wave amplitude [4], the surface of the material is first covered with a black electrical tape and then with a constraining water layer, as illustrated in figure 1. Under our generation conditions, it should be noted that the pressure level is below the Hugoniot Elastic Limits (HEL) of all the materials, so that wave propagation is in the elastic regime [5].

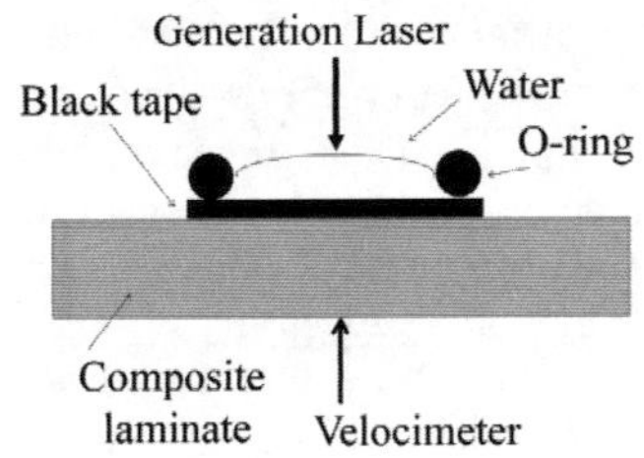


Figure 1: Schematic setup for laser shock generation.

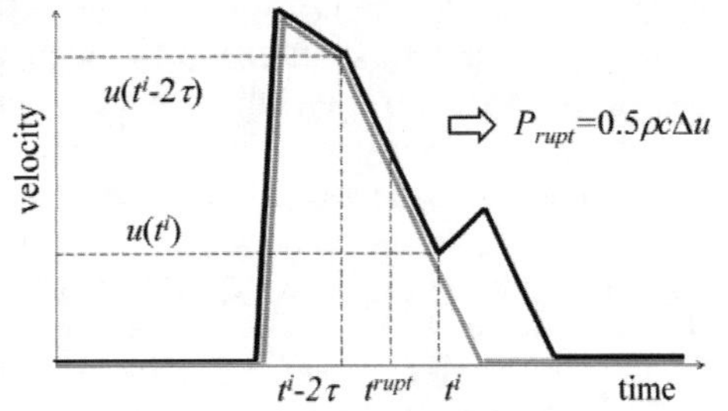


Figure 2: Schematic representation of velocity signals at the back surface. The black and grey signals represent respectively a signal slightly above and below damage threshold. Grey signal has been normalized to the black signal maximum.

The source size is a few times larger than the sample thickness, with the result that the waves generated at the top surface are mostly compressional plane waves. When propagating through the thickness of a homogeneous plate, they are reflected by the free back surface as a tensile wave. Close to this surface, almost no stress is imposed because the compression due to the end of the incoming wave is balanced by the beginning of the reflected tensile wave. If attenuation mechanisms and diffraction are neglected, the maximum tensile stress begins at a certain distance from the back surface, and propagates unchanged until the next reflection. For multi-layer structure, the propagation is altered by the transmission/reflection of the waves at the different interfaces. Only the tensile stress can induce failure within the laminate or at the adhesive bonded interfaces.

To quantitatively evaluate the stresses inside a sample, an optical velocimeter based on a Fabry-Perot solid state etalon was developed and used to monitor the back surface velocity $u(t)$ (see figure 1). A detailed description of the interferometer can be found in reference [6]. Then a simple relationship is used to relate this velocity to the stresses within the sample. In particular, the rupture threshold is given by the following equation [7]:

$$P_{rupt} = 0.5\rho c \left[ T_0^{-1}(z)\, u(t^i) - T_1(z)\, u(t^i - 2\tau) \right], \text{with } \tau = \int_0^{z_{rupt}} z/c(z)\, dz \quad (1)$$

where $t^i$ corresponds to the time on the velocity signal at which the damage is identified. It is delayed by $\tau$ from the time $t^{rupt}$ at which rupture occurs. $z_{rupt}$ is the distance of the damage from the back surface, measured by post shock laser ultrasonic inspection [8]. $c(z)$ is the propagation velocity through the sample in the $z$ direction and $\rho c$ is the acoustic impedance of the ultimate bottom layer. $T_0(z)$ and $T_1(z)$ stand for the product of all the pressure transmission coefficients between the free surface and the position $z_{rupt}$ and according of the wave propagation direction. Since shock propagation is elastic for the material studied, the velocity signals obtained with different laser energies should all be superimposed when normalized, except for the signal in which a disbond signature appears. The way to obtain $t^i$ is thus to increase gradually the laser energy until obtaining the disbond and to compare the normalized velocity signals. Figure 2 gives a schematic representation of the procedure.

## 4. Application to bonded laminates

To show that the technique is able to distinguish a strong joint from a weak one, bonded laminate assemblies (4-ply laminate on top of another 4-ply laminate) were prepared to present two areas with different adhesion strengths. All surfaces of the laminates were first cleaned by a solvent wipe. Then, only the half of the laminate's surfaces was treated by a corona discharge, a technique which is well known to improve adhesion [9]. Finally the laminates were assembled with their treated surface face to face. The dimensions of the samples were about 30 mm x 40 mm. The bond thickness all along the joint was measured to range between 90 and 160 µm.

Before laser shock testing, samples were non-destructively tested by laser-ultrasonics to detect delaminations or voids in the bond. Then, to distinguish the areas of different adhesive strengths, several shocks were applied over the entire surface, beginning with a "low" laser energy level and were followed by another laser-ultrasonic inspection of the whole sample. This procedure was repeated for increasing values of laser energy. Since the disbonds or delaminations occur from the reflection at the back surface, the laser-ultrasonic inspections were always made from this back surface. Figure 3 presents the B&C-scans representation of the signals obtained from the laser-ultrasonic inspections before and after shocks. The laser shock energy started at 400 mJ and was increased up to 1200 mJ by step of 200 mJ. No void is noticed in the pre-shock C-scan of the figure 3(a), and no disbond either is observed for 400 mJ (the C-scan at 400 mJ is identical to the one shown in figure 3(a)). The dashed rectangle delimits the weak area (without corona discharge). Disbonds are observed in this area at 600 mJ and becomes more evident when laser energy is increased to 1000 mJ, figure 3(b). B-scans indicate that the disbonds have occurred at the joint depth. All the shocks realized on the weakly bonded area have produced disbonds and almost no disbond is observed in the strong area. When laser energy reaches 1200 mJ, all the new delaminations observed are found to take place between the second and third laminate plies from the back surface (figure 3(c)). The epoxy matrix damage threshold is thus lower than the joint strength in the good bonded area.

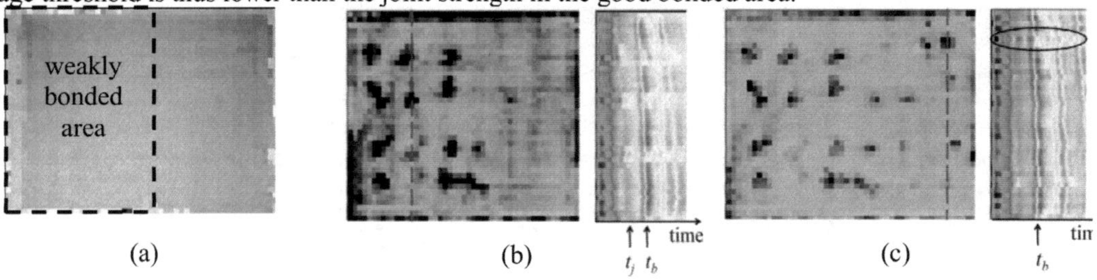


Figure 3. Typical C-scan obtained after shocks realized with laser energies (a) up to 400mJ, (b) from 600 mJ to 1000mJ, and (c) from 1200 mJ. Some C-scans are associated with their B-scans obtained along the dashed line. The lines pointed respectively by tb (tj) correspond to the reflection of the wave on the shock loading surface (on the disbonded joint interface). The encircled multiple echoes in (c) are reflections at the disbonded interface between the 2$^{nd}$ and 3$^{rd}$ ply.

Figure 4 shows the normalized velocity signals obtained below (400mJ), just above the weak joint damage threshold (600mJ) and just above the epoxy matrix damage threshold (1200mJ). In fact, the maximum of these signals are respectively about 50 m/s, 80 m/s and 150 m/s. Until the disbond signature, the signals are almost identical, which confirms again the elastic regime of wave propagation and that damage results from a brittle behavior of the material under high strain rate deformation. When applying the equation (1) to these signals, we obtain a value of about 340 MPa for the epoxy matrix strength and about 150 MPa for joint strength in the weak bonded area.

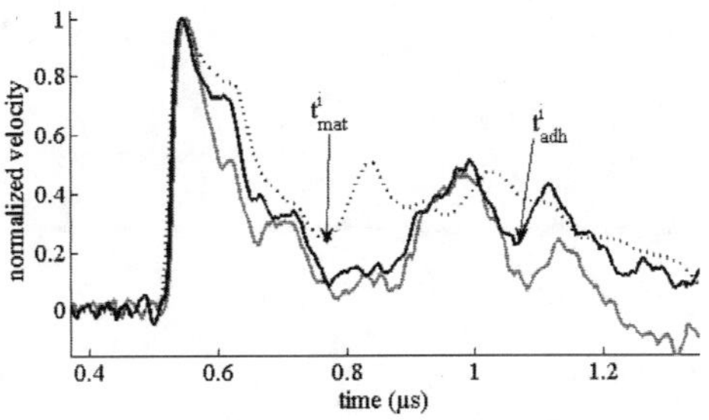


Figure 4. Normalized velocimeter signals produced at 400mJ (grey), at 600mJ (black) and at 1200mJ (point). The arrows indicate signatures of the matrix and adhesive ruptures.

## 5. Summary & Perspectives

A method based on laser shock waves combined with laser-ultrasonic inspection has been used to evaluate quantitatively the bond strength of carbon fibre composite laminates. It has been shown that the shock waves propagate only under the elastic regime, and that damage results of a brittle behavior of the material under high strain rate deformation. These observations confer to the method a quality of non invasive proof test and allow the adhesion strength to be measured quantitatively. The method requires first the identification of a damage signature in the back surface velocity signal and then the evaluation of the damage depth by laser-ultrasonic inspection. The inter-ply bond strength within the laminate was evaluated to about 340 MPa. Then two laminates were bonded in such way that half of their joint presents a weak strength and the other half a strong strength (comparing to the inter-ply bond strength). The test technique actually reveals the two different bonded areas. The strength of the strong bonded area was found to be at least equal than the one between the plies whereas the weakly bonded area showed a strength of about 150 MPa. The encouraging results provided by this work are an incentive for further development of the technique in view ultimately of its use for certifying adhesive bonding of primary aircraft structures.

**Acknowledgements**

The authors would like to thank Martin Lord and Christian Néron for their assistance in instrumentation for all aspects of this project. We also would like to thank Andrew Johnston, Richard Cole and Julieta Barroeta Robles of the Institute of Aerospace Research of NRC for providing the laminates and for useful information and discussions on composite materials and adhesive bonding. This work is part of the collaborative project SATAC between the National Research Council of Canada and the Centre National de la Recherche Scientifique of France.

**References**

[1] Vossen J. L., *ASTM Spec. Tech. Publ.*, **640**, (1978): 122-131.
[2] Gupta V. et al., *Mater. Sci. Eng* **A126**, (1990): 105–117.
[3] Bossi R. et al., "Laser Bond Testing", *Materials Evaluation*, **67**, (2009): 819-827.
[4] Fabbro R., et al., *J. Appl. Phys.* **68**, (1990): 775–781.
[5] Asay J. R. et al., *High-Pressure Shock Compression of Solids*, Springer, New York, 1993.
[6] Arrigoni M. et al., *Meas. Sci. Technol.* **20**, 015302 (7pp) (2009).
[7] Perton M. et al., Adhesive bond testing of carbon-epoxy composites by laser shockwave, to be published in Journal of Physics: Applied D.
[8] J.-P. Monchalin , "*Ultrasonic and Advanced Methods for Nondestructive Testing and Material Characterization*", C.H. Chen Ed., World Scientific Publishing Co., 2007.
[9] Kinloch A. J., *Adhesion and Adhesives*, Chapman and Hall, New York, 1987.

# AUTHOR INDEX

# AUTHOR INDEX